EXTRACTIVE INDUSTRY AND THE SUSTAINABILITY OF CANADA'S ARCTIC COMMUNITIES

T0132880

Extractive Industry and the Sustainability of Canada's Arctic Communities

Edited by

CHRIS SOUTHCOTT,
FRANCES ABELE,
DAVID NATCHER,
and BRENDA PARLEE

McGill-Queen's University Press
Montreal & Kingston • London • Chicago

© McGill-Queen's University Press 2022

ISBN 978-0-2280-1154-5 (cloth)
ISBN 978-0-2280-1155-2 (paper)
ISBN 978-0-2280-1347-1 (ePDF)

Legal deposit third quarter 2022
Bibliothèque nationale du Québec

Printed in Canada on acid-free paper that is 100% ancient forest free
(100% post-consumer recycled), processed chlorine free

Funded by the Government of Canada   Financé par le gouvernement du Canada | Canada   Canada Council for the Arts   Conseil des arts du Canada

We acknowledge the support of the Canada Council for the Arts.

Nous remercions le Conseil des arts du Canada de son soutien.

Library and Archives Canada Cataloguing in Publication

Title: Extractive industry and the sustainability of Canada's Arctic communities /
edited by Chris Southcott, Frances Abele, David Natcher, and Brenda Parlee.

Names: Southcott, Chris, editor. | Abele, Frances, editor. | Natcher, David C.,
1967– editor. | Parlee, Brenda, 1970– editor.

Description: Includes bibliographical references.

Identifiers: Canadiana (print) 20220175322 | Canadiana (ebook) 2022017539X
| ISBN 9780228011545 (hardcover) | ISBN 9780228011552 (softcover) | ISBN
9780228013471 (PDF)

Subjects: LCSH: Mineral industries—Canada, Northern. | LCSH: Mineral industries—
Economic aspects—Canada, Northern. | LCSH: Mineral industries—Social aspects—
Canada, Northern. | LCSH: Mineral industries—Environmental aspects—Canada,
Northern. | LCSH: Sustainable development—Canada, Northern. | LCSH: Indigenous
peoples—Canada, Northern—Social conditions. | LCSH: Canada, Northern—
Economic conditions. | LCSH: Canada, Northern—Social conditions. | LCSH: Canada,
Northern—Environmental conditions.

Classification: LCC HD9506.C22 E98 2022 | DDC 338.09719—dc23

Contents

Tables and Figures

TABLES

FIGURES

EXTRACTIVE INDUSTRY AND THE SUSTAINABILITY OF CANADA'S ARCTIC COMMUNITIES

Extractive Industry and the Sustainability of Canada's Arctic Communities: An Introduction

Chris Southcott, Frances Abele, David Natcher, and

Brenda Parlee

Canada's northern communities face important socio-economic disparities when compared to the rest of Canada (Canada 1996; Daley, Burton, and Phipps 2015; Southcott 2014). They have fewer employment opportunities, lower incomes, higher rates of poverty, inadequate housing, and lower rates of food security. These conditions contribute to lower measures of personal well-being for people living in this region (Adelson 2005; CCA 2014; Lauster and Tester 2010). Poor socio-economic conditions in northern communities have been linked to negative health indicators and a range of social pathologies that are consistently displayed in national and international media. International observers have long been critical of the conditions experienced by northern Indigenous communities and have called upon the Canadian government to take concerted action (United Nations 2009, 2012; Human Rights Watch, 2020).

For well over a century, extractive industries have been seen as part of the cause rather than the solution to these conditions. Yet in looking for socio-economic development choices, there are few immediate options other than mining and oil and gas development. Northern communities are facing continuing pressure to allow these types of industries to develop in their regions.[1] The cyclical nature of the industry and the fact that Arctic resources are more costly to develop than those of many other areas means that the pressure varies depending on international commodity pricing, but it is likely to grow as cheaper sources are depleted and new technologies reduce costs (AANDC 2015; Coates 2015; Gill and Sevigny 2015; GNWT 2013; Lang et al. 2014; NAEDB 2015, 2016). The prospect of extractive development is not an attractive one for most of these communities. While in many areas their experience with mining and oil and gas development is limited, for those that have had some experience, it has not been a pleasant one. In the past, there have been few lasting benefits for northern communities from extractive developments and often these projects have produced a legacy of negative impacts (Abele 1987; Berger 1977; Niezen 1993; Page 1999; Watkins 1977).

In the face of colonial barriers, the people of the region have fought for and succeeded in securing rights to control their own future (Dahl and Hicks 2000; Hamilton 1994; McPherson 2003; Saku 2002). New environmental regulations, political devolution, new comprehensive land claim agreements, and increased recognition and respect of Indigenous rights have changed the situation in Canada's North over the past forty years. There is a belief in many northern communities that it may be possible to control and mitigate the negative aspects of extractive development and ensure that a larger share of benefits from these activities can be used by the region to improve the long-term well-being and sustainability of these communities (Coates and Crawley, 2013; Dana, 2008). Yet northern communities acknowledge that the decisions they make now will have lasting ramifications for the future. For this reason, communities have sought information, based largely on the experience of others, to help them evaluate the benefits and costs of extractive resource development.

It was this desire on the part of northern communities for more research on these questions that gave rise to the Resources and Sustainable Development in the Arctic (RESDA) project. From a series of community research workshops starting in 2009, as part of an earlier project relating to the socio-economic development of northern communities, a project proposal was developed with community partners and researchers. That resulted in the RESDA project being funded by the Social Sciences and Humanities Research Council of Canada as part of their Major Collaborative Research Initiatives program.[2] Starting in 2011, the project organized a series of gap analyses to determine what research existed on the costs and benefits of extractive resource development in the contemporary context of Northern Canada and similar regions of the circumpolar world (Southcott et al. 2018). Using these gap analyses, communities and researchers organized a series of projects that examined different aspects of community-industry interaction. It is the findings of these projects, called subprojects during the life of the RESDA project, are the basis for this volume.

In presenting these findings we by no means claim to have arrived at a consensus on the relative value of extractive resource development for communities in Canada's North.

Researchers continue to warn that any increased dependence on extractive development under current conditions is unlikely to help communities increase their well-being or help them deal with a colonial heritage. It will instead likely continue to entrench it in the history of northern regions (Bernauer 2019; Hall 2013; Hoogeveen 2015). Some of the researchers involved in this project had spent most of their careers highlighting the negative aspects of extractive resource development on northern communities and as such were somewhat uneasy with the desire of some northern communities and organizations to engage with these activities. Yet the researchers involved in the project also recognized that the political situation had changed in Northern Canada and some other areas of the Arctic. These communities were now in a position to decide for themselves whether they wanted a mine or an oil and gas project to go ahead. A debate continues concerning whether northern communities have a veto over these developments. In practical terms, though, it would be difficult to see a project go ahead in the region if the communities did not support it.[3] The researchers involved in the project shared a desire to collaborate with communities so they can decide for themselves how or if to engage with extractive industries.

While the overall project was meant to be international in scope, in terms of the actual research, the focus was on the situation in Northern Canada. Extractive activities and communities in other areas of the Arctic were examined for comparative purposes. However, both researchers and community partners involved in RESDA quickly realized that resources could be best utilized by first concentrating our attentions exclusively on Canada. Following that study, they could partner with other projects to provide us with comparative data from other Arctic regions.[4] The chapters in this book are therefore focused on Northern Canada only. Likewise, the material is limited as far as some of the provincial norths are concerned. Projects limited themselves to the three territories, Nunavik, and Labrador. It is true that most conditions in the other provincial norths are similar to that of the regions covered by this research, but provincial regions are included in historic treaties (1-11) and have not been impacted by modern comprehensive land claims (Coates, Holroyd, and Leader 2015). It is the increased level of empowerment brought about by these modern treaties that coloured much of RESDA's research. Nonetheless, we hope that the findings from our project can be used by communities across the provincial north as they, too, grapple with some of these very same challenges.

INDUSTRIAL DEVELOPMENT AND CANADA'S NORTH

The resource curse debate which emerged in the 1990s has popularized the notion that extractive industrial development may have negative consequences on nations and regions (Sachs and Warner, 2001). This notion is not new to social science and humanities researchers looking at the Canadian North through the optics of staples theory (Huskey 2018; Huskey and Southcott 2016; Parlee 2015). While negative environmental impacts are often highlighted in the media, communities and researchers have long borne witness to the negative socio-economic and cultural impacts. The Berger Inquiry of the 1970s firmly established that industrial resource development projects could have substantial negative impacts on northern communities in these areas (Abele 1987; Berger 1977; Bone 2009; Southcott et al. 2018a).

While the report of the Berger Inquiry and its associated publications (Watkins 1977) were largely considered the first widely-known criticism of extractive resource development in northern communities, this theme has dominated social science research until recently (Angell and Parkins 2011). This started to change in the mid-1990s, as the literature began to highlight Indigenous peoples more as empowered participants (Dahl and Hicks 2000; Saku, Bone, and Duhaime 1998). The signing of modern land claim treaties in the years following the Berger Inquiry, combined with the processes of decolonization, gave communities an increased ability to control extractive resource projects (White 2002).

Northern communities continue to be very much aware of the risks of extractive resource development yet, given limited economic options, these projects are often pursued. Over the past seven years, the RESDA project, along with other researchers and organizations, have focused their attention to finding ways that communities could get more benefits from extractive resource development with fewer negative impacts (Southcott 2016). One key area of research has been a better understanding of social impacts of extractive development on communities and what to do about them (Cater and Keeling 2013; Czyzewski et al. 2014; Davison and Hawe 2012; Green 2013;

Ritsema et al. 2015; Rodon and Lévesque 2015, 2018; Schweitzer et al. 2016, 2018; Tester, Lambert, and Lim 2013). Another important area of research has been on finding ways of tracking and measuring impacts through developing new indicators and monitoring systems (Édouard and Duhaime 2013; NAEDB 2013; Petrov 2014, 2018; Petrov et al. 2018). The Environmental Impact Assessment (EIA) process has been recognized as an important means by which negative impacts can be considered and mitigated, and in what ways communities can be better engaged in decision-making. An important area of research in this regard is the barriers to the EIA system fulfilling these objectives and what can be done about it (Cox and Mills 2015; Kennedy Dalseg et al. 2018; Noble, Hanna, and Gunn 2018; Noble and Udofia 2015; Udofia, Noble, and Poelzer 2015).

In the past, very little of the resource rents from extractive projects stayed in northern regions. Yet with new land claims and devolution of powers to territorial and regional governments, northern communities are now in a better position to capture more of these resource revenues (Huskey and Southcott 2016; Huskey and Southcott 2018; O'Faircheallaigh and Gibson 2012; Thistle 2016). Our research has looked at the best ways for communities to manage and distribute these new revenues and avoid the resource curse that has characterized northern resource development in the past (Briones et al. 2014; Coates 2015; Coates and Crowley 2013; Huskey 2018; NAEDB 2015; O'Faircheallaigh 2013; Rodon, Lemus-Lauzon, and Schott 2018). Making sure extractive industry development does not negatively impact the environment is a central concern for northern communities. Research showing how communities can say no to those proposed projects that present too much risk is an important part of ensuring that benefits outweigh potential negative impacts (Bernauer 2010; Procter 2016). A new area of research relates to how communities can best deal with the environmental legacies of past extractive developments, and how they can use site remediation for local benefit and improve plans to stop or reverse damage (Dance 2015; Keeling et al. 2018; LeClerc and Keeling 2015; Parlee, Sandlos, and Natcher 2018; Sandlos and Keeling 2016).

Perhaps the most prominent ways northern communities have sought to secure benefits from extractive industry activities has been through the negotiation of Impact and Benefit Agreements (IBAS). Yet IBAS have also proven problematic for northern communities (Caine and Krogman 2010). Recent research has looked at improving IBAS to ensure that they more adequately address community needs (Bradshaw, Fidler, and Wright 2018; Gibson and O'Faircheallaigh 2015; Jones and Bradshaw 2015; O'Faircheallaigh 2018). The Berger Inquiry highlighted the fact that subsistence activities can be negatively impacted by industrial development in an era where communities exercise more control, and when the costs of subsistence activities are often prohibitive to many. However, several studies have also pointed out that extractive industry development is now a crucial support for the subsistence economy (Koke 2008; Kruse 1986, 1991). New research is trying to understand the conditions under which extractive industry could be used to ensure the long-term survival of subsistence activities (Natcher et al. 2016; Southcott and Natcher 2018).

Differing gender impacts is an important issue and recent research has tried to isolate these differences and suggest ways to deal with them (Cox and Mills 2015; Kennedy Dalseg et al. 2018; Mills, Dowsley, and Cox 2018; Natcher 2013; Staples and Natcher 2015; Stienstra 2015). Finally, it is difficult to determine the best way to maximize benefits of resource development to communities when there is an inadequate notion of well-being to both interpret results and guide future negotiations. Recent research has therefore tried to better understand what communities perceive as well-being in a manner that can contribute to better managing of extractive industry impacts and benefits (Édouard and Duhaime 2013; Jones and Bradshaw 2015; Parlee 2018; Parlee and Furgal 2012).

THE ReSDA PROJECTS

Much of the research discussed above includes some of the earlier publications and findings of the ReSDA projects. This particular volume adds to that earlier work by summarizing the central findings of the subprojects organized by ReSDA researchers. In so doing, the chapters here contextualize results published elsewhere and discuss findings not previously described. Before discussing the chapters themselves, it is important to understand the processes behind the formation of those projects. We found that it was not easy to decide on research projects and then see them carried out. Academics may have a logically coherent list of priority areas for research, but communities often have a different view of those priorities. It was a difficult process to arrive at a series of projects that communities valued and with which they would assist us as partners. This was combined with the difficulties of finding appropriate researchers with interests similar to potential partners. These challenges led to a list of research projects that cover a gamut of issues but which are not as comprehensive as some may have hoped. A host of important research questions surrounding extractive resource development and northern communities still exists to be dealt with, and members of the ResDA network hope to continue to deal with these issues in the future[5].

The first six chapters in this volume deal with what some would consider the "traditional" focus of extractive industry benefits and impacts – money, jobs, training, and how these are impacted by gender. As we have discussed elsewhere, these "wage-economic" issues have often been at the center of research and discussions concerning extractive industry impacts on communities (Southcott et al. 2018a). The ResDA projects have examined these issues, which still remain important, but from a newer perspective that attempts to integrate more authentic notions of well-being in northern communities (Parlee 2018). The inability of northern communities to capture a fair share of the monetary wealth from these projects has long been an important critique of extractive industry activities in the region. Indeed, it may be a reason that the resource curse has not been an important issue in the region (Huskey and Southcott 2016).

With the new structures outlined above, this is changing. Increasingly, northern communities are able to capture a larger share of this wealth though IBAs and other agreements. Yet as the resource curse debate has shown us, how these benefits are shared has an important impact on whether they promote well-being or not (Humphreys, Sachs, and Stiglitz 2007). Thierry Rodon led a team of community and academic researchers looking at the ways resource revenues captured by communities were shared in Canada's North and the challenges associated with different models of distribution. His findings point out that some models provide more long-term benefits and create less problems, but that the choice is often constrained by the social circumstance prevailing in that community. Overall, communities need to ensure that lost natural capital is adequately replaced and that communities don't become overly dependent on extractive industry revenues.

The next chapter examines the questions often associated with forward linkages of economic development. Stephan Schott led a team of researchers associated with ResDA in an ArcticNet funded project looking at the impacts of extractive industry developments on local and regional employment and businesses. To what extent are northern communities benefitting from mining in the region in terms of royalties, jobs, and local business development? They examined the case of mining in both the Nunavik and Nunatsiavut regions. Their findings show that while things are improving, there are still problems. Benefits regarding local employment are still relatively low and workforce turnover remains an important issue. Local businesses are still not capturing a large share of revenues and capacity in this area needs to be increased. More work needs to be done to find ways of ensuring local employment and local business development. Adequate data collection has to be put in place to better inform communities if progress is being made.

Andrew Hodgkins looked at the question of training and employment. Building on his previous work in the Northwest Territories and Northern Alberta, Hodgkins examined the relationship between mining, training, and sustainable employment in the North Baffin region of Nunavut. His research shows that important challenges remain in any attempt to use mining to support training leading to sustainable employment in Nunavut. Comparing the situation there to conditions in Northern Alberta, Hodgkins identifies a series of factors needed to ensure that northern communities do indeed benefit to a greater degree from training and employment opportunities in extractive industry projects.

The question of industrial employment and participation in traditional subsistence economies has been an important one in Northern Canada ever since the Berger Inquiry of the 1970s. While some see the two as incompatible, others see employment income as necessary to allow participation in increasingly costly subsistence activities (Southcott and Natcher 2018). As a result, the ResDA network supported a range of research looking at the impacts of resource development on subsistence activities. Much of this has been published elsewhere (Natcher 2018; Natcher et al. 2016; Parlee et al. 2018). The chapter in this volume examines one aspect of the research not published elsewhere, namely the strategies mining companies use to try to support subsistence activities. Rooke et al. examine these strategies to understand the value of each. They conclude that what is essential in ensuring successful strategies are "respectful relationships" that can build "meaningful collaboration."

Extractive industry employment in Northern Canada has been significantly transformed by the increasing use of long-distance commuting. Canada no longer builds semi-permanent resource towns in its North. Instead, companies rely on a fly-in/fly-out (FIFO) workforce. What does this mean for existing northern communities? While they may now be protected from some of the effects of the boom and bust economic cycle, this also means that it is increasingly difficult for these communities to benefit from jobs and associated economic activity. Building on her earlier work in the Russian North, Gerti Saxinger partnered with the First Nation of Na-Cho Nyäk Dun in the Yukon to look at the question of long-distance commuting impacts. She found an increased number of community members were part of this new mobile workforce and that this created new challenges and opportunities.

Extractive industry development impacts women differently than men. While there is an increasing availability of research on this reality, there is little done specific to Indigenous women. The RESDA network has tried to remedy this situation by looking at the gender impacts of extractive industry activities in Indigenous communities. The chapter by Mills and Simmons looks at the situation in two regions of the Canadian North: the Nunatsiavut region of Labrador and the Sahtú region of the Northwest Territories. Both case studies indicate the need for "gender-informed planning" to ensure that extractive activities lead to "healthy communities."

The next section of chapters represents a series of research projects that are not normally associated with traditional notions of extractive industry impacts. For some community partners involved in RESDA, any sort of new activity in their region should be looked at as a potential source of solution to long-standing problems. Housing, food security, and environmental damage are issues of importance to these communities. Is it inevitable that industrial development will make these problems worse? Given the long-standing inability of governments to help them deal with these issues, can extractive activities be used to help deal with them?

Many communities in the Canadian North are concerned about housing. Homelessness is increasing as housing shortages and overcrowding become more and more evident (Canada 2017; Minich et al. 2011). At the request of community partners, a project was organized led by Lisa Freeman and Julia Christensen that examined the relationship between extractive resource development and housing in Yellowknife, Northwest Territories. Their findings indicate that it may be possible to alter past negative impacts on housing and instead improve housing security through "collaborations between the non-profit sector, private rental housing providers and the territorial government."

The environmental legacies of past mining developments represent serious challenges for northern communities and of all types of impacts, these have received the most attention in the press. Increasingly, governments are attempting to ensure that the environmental degradations caused by extractive industries are properly remediated and monitored. This entails a considerable amount of expenditures and activity. Community partners were interested in finding out more about the process surrounding remediation and monitoring and how they are related to their well-being. The chapter by Dance et al. attempts to "unpack current Canadian mine remediation policies and practices" and shows that problems remain that require "northern involvement through employment, regulation, and the integration of TK (traditional knowledge) and Indigenous governance" so that "positive outcomes for local communities" can be prioritized.

Food security is another long-standing concern in northern communities (Campbell et al. 2014; CCA 2014). Partners wondered if industrial resource development could help communities deal with food insecurity. Researchers and communities in Labrador looked at the Muskrat Falls hydroelectric dam development and the possibility that waste from the project could be used to improve the region's food security. Through several concrete examples, their research showed that this was indeed possible.

The final chapter in this section builds on previous work done by many members of the network on social economy organizations in the region. Can these "civil society" organizations play a role in realizing the democratic decision-making potential of extractive resource governance in the region? Is it possible to ensure a better situation of well-being, more benefits, and a more sustainable type of development by allowing the involvement of a range of community organizations in the processes surrounding extractive industry development? Joshua Gladstone and Sheena Kennedy Dalseg conducted an initial investigation into this question by looking at the situation in Nunavut. Their findings show that while there is potential for communities to benefit from more active involvement through social economy organizations, up to now it has been limited – especially for smaller, locally-based groups. Their research suggests a number of ways organizations could become more actively involved in resource governance.

The final section of chapters deals with measuring extractive industry impacts. The question of whether or not the industry can contribute to the long-term sustainability of Arctic communities cannot be answered if there is no way to measure these impacts adequately. There is strong desire in the region to develop a dependable set of indicators that can look at all aspects of impacts and that can be both controlled by and reflect the particular conditions of these communities. The chapter by Brenda Parlee provides us with a background to previous work on measuring the social impacts of extractive activity on community well-being. Building on her earlier work on community-based monitoring, she shows the importance of developing indicators that can be specific to the community and that they can develop and manage themselves.

Even prior to the start of the RESDA project, the Inuvialuit Regional Corporation had begun a project that would enable them to have access to reliable and easily quantifiable data to measure extractive industry impacts. The chapter by Petrov et al. describes how RESDA partnered with the Inuvialuit to build on this project. A comprehensive set of baseline indicators were developed and used to evaluate past impacts. An important finding of this project was that the data could be greatly enhanced by more qualitative indicators that would allow for "community-based collection, management, and analysis of data by local actors."

The initial indicator work was continued in a new project that attempted to find ways that could more readily make data from indicators available to communities. The chapter by Wichmann summarizes this work and includes a discussion of a range of challenges faced by these attempts. While some data now exists that makes measurement of some impacts possible, much more work has to be done to develop indicators that are truly useful to communities trying to understand extractive industry impacts.

Not all the projects that RESDA put in place have been included as a chapter in this volume. Other projects were delayed for logistical and other reasons and their results are not included here. However, partial findings from these projects are discussed in the concluding chapter.

Readers of chapters in this volume will find that, while the projects have produced some information useful to northern communities considering extractive resource projects, much more needs to be done to provide further and clearer answers to the questions communities are asking. Some of the key conclusions and areas for future research are dealt with in the final chapter. Our findings do show that new possibilities exist that can lead to more sustainable relationships between northern communities and extractive industries. However, we have also learned that as relationships change, new challenges emerge. The "wicked" nature of northern development will likely continue and will require institutions that are responsive and committed to resolving future challenges. We hope that the findings from our research can inform those institutional responses and help to advance the sustainable development of Northern Canada.

NOTES

1   In addition to increased exploration activity (including oil and gas), since the 1990s the region has seen, among others, the opening of the diamond mines in the Northwest Territories and Nunavut, the Voisey's Bay Mine in Labrador, the Raglan Mine in Nunavik, the Meadowbank Gold Mine and the Mary River Mine in Nunavut, and the Minto and Eagle Gold Mines in the Yukon.

2   The earlier project, also funded by the Social Sciences and Humanities Research Council of Canada(SSHRC), was called the Social Economy Research Network for Northern Canada (SERNNOCA) and looked at the socio-economic possibilities for non-profit and non-state organizations in community development. See http://yukonresearch.yukoncollege.yk.ca/ sern/aboutsernnoca/. For the purposes of our SSHRC grant, our region was restricted to the territorial north of Canada, as well as Nunavik and Labrador. At the same time, research tried to include comparative studies from the provincial norths. For this reason, the project's results will be of interest to those interested in both the territories and the provincial norths.

3   See the work done on attempts to develop a uranium mine in the Kivalliq region of Nunavut as evidence of this (Bernauer 2010; Blangy and Deffner 2014).

4   Beginning in 2011, RESDA started negotiations with American researchers and the National Science Foundation to assist the development of an American counterpart to RESDA. The Arctic-FROST project received funding in 2013 and has assisted RESDA in developing more useful notions of sustainability for northern communities. It has also co-sponsored community workshops with RESDA in Alaska, Greenland, and Russia. The University of the Arctic Thematic Network on Extractive Industries at the University of Lapland, Finland, provided RESDA graduate students with

international collaborative courses. RESDA worked closely with the Institute of Social and Economic Research at the University of Alaska Anchorage on a number of projects. Nordforsk's Resource Extraction and Sustainable Arctic Communities (REXSAC) network project provides RESDA with comparative research in the Nordic region. These include the issues of the integration of Indigenous knowledge, the impacts of new assessment regimes, new forms of Impact and Benefit Agreements, and corporate social responsibility.

REFERENCES

AANDC. 2015. *Nunavut Mineral Exploration, Mining, and Geoscience Overview 2014*. Ottawa: Aboriginal Affairs and Northern Development Canada. http://cngo.ca/app/uploads/Exploration_Overview-2014-Magazine-English.pdf.

Abele, Frances. 1987. "Canadian Contradictions: Forty Years of Northern Political Development." *Arctic* 40, no. 4 (December): 310–320.

Adelson, Naomi. 2005. "The Embodiment of Inequity: Health Disparities in Aboriginal Canada." *Canadian Journal of Public Health/Revue Canadienne de Santé Publique* 96 (S2): S45–S61.

Angell, Angela C. and John R. Parkins. 2011. "Resource Development and Aboriginal Culture in the Canadian North." *Polar Record* 47, no. 1 (January): 67–79.

Berger, Thomas R. 1977. *Northern Frontier, Northern Homeland: The Report of the Mackenzie Valley Pipeline Inquiry* Volume 1. Ottawa: Mackenzie Valley Pipeline Inquiry.

Bernauer, Warren. 2010. "Mining, Harvesting and Decision Making in Baker Lake, Nunavut: A Case Study of Uranium Mining in Baker Lake." *Journal of Aboriginal Economic Development* 7, no. 1: 19–33.

– 2019. "The Limits to Extraction: Mining and Colonialism in Nunavut." *Canadian Journal of Development Studies-Revue Canadienne d'Études Du Développement* 40, no. 3 (July): 404–422. doi: 10.1080/02255189.2019.1629883.

Blangy, Sylvie and Anna Deffner. 2014. "Impacts du développement minier sur les hommes et les caribous à Qamani'tuaq au Nunavut: approche participative." Études/*Inuit*/Studies 38, no. 1–2: 239–265.

Bone, Robert M. 2009. *The Canadian North: Issues and Challenges*. 3rd ed. Don Mills: Oxford University Press.

Bradshaw, Ben, Courtney Fidler, and Adam Wright. 2018. "Impact Benefit Agreements and Northern Resource Governance." In *Resources and*

Sustainable Development in the Arctic, edited by Chris Southcott, Frances Abele, Dave Natcher, and Brenda Parlee, 204–218. London: Routledge.

Briones, Jesika, Sarah Daitch, Andre Dias, Martin Lajoie, Julia Fan Li, and Alyssa Schwann. 2014. *A Question of Future Prosperity: Developing a Heritage Fund in the Northwest Territories.* Ottawa: Action Canada. http://www.actioncanada.ca/project/ question-future-prosperity-developing-heritage-fund-northwest- territories.

Caine, Ken J. and Naomi Krogman. 2010. "Powerful or Just Plain Power-Full? A Power Analysis of Impact and Benefit Agreements in Canada's North." *Organization & Environment* 23, no. 1 (March): 76–98. <Go to ISI>://WOS:000274556500004.

Campbell, Megan, Lara Honrado, Brian Kingston, Alika Lafontaine, Leslie Lewis, and Kate Muller. 2014. *Hunger in Nunavut: Local Food for Healthier Communities.* Ottawa: Action Canada. http://www.actioncan- ada.ca/project/hunger-nunavut-local-food-healthier-communities/.

Canada. 1996. *Report of the Royal Commission on Aboriginal Peoples.* Ottawa: Canada. https://www.bac-lac.gc.ca/eng/discover/aboriginal-her- itage/royal-commission-aboriginal-peoples/Pages/final-report.aspx.

– 2017. *WE CAN DO BETTER: Housing in Inuit Nunangat: Report of the Standing Senate Committee on Aboriginal Peoples.* Ottawa: Senate Canada. https://homelesshub.ca/sites/default/files/APPA-RPT-Northern- Housing-Report-2017-02-28.pdf.

Cater, Tara and Arn Keeling. 2013. ""That's Where Our Future Came From": Mining, Landscape, and Memory in Rankin Inlet, Nunavut." *Études/Inuit/Studies* 37, no. 2: 59–82.

CCA. 2014. *Aboriginal Food Security in Northern Canada: An Assessment of the State of Knowledge.* Council of Canadian Academies.

Coates, Ken. 2015. *Sharing the Wealth: How Resource Revenue Agreements can Honour Treaties, Improve Communities, and Facilitate Canadian Development.* Ottawa: Macdonald-Laurier Institute. http:// www.macdonaldlaurier.ca/files/pdf/MLIresourcerevenuesharingweb.pdf.

Coates, Ken and Brian Crowley. 2013. *New Beginnings: How Canada's Natural Resource Wealth Could Re-shape Relations with Aboriginal People.* Ottawa: Macdonald-Laurier Institute. http://www.macdonaldlaurier.ca/files/pdf/2013.01.05-MLI-New_ Beginnings_Coates_vWEB.pdf.

Coates, Ken, Carin Holroyd, and Joelena Leader. 2015. "Managing the Forgotten North: Governance Structures and Administrative Operations of Canada's Provincial Norths." *The Northern Review* 38: 6.

Cox, David and Suzanne Mills. 2015. "Gendering Environmental Assessment: Women's Participation and Employment Outcomes at Voisey's Bay." *Arctic* 68, no. 2 (June): 246–260. Doi: 10.14430/arctic4478.

Czyzewski, Karina, Frank Tester, Nadia Aaruaq, and Sylvie Blangy. 2014. *The Impact of Resource Extraction on Inuit Women and Families in Qamani'tuaq, Nunavut Territory.* Vancouver: School of Social Work, University of British Columbia. http://pauktuutit.ca/wp-content/blogs.dir/1/assets/Final-mining-report-PDF-for-web.pdf.

Dahl, Jens and Jack Hicks. 2000. *Nunavut: Inuit Regain Control of Their Lands and Their Lives.* Copenhagen, Denmark: IWGIA, International Work Group for Indigenous Affairs.

Daley, Angela, Peter Burton, and Shelley Phipps. 2015. "Measuring Poverty and Inequality in Northern Canada." *Journal of Children and Poverty* 21, no. 2 (July): 89–110.

Dana, Léo-Paul. 2008. Oil and gas and the Inuvialuit people of the Western Arctic. *Journal of Enterprising Communities: People and Places in the Global Economy* 2(2), 151–167.

Dance, Anne. 2015. "Northern Reclamation in Canada: Contemporary Policy and Practice for New and Legacy Mines." *The Northern Review* 41: 41–80. <Go to ISI>://WOS:000384650500003.

Davison, Colleen M. and Penelope Hawe. 2012. "All That Glitters: Diamond Mining and Tåîchô Youth in Behchokö, Northwest Territories." *Arctic* 65, no. 2: 214–228.

Édouard, Roberson and Gérard Duhaime. 2013. "The Well-Being of the Canadian Arctic Inuit: The Relevant Weight of Economy in the Happiness Equations." *Social Indicators Research* 113, no. 1 (August): 373–392.

Gibson, Ginger and Ciaran O'Faircheallaigh. 2015. *IBA Community Toolkit: Negotiation and Implementation of Impact and Benefit Agreements.* Summer 2015 ed. Toronto: Walter & Duncan Gordon Foundation.

Gill, Alan and David Sevigny. 2015. *Sustainable Northern Development: The Case for an Arctic Development Bank.* Waterloo, ON: CIGI. https://www.cigionline.org/publications.

GNWT. 2013. *What We Heard and Recommendations: Report of the NWT Economic Opportunities Strategy Advisory Panel.* Yellowknife: GWNT. https://www.ntassembly.ca/sites/assembly/files/13-05-30td76-174.pdf.

Green, Heather. 2013. "State, Company, and Community Relations at the Polaris Mine (Nunavut)." *Études/Inuit/Studies* 37, no. 2: 37–57.

Hall, Rebecca. 2013. "Diamond Mining in Canada's Northwest
    Territories: A Colonial Continuity." *Antipode* 45, no. 2 (March):
    376–393. Doi: 10.1111/j.1467-8330.2012.01012.x.

Hamilton, John D. 1994. *Arctic Revolution: Social Change in the
    Northwest Territories, 1935–1994.* Toronto: Dundurn Press.

Hoogeveen, Dawn. 2015. "Sub-surface Property, Free-entry Mineral
    Staking and Settler Colonialism in Canada." *Antipode* 47, no. 1
    (January): 121–138. <Go to ISI>://WOS:000347695500007.

Human Rights Watch. 2020. World Report 2020. https://www.hrw.org/sites/
    default/files/world_report_download/hrw_world_report_2020_0.pdf.

Humphreys, Macartan, Jeffrey D. Sachs, and Joseph E. Stiglitz, eds. 2007.
    Escaping the Resource Curse. New York: Columbia University Press.

Huskey, Lee. 2018. "An Arctic Development Strategy? The North Slope
    Inupiat and the Resource Curse." *Canadian Journal of Development
    Studies / Revue Canadienne d'études du développement* 39, no. 1
    (January): 89–100. Doi: 10.1080/02255189.2017.1391067.

Huskey, Lee and Chris Southcott. 2016. ""That's Where My Money
    Goes": Resource Production and Financial Flows in the Yukon
    Economy." *The Polar Journal* 6, no. 1 (January): 11–29. http://dx.doi.
    org/10.1080/2154896X.2016.1171002.

– 2018. "Resource Revenue Regimes Around the Circumpolar North." In
    Resources and Sustainable Development in the Arctic, edited by Chris
    Southcott, Frances Abele, Dave Natcher, and Brenda Parlee, 156–174.
    London: Routledge.

Jones, Jen and Ben Bradshaw. 2015. "Addressing Historical Impacts Through
    Impact and Benefit Agreements and Health Impact Assessment." *The
    Northern Review* 41: 81–109. <Go to ISI>://WOS:000384650500004

Keeling, Arn, John Sandlos, Jean-Sébastien Boutet, and Hereward Longley.
    2018. "Knowledge, Sustainability and the Environmental Legacies of
    Resource Development in Northern Canada." In *Resources and
    Sustainable Development in the Arctic,* edited by Chris Southcott,
    Frances Abele, Dave Natcher, and Brenda Parlee, 187–203. London:
    Routledge.

Kennedy Dalseg, Sheena, Rauna Kuokkanen, Suzanne Mills, and Deborah
    Simmons. 2018. "Gendered Environmental Assessments in the
    Canadian North: Marginalization of Indigenous Women and
    Traditional Economies." *The Northern Review* 47: 135–166.

Koke, Paul E. 2008. "The Impact of Mining Development on Subsistence
    Practices Of Indigenous Peoples: Lessons Learned From Northern
    Quebec And Alaska." Master of Arts, University of Northern British
    Columbia, Prince George.

Kruse, John. 1986. "Subsistence and the North Slope Inupiat: The Effects of Energy Development." In *Contemporary Alaskan Native Economies,* edited by Steve Langdon, 121–152. Lanham, MD: University Press of America.

– 1991. "Alaska Inupiat Subsistence and Wage Employment Patterns: Understanding Individual Choice." *Human Organization* 50, no. 4 (December): 317–326.

Lang, Paul, Antonia Maioni, Jaimie Boyd, Mélanie Loisel, Alexandra Laflamme-Sanders, and Ian Anderson. 2014. *Breaking Ground in Nunavut: Assessing Transportation Infrastructure Proposals for Resource Development.* Ottawa: Action Canada. http://www.actioncanada.ca/project/breaking-ground-nunavut-assessing-transportation-infrastructure-proposals-resource-development/.

Lauster, Nathanael and Frank Tester. 2010. "Culture as a Problem in Linking Material Inequality to Health: On Residential Crowding in the Arctic." *Health & Place* 16, no. 3 (May): 523–530. Doi: 10.1016/j.healthplace.2009.12.010.

LeClerc, Emma and Arn Keeling. 2015. "From Cutlines to Traplines: Post-Industrial Land Use at the Pine Point Mine." *Extractive Industries and Society-an International Journal* 2, no. 1: 7–18. <Go to ISI>://WOS:000363542800003.

McPherson, Robert. 2003. *New Owners in their Own Land: Minerals and Inuit Land Claims.* Calgary: University of Calgary Press.

Mills, Suzanne, Martha Dowsley, and David Cox. 2018. "Gender in Research on Northern Resource Development." In *Resources and Sustainable Development in the Arctic,* edited by Chris Southcott, Frances Abele, Dave Natcher, and Brenda Parlee, 251–270. London: Routledge.

Minich, Katherine, Helga Saudny, Crystal Lennie, Michele Wood, Laakkuluk Williamson-Bathory, Zhirong Cao, and Grace M. Egeland. 2011. "Inuit Housing and Homelessness: Results from the International Polar Year Inuit Health Survey 2007–2008." *International Journal of Circumpolar Health* 70, no. 5 (February): 520–531.

NAEDB. 2013. *The Aboriginal Economic Benchmarking Report Core Indicator 3: Wealth and Well-Being.* Ottawa: NAEDB. http://www.naedb-cndea.com.

– 2015. *Enhancing Aboriginal Financial Readiness for Major Resource Development Opportunities.* Ottawa: NAEDB. http://www.naedb-cndea.com.

– 2016. *Recommendations on Northern Infrastructure to Support Economic Development.* Ottawa: NAEDB. http://www.naedb-cndea.com.

Natcher, Dave. 2013. "Gender and Resource Co-Management in Northern Canada." *Arctic* 66, no. 2 (June): 218–221. <Go to ISI>://WOS:000321092000010.

– 2018. "Normalizing Aboriginal Subsistence Economies in the Canadian North." In *Resources and Sustainable Development in the Arctic,* edited by Chris Southcott, Frances Abele, Dave Natcher, and Brenda Parlee, 219–233. London: Routledge.

Natcher, Dave, Shea Shirley, Thierry Rodon, and Chris Southcott. 2016. "Constraints to Wildlife Harvesting Among Aboriginal Communities in Alaska and Canada." *Food Security* 8, no. 6: 1153–1167.

Niezen, Ronald. 1993. "Power and Dignity: The Social Consequences of Hydro-Electric Development for the James Bay Cree." *Canadian Review of Sociology and Anthropology* 30, no. 4 (November): 510–529. <Go to ISI>://WOS:A1993MJ89000005.

Noble, Bram, Kevin Hanna, and Jill Gunn. 2018. "Northern Environmental Assessment: A Gap Analysis and Research Agenda." In *Resources and Sustainable Development in the Arctic,* edited by Chris Southcott, Francis Abele, Dave Natcher, and Brenda Parlee, 65–87. London: Routledge.

Noble, Bram and Aniekan Udofia. 2015. *Protectors of the Land: Toward an EA Process that Works for Aboriginal Communities and Developers.* Ottawa: Macdonald-Laurier Institute. http://www.macdonaldlaurier.ca/files/pdf/Noble-EAs-Final.pdf.

O'Faircheallaigh, Ciaran. 2013. "Extractive Industries and Indigenous Peoples: A Changing Dynamic?" *Journal of Rural Studies* 30: 20–30. <Go to ISI>://WOS:000317451900003.

– 2018. "Using Revenues from Indigenous Impact and Benefit Agreements: Building Theoretical Insights." *Canadian Journal of Development Studies/Revue canadienne d'études du développement* 39, no. 1 (January): 101–118.

O'Faircheallaigh, Ciaran and Ginger Gibson. 2012. "Economic Risk and Mineral Taxation on Indigenous Lands." *Resources Policy* 37, no. 1 (March): 10–18. <Go to ISI>://WOS:000304635600002

Page, Robert. 1999. *Northern Development: The Canadian Dilemma.* Toronto: McClelland & Stewart.

Parlee, Brenda. 2015. "Avoiding the Resource Curse: Indigenous Communities and Canada's Oil Sands." *World Development* 74: 425–436. <Go to ISI>://WOS:000358468700031.

– 2018. "Resource Development and Well-being in Northern Canada." In *Resources and Sustainable Development in the Arctic,* edited by Chris

Southcott, Frances Abele, Dave Natcher, and Brenda Parlee, 132–155. London: Routledge.

Parlee, Brenda and Chris Furgal. 2012. "Well-Being and Environmental Change in the Arctic: A Synthesis of Selected Research from Canada's International Polar Year Program." *Climatic Change* 115, no. 1 (November): 13–34. <Go to ISI>://WOS:000309866900002.

Parlee, Brenda, John Sandlos, and Dave Natcher. 2018. "Undermining Subsistence: Barren-ground Caribou in a "Tragedy of Open Access"". *Science Advances* 4, no. 2 (February). Doi:10.1126/sciadv.1701611.

Petrov, Andrey. 2014. *Measuring Impacts: A Review of Frameworks, Methodologies, and Indicators for Assessing Socio-Economic Impacts of Resource Activity in the Arctic: Gap Analysis 3.* Whitehorse: ReSDA. www.resda.ca.

– 2018. "Inuvialuit Social Indicators: Applying Arctic Social Indicators Framework to Study Well- Being in the Inuvialuit Communities." *The Northern Review* 47: 167–185.

Petrov, Andrey, Jessica Graybill, Matthew Berman, Philip Cavin, Vera Kuklina, Rasmus O. Rasmussen, and Matthew Cooney. 2018. "Measuring Impacts: A Review of Frameworks, Methodologies, and Indicators for Assessing Socio-Economic Impacts of Resource Activity in the Arctic." In *Resources and Sustainable Development in the Arctic,* edited by Chris Southcott, Frances Abele, Dave Natcher, and Brenda Parlee, 107–131. London: Routledge.

Procter, Andrea. 2016. "Uranium and the Boundaries of Indigeneity in Nunatsiavut, Labrador." *Extractive Industries and Society-an International Journal* 3, no. 2 (April): 288–296. <Go to ISI>:// WOS:000375104100004

Ritsema, Roger, Jackie Dawson, Miriam Jorgensen, and Brenda Macdougall. 2015. ""Steering Our Own Ship?" An Assessment of Self-Determination and Self-Governance for Community Development in Nunavut." *The Northern Review* 41: 205–228.

Rodon, Thierry, Isabel Lemus-Lauzon, and Stephan Schott. 2018. "Impact and Benefit Agreement (IBA) Revenue Allocation Strategies for Indigenous Community Development." *The Northern Review* 47: 9–29.

Rodon, Thierry and Francis Lévesque. 2015. "Understanding the Social and Economic Impacts of Mining Development in Inuit Communities: Experiences with Past and Present Mines in Inuit Nunangat." *The Northern Review* 41: 13–39. <Go to ISI>://WOS:000384650500002

– 2018. "From Narrative to Evidence: Socio-Economic Impacts of Mining in Northern Canada." In *Resources and Sustainable Development in the*

Arctic, edited by Chris Southcott, Frances Abele, Dave Natcher, and Brenda Parlee, 88–106. London: Routledge.

Sachs, Jeffrey D. and Andrew M. Warner. 2001. "The Curse of Natural Resources." *European Economic Review* 45, no. 4–6: 827–838. Doi: 10.1016/S0014-2921(01)00125-8.

Saku, James C. 2002. "Modern Land Claim Agreements and Northern Canadian Aboriginal Communities." *World Development* 30, no. 1: 141–151.

Saku, James C., Robert M. Bone, and Gérard Duhaime. 1998. "Towards an Institutional Understanding of Comprehensive Land Claim Agreements in Canada." *Études/Inuit/Studies* 22, no. 1: 109–121.

Sandlos, John and Arn Keeling. 2016. "Aboriginal Communities, Traditional Knowledge, and the Environmental Legacies of Extractive Development in Canada." *Extractive Industries and Society-an International Journal* 3, no. 2 (April): 278–287. <Go to ISI>://WOS:000375104100003

Schweitzer, Peter, Florian Stammler, Cecilie Ebsen, Aytalina Ivanova, and Irina Litvina. 2016. *Social Impacts of Non-Renewable Resource Development on Indigenous Communities in Alaska, Greenland and Russia: Gap Analysis 2A.* Whitehorse: RESDA. www.resda.ca.

– 2018. "Social Impacts of Non-Renewable Resource Development on Indigenous Communities in Alaska, Greenland, and Russia." In *Resources and Sustainable Development in the Arctic,* edited by Chris Southcott, Frances Abele, Dave Natcher, and Brenda Parlee, 42–64. London: Routledge.

SERNNOCa. 2010. *Nunavut Summit on the Social Economy Proceedings.* Iqaluit: SERNNOCa. http://yukonresearch.yukoncollege.yk.ca/frontier/files/sernnoca/NunavutSummitontheSocialEcon.pdf.

Southcott, Chris. 2014. "Socio-Economic Trends in the Canadian North: Comparing the Provincial and Territorial Norths." *The Northern Review* 38, no. 2: 159–177.

– 2016. "Est-il possible d'échapper à la malédiction de ressources dans l'Arctique?" In *Ressources naturelles, gouvernance et collectivités: Refonder le développement des territoires,* edited by Marie-Josée Fortin, Guy Chiasson, Maude Flamand-Hubert, Yann Fournis, and François L'Italien, 9–27. Rimouski, Que.: Éditions de GRIDEQ.

Southcott, Chris, Frances Abele, Dave Natcher, and Brenda Parlee. 2018a. "Beyond the Berger Inquiry: Can Extractive Resource Development Help the Sustainability of Canada's Arctic Communities?" *Arctic* 71, no. 4 (December): 393–406.

– 2018b. *Resources and Sustainable Development in the Arctic*. London: Routledge.

Southcott, Chris, and Dave Natcher. 2018. "Extractive Industries and Indigenous Subsistence Economies: A Complex and Unresolved Relationship. *Canadian Journal of Development Studies / Revue canadienne d'études du développement* 39, no. 1 (January): 137–154. Doi: 10.1080/02255189.2017.1400955.

Staples, Kiri, and Dave Natcher. 2015. "Gender, Decision Making, and Natural Resource Co-management in Yukon." *Arctic* 68, no. 3 (September): 356–366. <Go to ISI>://WOS:000362696900009.

Stienstra, Deborah. 2015. "Northern Crises: Women's Relationship and Resistances to Resource Extractions." *International Feminist Journal of Politics* 17, no. 4 (October): 630–651. <Go to ISI>:// WOS:000364552100007.

Tester, Frank, Drummond Lambert, and Tee Lim. 2013. "Wistful Thinking: Making Inuit Labour and the Nanisivik Mine Near Ikpiarjuk (Arctic Bay), Northern Baffin Island." Études/Inuit/Studies 37, no. 2: 15–36.

Thistle, John. 2016. "Forgoing Full Value? Iron Ore Mining in Newfoundland and Labrador, 1954–2014." *Extractive Industries and Society-an International Journal* 3, no. 1 (January): 103–116. <Go to ISI>://WOS:000373083000013.

Udofia, Aniekan, Bram Noble, and Greg Poelzer. 2015. "Community Engagement in Environmental Assessment for Resource Development: Benefits, Enduring Concerns, Opportunities for Improvement." *The Northern Review* 39: 98–110. <Go to ISI>://WOS:000384648100009

United Nations. 2009. *Report of the Special Rapporteur on Adequate Housing as a Component of the Right to an Adequate Standard of Living, and on the Right to Non-discrimination in this Context, Miloon Kothari: Mission to Canada*. New York: UNHRC.

– 2012. *Report of the Special Rapporteur on the Right to Food, Olivier De Schutter: Mission to Canada*. New York: UNHRC.

Watkins, Mel, ed. 1977. *Dene Nation: The Colony Within*. Toronto: University of Toronto Press.

White, Graham. 2002. "Treaty Federalism in Northern Canada: Aboriginal-Government Land Claims Boards." *Publius-the Journal of Federalism* 32, no. 3 (July): 89–114.

2

Resource Revenue Allocation Strategies and Indigenous Community Sustainable Development

Thierry Rodon, Isabel Lemus-Lauzon, Jean-Marc Séguin,
and Stephan Schott

As pointed out in the introduction, in Canada, Indigenous peoples are increasingly able to capture a share of the revenue from resource development that takes place in their territory. The different court cases have made it clear that Indigenous peoples need to be consulted and that, in some cases, their consent might be necessary (Papillon and Rodon 2017). The development of non-renewable resources in Northern Canada has led many mining, oil extraction, and hydroelectric companies to negotiate resource revenue-sharing agreements through Impact and Benefit Agreements (IBA) or similar arrangements with Indigenous communities (Prno et al. 2010). The negotiation of these agreements greatly reduces uncertainties over the legality and the legitimacy of a given project, considering the legal context surrounding resource development and Indigenous land rights in Canada (Papillon and Rodon 2017; Bradshaw and McElroy 2014). These agreements are also seen by many project proponents as a way to facilitate the acceptability of development projects on Indigenous lands since they establish guidelines for Indigenous employment, community development, and often include provisions for profit sharing.

Resource revenue can be shared in many ways, including arrangements for sharing profits, the value of production, or production rates. There can also be fixed payments that will allow communities to cover their IBA administration expenses, such as hiring an IBA administrator or "enforcer" (Gibson and O'Faircheallaigh 2010). The money from profit-sharing revenues is generally paid to the organization and/or community that signed the agreement with the company. In turn, this organization is responsible for distributing the sums paid by the company. There are no uniform ways to distribute these sums. For example, in some cases, revenue may be distributed directly to individuals, while in others, it may be used for community projects or invested in resource trust funds.

In fact, there are only a few studies about how the different modes of allocating revenues and profit shares affect the communities (Altman and Levitus 1999; NWT 1989; O'Faircheallaigh 2010, 2012). However, how these benefits are shared can have an important impact on whether they promote well-being or not: What are the social and economic impacts of the various modes of distribution and investment? Are there modes that can better mitigate the impacts of a resource development on Indigenous communities? Which modes of distribution or investment best benefit communities? Which models provide the most intragenerational and intergenerational equity and sustainability? Balancing the needs of households and communities today with the needs and lost opportunities of future generations is definitively a challenging task.

This chapter is divided into three sections. First, we will review the literature of the different models for distributing resource revenues and their positive and negative impacts for communities. We will also assess their level of sustainability according to the criteria developed by Gibson (2006). In the second section, we present the results of a survey and follow-up interviews conducted with representatives from twenty-one Canadian First Nations that have signed a profit-sharing agreement. Finally, we present three case studies of resource revenue allocation: Raglan Mine in Nunavik, Quebec, Red Dog Mine in Alaska, and Musselwhite Mine in Ontario.

CHARACTERISTICS OF RESOURCE REVENUE ALLOCATION

Non-renewable resources are a depletable asset by definition. Any exploitation diminishes availability of the resource to future generations unless the returns or rents[1] from its extraction are reinvested in other assets that benefit future generations. Sustainability principles, therefore, should be at least based on Hartwick's rule (Hartwick 1977,

1978), which states that, at every point in time, the total rent arising in the resource extraction industry must be saved and invested in reproducible capital. This will at least guarantee that continuing consumption levels for future generations benefit from the returns of the transformed levels of capital, and that weak sustainability criteria are satisfied (see also Chapter 3 by Schott et al. on applications to the mining industry in the Canadian Eastern Subarctic). For this reason, some economists (Markus Herrmann in Northern Sustainable Development Research Chair 2013, 14–15) advise that resource revenue should be invested in trust funds, and that only the interest generated should be spent. This would ensure that the capital is not used up and can benefit future generations. In this section, drawing on the existing literature, we will discuss the different resource revenue allocation models and their negative and positive outcomes for Indigenous communities. The review presented here is not exhaustive but examines the four more widely used models: direct payments to individuals, trust funds, social programs and public services, and infrastructure investments.

Direct Payments to Individuals

The Direct Distribution Model (DDM) transfers resource revenues directly to individuals. Different operational definitions exist such as lump-sum payments, conditional cash transfers, and dividends (see Segal 2012b for an in-depth discussion of DDM). This system is likely the simplest allocation model to implement. It presents the potential to reduce social and revenue inequalities, and therefore could provide greater social and income security to the most vulnerable (Gupta et al. 2014; G. Standing 2008; A. Standing 2014; DFID 2011; Segal 2011, 2012b; Gibson and O'Faircheallaigh 2010; O'Faircheallaigh 2010, 2012; Moss 2010; Soares et al. 2007; Cornell et al. 2007; Schubert and Huijbregts 2006; Morley and Coady 2003; Bunting and Trulove 1970). Moreover, the DDM provides beneficiaries with the freedom to decide personally how to spend the money, whether it be for immediate urgent needs or long-term investments (O'Faircheallaigh 2010, 2012; Moss 2010; Gibson and O'Faircheallaigh 2010; Cornell et al. 2007; Weinthal and Jones Luong 2006; Palley 2003). This model allows those citizens who live outside the region to benefit from the nation's success (Cornell et al. 2007). Direct payments could certainly contribute to reducing corruption and rent-seeking behaviour, leading to a more responsible and transparent governance of the funds (Gupta, Segura-Ubiergo, and Flores 2014; Segal 2012b; Moss 2010; Cornell et al. 2007; Weinthal and Jones Luong 2006; Palley 2003). Additionally, this model could have a positive impact on political participation and stronger institutions since it would increase the interest of citizens in the management of the resource revenues, as they have a personal stake in the profit (Gupta et al. 2014; A. Standing 2014; Devarajan et al. 2011; Gillies 2010; Moss 2010; Weinthal and Jones Luong 2006; Mahon 2005; Birdsall and Subramanian 2004; Palley 2003; Sala-i-Martin and Subramanian 2003).

If direct payments offer many benefits, they can also provoke tensions between beneficiaries and non-beneficiaries (O'Faircheallaigh 2013; Gibson and O'Faircheallaigh 2010), besides fostering disputes over the borders of tribal territories (O'Faircheallaigh 2016) and Indigenous citizenship (Cornell et al. 2007). Moreover, direct payments are associated with a disincentive to work and with destructive behaviours such as substance abuse and domestic violence (Hill 2012; Cornell et al. 2007; Bunting and Trulove 1970). In addition, this distribution system assumes individuals will, to some extent, invest revenues in the form of savings or investments for their children or grandchildren. Given the limited incomes of Indigenous households and the generally low saving rates among them (O'Faircheallaigh 2010, 2012), this type of redistribution will not guarantee that assets and savings for the future will be generated with the result of declining capital for the region or community.

Trust Funds

Trust funds are another popular means to manage natural resource revenues. Their definitions and legal structures greatly vary. For example, *stabilization funds* essentially reduce potential impacts due to their lack of volatility and help to stabilize the pace of government spending, while *trusts* are saving funds that transfer a portion of the revenues to future generations (Baena et al. 2012). A common characteristic is that trust funds are (in theory) managed by an independent institution and are separated from the current operations (Baena et al. 2012; World Bank 2011). ·

It is quite clear from the literature that trust funds can represent an opportunity to accumulate an asset base from mining payments.

In comparison to direct payments, this strategy has the potential to generate stable, long-term, and sustainable revenue for the future, thus promoting intergenerational equity (Söderholm and Svahn 2015; Tsani 2013; Baena et al. 2012; O'Faircheallaigh 2010, 2012; Gibson and O'Faircheallaigh 2010; Fischer 2007; Poole et al. 1992).

Critics argue that the actual resource revenue that is invested in trust funds is therefore unavailable to address other needs of the community, and neither are they available for direct distribution (O'Faircheallaigh 2010, 2012; Gibson and O'Faircheallaigh 2010; Cornell et al. 2007). Thus, for communities with rapid growth, and those with significant poverty, in the short term, saving less and spending more in the present time could be a better strategy for current, and even future, generations (Segal 2012a).

Social Programs and Public Services

Resource development often exists in areas where government may be weak or simply nonexistent. When public services are inadequate, governments may promote the spending of resource revenue to fill gaps in public service or to extend the scope of services being provided. For instance, the investments might be made in cultural programs, education, and health care. On the positive side, this strategy allows Indigenous peoples to have a greater control over the priorities and decisions about these services (O'Faircheallaigh 2004). Such a use of funds may also offer them opportunities to develop and improve their administrative skills and their capacity for self-governance (Cornell 2006; O'Faircheallaigh 2004). The latter could benefit future generations by making them more autonomous and giving them more control of their own economic, social, and cultural direction. By choosing to invest in public services, the community could experience social returns that are greater than just financial (Segal 2012b; Collier et al. 2009).

However, the investment of mining revenue in essential public services (health, education, etc.) or community programs, such elder and hunter support, can result in the disengagement of the central government that should be the agency responsible for providing these services in the first place. This risks establishing a dependency of the institutions on development revenues (Gibson and O'Faircheallaigh 2010; O'Faircheallaigh 2004, 2010, 2012, 2017).

Infrastructure Investments

Investments are also made in community infrastructure. The advantage is that it can benefit the community at large, making assets available for longer time periods if properly maintained. The downside is that infrastructure needs to be maintained and that requires services. So, it is important that the cost of upkeep is built into the finance model, perhaps by continuous support of the extractive industry during the life cycle of their operations. It then could have spinoff effects, as new economic operations and jobs might be created through the infrastructure construction and operation.

*

The four revenue distribution and investment models presented above—direct payments to individuals (short-term private handouts); trust funds (long-term investments); investments in social programs and public services; and infrastructure investments (short-, mid- and long-term public investments)—all present benefits and drawbacks for Indigenous communities.

ASSESSING THE REVENUE ALLOCATION STRATEGIES ACCORDING TO SUSTAINABILITY CRITERIA

In order to go further in assessing the merit of each strategy, we use specific sustainability criteria. Gibson (2006) has developed an ambitious framework to assess sustainability in the context of resource development, but in the case of resource revenue allocation, the main criteria are intragenerational and intergenerational equity. Intragenerational equity measures the equity of revenue distribution or investment in the region, community, and among different groups; intergenerational equity refers to the equity between different generations and the maintenance of overall asset values. Some authors have insisted on the interconnection of both forms of equity (Weiss 1992; Godden 2009), but others (Dobson 1999; Gosseries 2008) argue that these two kinds of equities are often in conflict. Since mines are depletable assets, intergenerational equity is a crucial issue in this sector—although the lack of intragenerational equity can be a major source of social conflict within Indigenous communities, making it difficult or impossible to address issues of intergenerational equity. Finally, as we will see later, the lack of intergenerational equity is clearly a major issue for some of this study's respondents.

Direct payments to individuals score low on intergenerational equity but can be high on intragenerational equity since the cash is usually distributed equally amongst community members near the development. Furthermore, if cash transfers can address urgent needs, they cannot secure lasting benefits. Thus, this model fails to address other causes of poverty, such as access to education and civil rights, and is not sufficient to ensure economic well-being in the long run, therefore lacking intergenerational equity (Gupta et al. 2014; Cornell et al. 2007; Bunting and Trulove 1970; Hartwick 1977, 1978).

Regarding investments in public services and community programs, the volatility and unpredictability of mineral markets make them vulnerable to the mining boom and bust cycles. One can argue that funds should preferably address the negative outcomes of mining and create opportunities for improving the economic situation. The funds could therefore act as automatic stabilizers that invest into programs, services, and spinoff activities in boom periods, and that later can make up for some of the downturns in bust periods. Programs and services can address issues of intragenerational equity through policies designed to support neglected or undersupplied areas (e.g., through income support, advancement of services in health, education, and so on, and additional retirement support, to name a few).

Investment in infrastructure seems like a good way to use resource revenue, although it is highly dependent on the type of infrastructure that is built. In general, Indigenous communities use the money for collective infrastructure such as gathering places, arenas, or swimming pools, and sometimes social housing. These investments help to improve well-being and social cohesion, two elements that are often threatened by resource development. However, in the case of infrastructure that needs constant investment for maintenance and upkeep in order to function, for example, swimming pools, their usefulness often doesn't go beyond the resource development lifetime. As for services, it also allows governments to avoid investing, especially in the case of social housing. The latter can create a false sense of security for the communities and regions, and can result in loss of support such as transfer payments from central governments. This type of investment can contribute both to intra- and intergenerational equity, but only if the infrastructure is well designed and maintained, which is often not the case.

Table 2.1
Assessing the different modes of distribution and investment

	Individual payment	Programs and services	Infrastructure	Trust funds
Advantages	Individual Choices Some aspect of Intragenerational Equity Contribute to income and food security	Local control on program and services Development of human capital and capacity building	Address Collective needs	Intergenerational equity
Disadvantages	Negative effects (disincentive to work, substance abuse) Lack of interregional equity Lack of long-term improvement	Other levels of Government tends to lower their transfer	High cost Maintenance cost can be prohibitive	Doesn't deal with urgent needs
Intragenerational equity	High	medium	medium	Low
Intergenerational equity	Low	Low	High	High
Sustainability	Low	Low	Low to High	High

Trust funds are, in essence, geared towards intragenerational equity when the capital is invested and only the interest is spent. However, investments, especially in the marketplace, can be risky; during the last financial crisis, the trust funds of many communities lost a good deal of their value. Trust funds, therefore, must be treated as long-term investments that will not deliver returns for some time. They cannot be relied on for constant streams of revenues. The governance of the trust funds is also a key element. In most Indigenous communities that were surveyed, the trust funds were administered by the political authority, which makes these funds susceptible to political motivation, and therefore risky investments. It is also difficult to convince community authorities to establish such a fund since their community is often in immediate need of returns to relieve poverty. A future vision for how these funds will contribute to future investments and expenditures is essential in persuading both the public and the community authorities.

Table 2.1 summarizes the advantages and disadvantages of each allocation mode and assesses their relative sustainability. In many cases, Indigenous communities use a mix of distribution or investment strategies, which allows them to address urgent needs, such as poverty relief, housing, and food security, but also lets them keep some funds for medium- and long-term needs. Many Indigenous communities face both urgent immediate needs and long-term objectives such as human capital development, economic diversification, road infrastructure improvements, and increased energy self-sufficiency (based on alternative energies that can displace dependence on diesel generators).

The success of revenue allocation appears to be associated with collective investments (short-term or long-term) characterized by a transparent and diversified investment strategy. Most importantly, strong and transparent institutions are key to maximizing the use of resource revenue and mitigating negative impacts. However, the key element, as shown in different research, is the presence of a vision of community development that allows for choosing the best strategies (Cornell and Kalt 2007; Rodon and Schott 2013; O'Faircheallaigh 2016).

SURVEY ON REVENUE ALLOCATION STRATEGIES OF CANADIAN IBA SIGNATORIES

Data on modes of allocating resource revenues and profit shares was collected through phone surveys targeting Indigenous communities in Canada that had signed a profit-sharing agreement. This component aimed to provide a community perspective on the various models used to distribute revenues from resource development, and the positive and negative impacts of each method. To do so, we developed an online questionnaire that addressed four main themes: the modes of distribution or investment chosen, the decision-making process, the administration and management of the funds, and the positive and negative impacts of the allocation mode.

In 2017, there were over 400 active agreements in Canada between a mining company and an Indigenous community, including memorandums of understanding, surface lease agreements, participation agreements, socio-economic agreements, and Impact and Benefit Agreements (Natural Resources Canada 2017). In 2015, at the time the survey was conducted, forty-two Indigenous communities and/or regional organizations had an active IBA and all of them were contacted to participate in this study.[2] IBA coordinators or negotiators were approached to fill out the survey, but other community members with a "position," such as social and health program coordinators, were also included to get a diverse perspective on the social impacts of revenues. Out of these forty-two communities, twenty-one representatives agreed to participate in the survey, but all of those did not necessarily answer all questions.[3]

This mixed success in recruiting and engaging community participants in the project could be explained by the sensitive, and the confidential and private nature of these IBA agreements. Although our questions did not address confidential aspects, participants may have felt uncomfortable answering and unsure as to what they could and could not say. Additionally, the difficulties could relate to the methodology itself, which did not permit face-to-face encounters. This highlights the importance of research relationships, particularly when working with Indigenous communities. Despite these obstacles, the results provide an overview of the challenges and outcomes of revenue distribution and investment for Indigenous communities. The respondents had the option to keep their community anonymous. A total of eleven participants chose to identify their community—of these, eight were based in Ontario, one in the Yukon, one in British Columbia, and one in New Brunswick.

DISTRIBUTION AND INVESTMENT MODES

The allocation strategies that were described by community and organization representatives fell under four main categories: community programs and services, resource or trust funds, infrastructure, and direct payments (Figures 2.1 and 2.2). The most common mode of distribution or investment was funding community services and projects, which included education and training opportunities, employment and business development initiatives, and cultural and social programming, such as Elder and youth initiatives and community events. Housing and environmental monitoring programs were also funded. Moreover, one community even used revenue to cover medical care fees that were not covered by the public regime.

Resource or trust funds were the second most common option chosen by respondents, with a large range in operating modes of these funds (administration, spending, and investment types). Revenues were also used for community infrastructure, such as arenas, road maintenance, and the improvement of internet networks. Direct payments to individuals or households were the least common mode of distribution, but were made in a few cases.

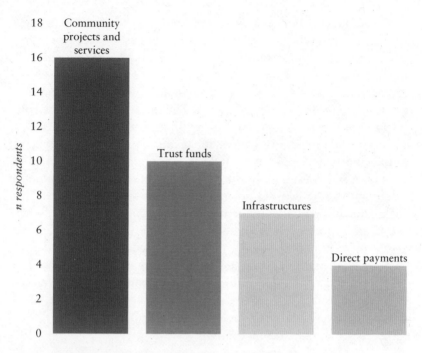

Figure 2.1 Models of resource revenue distribution adopted by Canadian Indigenous communities

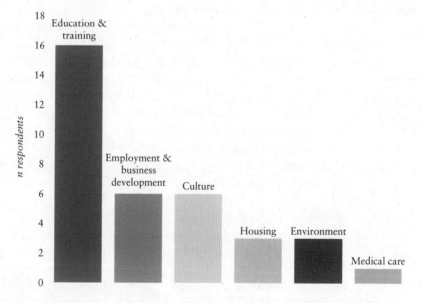

Figure 2.2 Main categories of community services and projects funded through resource revenues

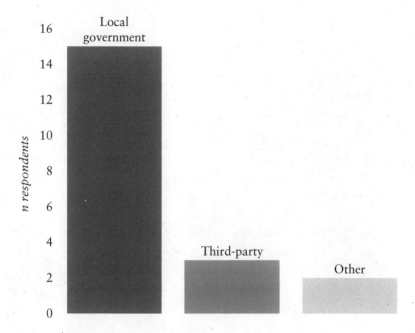

Figure 2.3 Resource revenues management

Decision-Making Process and Funds Management

During the process of choosing a mode of distribution or investment, most communities and organizations involved legal and financial advisors. The Assembly of First Nations was also involved in one case and facilitated information sharing between communities. Standard practice was then to consult community members and local organizations on which distribution mode to choose and what services and projects to prioritize. Specific groups, such as hunters, trappers, and gatherers' associations and Elder committees, were targeted for the consultation process, in order to bring forward their respective concerns, needs, and expectations. However, one informant mentioned that women's associations and youth organizations were not consulted in the decision-making process regarding the distribution mode and the projects that would be prioritized. Nevertheless, it was expected that revenue distribution would positively impact all social groups in the community.

The fund management and distribution were predominantly done through the local government (Figure 2.3), which in most cases was the band council executive. A minority used a third party such as a

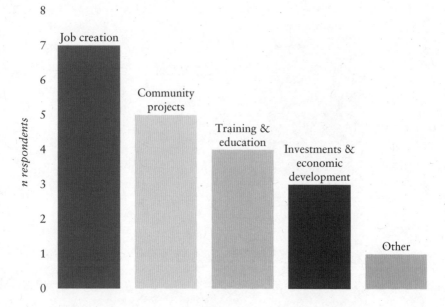

Figure 2.4 Positive impacts of resource revenues distribution on economic development

trustee board, a corporation, or an association. The Assembly of First Nations also acted as fund manager in one case.

Economic and Social Impacts of the Allocation Modes

When asked about the positive impacts of revenue distribution or investment on community economic development, respondents predominantly mentioned employment (Figure 2.4). Thus, in addition to direct employment, mining activities can support the parallel contracting of local companies in fields such as catering, housekeeping, and security services. Moreover, the use of revenue for community projects, services, and infrastructure can also lead to the creation of local jobs, which, considering the lack of work opportunities in these often remote locations, presents the potential to galvanize the local economy. However, one informant pointed out that skilled jobs (e.g., mechanics) are mostly awarded to workers from outside the community.

The importance of mining revenues as additional funding for community projects and opportunities for education and training development was also stressed. As specified by a respondent, "the funds used to improve programs, and in particular the post-secondary program, had a significant impact. Each community was able to fund ten to twenty more students for school." More education and training opportunities could ultimately result in an increase of the local qualified workforce.

One community representative also mentioned the investments in a hydropower plant as a major economic impact of resource revenue, as this alternative energy source will allow for reducing the energy costs for community members and could facilitate creating new businesses, thus contributing to long-term sustainability through transforming a non-renewable asset into a renewable asset.

A recurring concern of several respondents referred to the lack of awareness, transparency, and accountability in the distribution process, sometimes causing distrust and a sense of unfairness and inequality. Thus, about a third of respondents thought that community members did not all benefit equally from revenue, and half of all respondents said that not all communities in a region benefit in the same way from resource development, clearly denoting an issue of intragenerational equity. Indeed, the revenue sharing formula between affected communities is often based on the level of impacts measured by the proximity to the mine and not on the needs of the community. Another concern referred to off-reserve members who did not benefit from revenue distribution except in the case of direct payments to individuals that are distributed to all members of the community in which they live. One respondent commented that revenue distribution failed to address urgent needs, and that was why more direct payments were needed.

Direct payments were also criticized, however, by a respondent who mentioned that they had little to no economic impact on the community. This was echoed by another respondent who explained the difficulty in retaining the money from payments locally, as it was spent on goods fabricated elsewhere. One respondent criticized the trust fund strategy, stating that the investments were not productive and that administrative costs were too high. Moreover, the money was not available to address community priorities and urgent needs.

Respondents were less vocal about the impacts of revenue distribution on social community development. Positive impacts were mainly seen around improved social and health programming that had improved overall community well-being. Education and training opportunities were also said to have opened up new possibilities for community members. Negative effects of mining revenues were mentioned, such as the rise in alcohol and drug consumption, which would sometimes result from direct payments to individuals.

Discussion of Survey Results

The data collected amongst respondents show that resource revenue use and distribution is a complex endeavour without a clear direction. It seems evident that Indigenous households that are affected by mining projects are in need of both short-term direct transfers (to address intragenerational equity) and long-term investment into replacement assets for depleted non- renewable resources (to address intergenerational equity). In addition, almost all of the Indigenous communities need to provide more public infrastructure and services. Using resource revenue for the latter speeds up the process of acquiring important public goods, such as arenas and swimming pools, but also introduces additional costs for maintenance and service that can use up some of the future revenues. Without a solid revenue base (from property and income taxation) communities run the risk of dependence on the boom and bust cycles of the mining industry. Furthermore, many of the infrastructure and public goods should be provided by other levels of government. In the short- and medium-term, communities need to provide training and education for local workers and entrepreneurs to fully benefit from higher-level jobs and business opportunities. This should be the joint responsibility of central governments, communities, and extractive industries.

The governance, management, and administration of resource revenues are another major challenge. For example, in the vast majority of the communities surveyed, the resource revenues are managed directly by the political authority, which is, in most cases, the band council. This goes against the practice of trust funds managed by an arm's-length body that is immune to political influence and political cycles. An arm's-length body would insulate fund management from the political short-term needs and would thus by more inclined to tend to foster intergenerational equity. However, it is reassuring that the majority of resource revenues are invested into community programs and services and trust funds, and only a few communities are using direct payment. On the other hand, this could also be evidence that central governments are failing to provide much needed services, public goods, and infrastructure in due time.

CASE STUDIES

In order to complete the phone surveys and get more in-depth information on resource revenue allocation, three case studies were carried out: 1) the Raglan Mine and the communities of Salluit and Kangiqsujuaq in Nunavik, 2) the Red Dog Mine and the NANA Corporation in Alaska, and 3) the Musselwhite Mine and the community of Kingfisher Lake, Northern Ontario. The three case studies, based on semi-directed interviews and fieldwork (direct observation), present different institutional and governance contexts and revenue allocation strategies.

In the case of the Raglan Mine, data was collected through forty-six semi-structured interviews with the residents of the two closest communities (Salluit and Kangiqsujuaq), local and regional leadership, and members of Nunavik's health and social services (Blais 2015). For the Red Dog Mine, twelve semi-structured interviews were conducted with officials from NANA Corporation, the department heads and staff in Anchorage and Kotzebue, and with members of the social and education services in Kotzebue. Finally, in the case of the Musselwhite Mine, eight semi-structured interviews were conducted with local leaders and members of the education, health, and social service sectors.

Raglan Mine, Nunavik, Quebec

The Raglan Mine is located in Nunavik in the province of Quebec, a territory covered by the James Bay and Northern Quebec Agreement (JBNQA), the first modern treaty signed in 1975. It is located on category III land as defined in the JBNQA as public land. In 1995, the Société Minière Raglan du Québec Limitée (a Falconbridge subsidiary) signed the first Canadian IBA with the Makivik Corporation, the organization representing the Inuit of Nunavik, and with the two neighbouring villages of Salluit and Kangiqsujuaq, and their respective landholding corporations. The IBA aimed to reduce environmental impacts and define the parties' responsibilities and benefits (Benoît 2004). Among other things, it assigns 4.5 per cent of the mine's profits distributed according to this formula: 45 per cent for Salluit, 30 per cent for Kangiqsujuaq, and 25 per cent for the Nunavik region (Kativik Regional Government 2007). This provision has led to significant financial benefits. From 1997 to 2011, Raglan has paid out over \$100 million in money transfers or profit sharing to Nunavik (George 2012).

In the case of the first village, Salluit, the decision to allocate the revenue from Raglan is mostly taken by referendum. Over the years, they have generally opted for individual cash distribution. In the second village, Kangiqsujuaq, the Raglan payments are managed by a board of directors and have mostly been invested in community infrastructure and projects (SNC-Lavalin 2015). The 25 per cent allocated to the Nunavik region has been invested in the Raglan Trust, with an important proportion being redistributed to each village in Nunavik every year so they can decide for themselves how to allocate the revenue in their communities. Sometimes every adult individual in Nunavik has received a small amount in cash or in food coupons from this fund, but the amounts are much smaller than in Salluit or Kangiqsujuaq.

There are a few reports on the negative social and economic impact of large individual payment in the first village, Salluit, especially in 2008, where record amounts of cash were distributed, with some families receiving up to \$50,000 (Nunatsiaq News 2009). With regards to how the money is shared, some participants indicated not wanting individuals to receive big cheques, as was the case in the past. In 2007-08, big cheques were given out and the observed effects were negative. Some Elders are still afraid and leave town in July when the money normally arrives. A few participants indicated being against individual cheques mostly because it enhances alcohol and drug abuse. Vandalism and violence also become important issues. In addition, individuals went into debt to pay the tax on these revenues. Another negative side effect would be the increase of prices in the stores when the cheques are distributed. However, it was also recognized that the individual cheques were a way for community members to buy equipment (SNC-Lavalin 2015).

In addition, this influx of money resulted in staff shortages in most local services with people not showing up for work (George 2008). After this episode, it was suggested by Makivik that the payment would be made with food coupons in order to reduce the social impacts (Rogers 2010).

In contrast, the community investment strategy chosen by the second village, Kangiqsujuaq, seems to have been more effective in terms of positive economic and social impacts: "[They] expressed their pride that the profit sharing is spent collectively for the benefit of the community. In this way, it is felt that the profit sharing helps the community (swimming pool, gym, renovations of the arena, emergency boat, garage, Elder's home, etc.). A minority of respondents (two) would prefer individual cheques to the collective sharing of the money" (SNC-Lavalin 2015).

The Raglan case allows us to compare two different modes of revenue allocation and their social impacts. The first village has usually decided to provide individual payments to each adult and, even if this contributes to improving living conditions by allowing people to buy materials (household appliances, vehicles, and boats) that they could not afford otherwise, it doesn't contribute to the economic development of the community, since all these products are bought in the South. However, direct allocations can also cause negative behaviours, such as absenteeism, substance abuse, and domestic violence. The second village has mostly chosen to invest the revenue in community projects and has thus contributed to strengthening the community cohesion. However, in both cases, no trust funds have been established and therefore there is no plan for a post-mining future. It should be noted that the first village is now increasing the amount invested in community programs and infrastructure.

Red Dog Mine, Alaska

In Alaska, as a result of the signature of the Alaska Native Claims Settlement Act (ANCSA) in 1971, the NANA Corporation was entitled to approximately 2.2 million acres of surface and subsurface land, including the rich Red Dog zinc deposit. NANA, who represents 14 000 Iñupiat, signed a joint-venture agreement with the mining company, Teck, in 1982, for the development of the Red Dog Mine.

NANA started receiving 4.5 per cent of net smelter returns until 2007, after which NANA and Teck started sharing the net proceeds of the mine. Profit sharing began at 25 per cent for NANA and increases every five years until the profits are equally shared between the two parties. In 2017, NANA received 35 per cent of net proceeds (Lasley 2019). NANA has retained approximately $750 million of the $2 billion it has received from net proceeds from the mine and has distributed the rest, approximately $1.25 million, to other regions and at-large shareholders, according to the sharing provisions of the ANCSA (Lasley 2019).

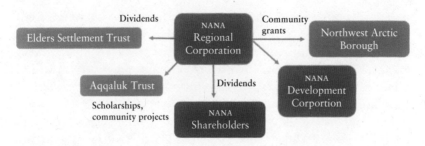

Figure 2.5 Dividend sharing for the Red Dog Mine

With the money the corporation retained, NANA has opted for a mixed strategy (see Figure 2.5), divided between individual payments (46 per cent of the dividends) and investment in trust funds for Elders, scholarships, and community projects (54 per cent of the dividends).

The allocation system raises some questions, since 50 per cent of the shareholders receiving individual payments live outside the region and therefore, these payments don't benefit the local communities. According to Wayne Westlake, NANA President and CEO: "We have shareholders all over the country. They have economic opportunities, they have a water and sewer system, we have communities that don't have that yet."

The question of the sustainability of this allocation formula is also especially of concern for the Elders. According to Westlake: "There's a number of Elders [...] they come to me and say: one day that mine is going to be gone, and we're not seeing enough being done within our communities to sustain them from an economic standpoint or even from an infrastructure standpoint once that mine is gone. What's going to happen? [...] We haven't come up with a really good answer..."

The NANA Corporation has chosen a diversified strategy for the Red Dog Mine resource revenue allocation. Nearly 50 per cent is distributed as dividends to shareholders, while the rest is put in trusts or invested in community development. This provides revenue to shareholders, but since 50 per cent of them live outside the region, the economic leakage is very important. Thus, even with some trust funds in place, the leaders of the NANA Corporation don't have a plan for a future without a mine.

Musselwhite Mine, Ontario

The Musselwhite gold mine, owned by Goldcorp, opened in 1997 and is located about 480 kilometres north of Thunder Bay. An all-weather road connects the mine to Pickle Lake, a community located 200 kilometres south (Musselwhite Environmental Working Committee 2012). The Musselwhite Mine has a revenue-sharing agreement with four First Nations communities and two First Nations Councils:

- North Caribou Lake and Cat Lake, which are represented by the planning board of the Windigo First Nations Council
- Wunnumin Lake and Kingfisher Lake, which are represented by the planning board of the Shibogama First Nations Council

Negotiations with the Kingfisher Lake First Nation started in the 1990s. These involved ten Elders from Kingfisher Lake, a tribal council worker, and representatives from other communities. The original agreement was signed in 1992 and amended in 1996 (Musselwhite Environmental Working Committee 2012). It was then renewed in 2001 and a portion of it was modified five years ago.

Kingfisher Lake First Nation receives royalties from the Musselwhite Agreement. According to the interviews conducted with reserve members, the community gets a smaller amount of money than the other communities, as the mine is not located on their traditional land but has impacts on their river system. However, the amount is not public. These royalties are the main source of resource revenue.

The money is managed by the band council. At the end of the fiscal year, the council decides about the Musselwhite Mine resource revenue allocation. The population is not consulted on that matter, but is informed about it. According to the survey results presented earlier, we can see that this way of making decisions without community consultations is unusual compared to other communities. The royalties are considered as funding for the entire community; therefore, there are no direct payments to individuals or families, and the funds are allocated as follows:

- Community projects (40 per cent)
- Economic development (20 per cent)
- Education (20 per cent)
- Infrastructure (20 per cent)

In the interviews conducted in the community, respondents were positive about the outcomes of the use of the revenue, but some have expressed their worries regarding the sustainability of mining for the territory. Local decision makers also expressed concerns regarding both the fluctuating prices of the ore, as well as the uncertainty that a dependency of community development on the mining sector creates.

In case of Kingfisher Lake, the income generated by the revenue sharing agreement is entirely managed by the local government, without any third-party involvement or consultation with the community members at large. These revenues are benefitting the community through community services and projects, economic development, education, and infrastructure. As the money is considered as a reserve fund, it is not dedicated to one particular project and thus is used to complete investments where they are needed. Therefore, we cannot measure one distinctive impact that would be due to one specific project. Although the life expectancy of the mine has been extended several times, the sustainability of this revenue over the long term is questionable.

*

The three cases presented show that the governance, distribution, and investment of resource revenues are key factors in building a better and more sustainable development strategy. Referendums seems to be the more democratic way of deciding but, in the case of revenue allocation decision, it tends to privilege individual benefits over collective benefits, limiting the positive economic and social impacts of mining revenues. One of the Nunavik villages, the NANA Corporation, and the Kingfisher Lake Band Council have been able to put more emphasis on collective investments, but none of the three cases presented have developed long-term strategies. They are still dependent on the recurrence of mining royalties, having therefore a very limited long-term sustainability.

CONCLUSION

In trying to answer the central question of this book—Is it possible for extractive resource development to improve the sustainability of communities in the North?—we have seen that the distribution models could contribute to improved community sustainability, but that the social circumstances often constrained their choices. Indeed, many

Indigenous communities are contending with both urgent immediate needs and plans for long-term development. As outlined in the chapter above, some communities use a combination of distribution and investment strategies, which allow them to address urgent needs, such as poverty relief, housing, and food security, while also enabling them to preserve some funds for medium- and long-term needs. However, communities need to determine the proportion of money they should distribute directly, to invest in trust funds, and to use these funds for education, training, business development, public goods, and infrastructure. This is clearly a political process, and in order to make this choice, communities need to develop a vision that will allow them to weigh preferences and initiate public debate. However, referendums tend to reflect individual choices and do not allow for a real deliberative process.

This being said, so long as many Canadian Indigenous communities, especially those in the North, are faced with poverty, difficult social circumstances, and poor access to public services—in addition to issues of housing shortages and lack of clean drinking water—it will be difficult to develop a vision given the magnitude of urgent needs to address.

We have also seen that in most of the communities surveyed, trust funds are directly managed by political authorities, thus providing no insulation from political influence. Nonetheless, it is difficult, especially for smaller communities, to set up independent boards to manage trust funds. One solution would be to create intercommunity trust funds or trust funds managed by independent authorities who would include a board of directors consisting of representatives from the community and other responsible authorities.

The survey also shows that some Indigenous communities are investing in infrastructure, and environmental and social programs, including medical care and social housing. All these programs should normally be funded by the federal and provincial/territorial governments, and this attests to the chronic lack of investment in Indigenous communities in Canada. The mining companies could also contribute more to community infrastructure, housing, and maintenance, and should cover all environmental costs created by their mining operations.

Finally, whatever the choice of distribution or investment mode, in order to ensure sustainable development, communities need to ensure that the lost natural capital (including non-renewable resource depletion, social capital in the form of community vitality, traditional knowledge transfer and retention of traditional languages, and the environmental damage created by the operations) will be replaced for future generations. This can be done in the form of long- term financial investments, investment in renewable resources (e.g., energy, food security), human capital, Elder and hunter support programs, or in long-term physical assets that contribute to community development. But here again, this ideal will be easier to attain once the basic needs of Canadian Indigenous communities are met first. Once basic needs are covered, Indigenous peoples will be better positioned to devise sustainable development visions and strategies, and in turn, to negotiate more favourable revenue sharing agreements to be used for investments in alternative assets and rent distributions that ensure sustainable livelihoods for many generations to come. However, mining has the potential to create a false sense of permanent economic opportunity, especially for the generations that have grown up with mining and who perceive it as an everlasting industry.

ACKNOWLEDGEMENTS

The authors wish to acknowledge the contribution of Juliette Bastide, Remy Darit Chhem, and Ève Harbour- Marsan who provided expert research assistance for this project. We also want to thank the members of Indigenous organizations and communities who agreed to share their experiences about Impact and Benefit Agreements. However, the content, as well as any factual errors, are the sole responsibility of the authors. This research was funded by the Resources and Sustainable Development in the Arctic network (ResDA) and by Mining Encounters and Indigenous Sustainable Livelihoods (MinErAL), a SSHRC (grant number 890-2012- 0111).

NOTES

1 The rent is the difference between the market price for an additional unit of the resource and the cost of extracting it.
2 In 2017, the number of active Impact and Benefit Agreements in Canada rose to fifty.
3 In two cases, two representatives from the same community or regional organization completed the survey, meaning that eighteen communities/ organizations are represented in total.

REFERENCES

Altman, Jon C., and Robert Levitus. 1999. "The Allocation and Management of Royalties Under the Aboriginal Land Rights (Northern Territory) Act: Options for Reform." CAEPR Discussion Paper 191. Canberra, Australia: Centre for Aboriginal Economic Policy Research.

Arnold, Catherine, Tim Conway, and Matthew Greenslade. 2011. *Cash Transfers Evidence Paper*. London, UK: Department for International Development [DFID].

Baena, César, Benoît Sévi, and Allan Warrack. 2012. "Funds From Non-Renewable Energy Resources: Policy Lessons from Alaska and Alberta." *Energy Policy* 51: 569–677.

Benoît, Catherine. 2004. "L'entente Raglan: Outil efficace pour favoriser la formation et l'emploi Inuit? Évaluation et documentation de la situation de l'emploi des Inuit à la mine Raglan, au Nunavik, dans le cadre de l'entente sur les impacts et bénéfices." Master's thesis. Université du Québec à Montréal.

Birdsall, Nancy, and Arvind Subramanian. 2004. "Saving Iraq From Its Oil." *Foreign Affairs* 83, no. 4 (July): 31–65.

Blais, Jonathan. 2015. "Les impacts sociaux de la mine Raglan auprès des communautés inuit de Salluit et de Kangiqsujuaq." Master's thesis. Université Laval: Quebec City.

Bradshaw, Ben, and Caitlin McElroy. 2014. "Company-Community Agreements in the Mining Sector." In *Critical Studies on Corporate Responsibility, Governance and Sustainability*. Vol. 7, *Socially Responsible Investment in the 21st Century: Does it Make a Difference for Society?*, edited by Céline Louche and Tessa Hebb, 173-193. Bingley, UK: Emerald Group Publishing Limited. https://doi.org/10.1108/S2043-905920140000007007.

Bunting, D., and W.T. Trulove. 1970. "Some Experiences with Guaranteed Incomes and Lump Sum Payments: The Case of the Klamath Indians." Presentation at the 137th meeting of the American Association for the Advancement of Science. Chicago, IL, December 26, 1970.

Collier, Paul, Rick van der Ploeg, Michael Spence, and Anthony J. Venables. 2010. "Managing Resource Revenues in Developing Economies." *IMF Staff Papers* 57, no. 1 (January): 84–118.

Cornell, Stephen 2006. "What Makes First Nations Enterprises Successful? Lessons from the Harvard Project." *Joint Occasional Papers on Native Affairs (JOPNA)* 2006-01. Cambridge: The Harvard Project on American Indian Economic Development.

Cornell, Stephen, Miriam Jorgensen, Stephanie Carroll Rainie, Ian Record, Ryan Seelau, and Rachel R. Starks. 2007. "Per Capita Distributions of American Indian Tribal Revenues: A Preliminary Discussion of Policy Considerations." *Joint Occasional Papers on Native Affairs (JOPNA)*. 2008-02. Tucson and Cambridge: Udall Center for Studies in Public Policy and Harvard Project on American Indian Economic Development.

Cornell, Stephen, and Joseph P. Kalt. 2007. "Two Approaches to the Development of Native Nations: One Works, the Other Doesn't." In *Rebuilding Native Nations: Strategies for Governance and Development*, edited by Miriam Jorgensen, 3–32. Tucson: University of Arizona Press.

Devarajan, Shantayanan, Hélène Ehrhart, Tuan M. Le, and Gaël Raballand. 2011. "Direct Redistribution, Taxation, and Accountability in Oil-Rich Economies: A Proposal." Working Paper 281. Washington, DC: Center for Global Development.

Fischer, Carolyn. 2007. *International Experience with Benefit-Sharing Instruments for Extractive Resources*. Washington, DC: Resources for the Future.

George, Jane. 2008. ""It's like winning the lottery" Mining windfall sparks spending spree in two Nunavik towns." *Nunatsiaq News*. August 8, 2008. https://nunatsiaq.com/stories/article/Its_like_winning_the_lottery/

Gibson, Ginger, and Ciaran O'Faircheallaigh. 2010. *IBA Community Toolkit: Negotiation and Implementation of Impact and Benefit Agreements*. Summer 2015 ed. Toronto: The Walter & Duncan Gordon Foundation.

Gibson, Robert B. 2006. "Sustainability Assessment: Basic Components of a Practical Approach." *Impact Assessment and Project Appraisal* 24, no. 3 (September): 1–13.

Gillies, Alexandra. 2010. "Giving Money Away? The Politics of Direct Distribution in Resource-Rich States." Working Paper 231. Washington, DC: Center for Global Development.

Godden, Lee. 2009. "Towards a New Ethic in Australian Water Law and Policy." In *Climate Change on for Young & Old*, edited by Helen Sykes, 46–60. Albert Park, VIC: Future Leaders.

Gosseries, Axel. 2008. "Theories of Intergenerational Justice: A Synopsis." *Surveys and Perspectives Integrating Environment and Society* 1.1.

Gupta, Sanjeev, Alex Segura-Ubiergo, and Enrique Flores. 2014. "Direct Distribution of Resource Revenues: Worth Considering?" *IMF Staff Discussion Note* 14/05. Washington, DC: International Monetary Fund.

Hartwick, John M. 1977. "Intergenerational Equity and the Investing of Rents from Exhaustible Resources." *American Economic Review* 67, no. 5 (December): 972–974.

– 1978. "Substitution Among Exhaustible Resources and Intergenerational Equity." *The Review of Economic Studies* 45, no. 2 (June): 347–354.

Hill, James R., and Peter A. Groothuis. 2012. "The Effects of Per Capita Tribal Payments on the Fertility, Education, and Labor Force Participation of Tribal Members." *Business and Economic Research* 2, no. 2: 106–118.

IBA Research Network. 2012. *List of Known IBAS*. University of Guelph. http:// www.impactandbenefit.com/IBA_Database_List/.

Kativik Regional Administration. 2007. "Cadre Socioéconomique, Kuujjuaq: Service des ressources renouvelables, de l'environnement et de l'aménagement du territoire."

Lasley, Shane. 2019. "NANA – "Two worlds, one spirit."" *North of 60 Mining News*. March 1, 2019. https://www.miningnewsnorth.com/ story/2019/03/01/in-depth/nana-two-worlds-one- spirit/5630.html.

Mahon, James. 2005. "Liberal States and Fiscal Contracts: Aspects of the Political Economy of Public Finance." Paper presented at *Annual Meeting of the American Political Science Association, September 1, Washington, DC.*

Morley, Samuel, and David Coady. 2003. *From Social Assistance to Social Development: Targeted Education Subsidies in Developing Countries.* Washington, DC: Center for Global Development and the International Food Policy Research Institute.

Moss, Todd. 2011. "Oil to Cash: Fighting the Resource Curse through Cash Transfers." Working Paper 237. Washington, DC: Center for Global Development.

Natural Resources Canada. 2017. *The Atlas of Canada: Indigenous Mining Agreements.* Government of Canada. http://atlas.gc.ca/imaema/en/.

Northern Sustainable Development Research Chair. 2013. *Actes du colloque, 81ᵉ congrès de l'ACFAS, Université Laval: Développement minier et communautés inuit et cries: comment rendre le développement minier plus durable dans le Nord?, 8 mai 2013.* Québec, QC. https://www. chairedeveloppementnord.ulaval.ca/sites/chairedeveloppementnord.ula- val.ca/files/acfasactes_du_colloque.pdf .

Nunatsiaq News. 2009. "A big cash windfall, murder, mayhem and much more: Nunavik 2008 in review." January 6, 2009. https://nunatsiaq.com/ stories/article/A_big_cash_windfall_murder_mayhem_and_much_more/.

NWT (Northwest Territories Legislative Assembly). 1989. *Coping with the Cash: A Financial Review of Four Northern Land Claims Settlement.* Yellowknife.

O'Faircheallaigh, Ciaran. 2004. "Denying Citizens Their Rights? Indigenous People, Mining Payments and Service Provision." *Australian Journal of Public Administration* 63, no. 2 (June): 42–50.

– 2010. "Aboriginal Investment Funds in Australia." In *The Political Economy of Sovereign Wealth Funds,* edited by Xu Yi-chong and Gawdat Bahgat, 157–176. London: Palgrave Macmillan.

– 2012. "Curse or Opportunity? Mineral Revenues, Rent Seeking and Development in Aboriginal Australia." In *Community Futures, Legal Architecture: Foundations for Indigenous People in the Global Mining Boom,* edited by M. Langton and Judy Longbottom, 45–58. Abingdon and New York: Routledge.

– 2013. "Community Development Agreements in the Mining Industry: An Emerging Global Phenomenon." *Community Development* 44, no. 2 (May): 222–238.

– 2016. *Negotiations in the Indigenous World: Aboriginal Peoples and the Extractive Industry in Australia and Canada.* New York: Routledge.

– 2018. "Using Revenues from Indigenous Impact and Benefit Agreements: Building Theoretical Insights." *Canadian Journal of Development Studies/Revue canadienne d'études du développement* 39, no. 1 (January): 101-118. Doi: 10.1080/02255189.2017.1391068.

Palley, Thomas. I., 2003. *Combating the Natural Resource Curse with Citizen Revenue Distribution Funds: Oil and the Case of Iraq.* Foreign Policy in Focus (FPIF) Special Report. http://www.fpif. Org/papers/ordf2003.html.

Papillon, Martin, and Thierry Rodon. 2017. "Proponent-Indigenous Agreements and the Implementation of the Right to Free, Prior, and Informed Consent in Canada." *Environmental Impact Assessment Review* 62 (January): 216–224. https://doi. Org/10.1016/j.eiar.2016.06.009.

Poole, Graham. R., Michael Pretes, and Knud Sinding. 1992. "Managing Greenland's Mineral Revenues. A Trust Fund Approach." *Resources Policy* 18, no. 3: 191–204.

Prno, Jason, Ben Bradshaw, and Dianne Lapierre. 2010. "Impact and Benefit Agreements: Are They Working?" Paper presented at *Canadian Institute of Mining, Metallurgy and Petroleum Annual Conference, Vancouver, BC, January 1, 2010.* Canadian Institute of Mining, Metallurgy and Petroleum. http://www.impactandbenefit.com/UserFiles/Servers/Server_625664/File/IBA%20PDF/CIM%202010%20 Paper%20 -%20Prno,%20Bradshaw%20and%20Lapierre.pdf.

Rodon, Thierry, and Stephan Schott. 2014. "Towards a Sustainable Future for Nunavik." *Polar Record* 50, no. 3 (July): 260– 276. https://doi. org/10.1017/S0032247413000132.

Rogers, Sarah. 2010. "Raglan's Inuit Royalty Cash to Go Out in Coupons." *Nunatsiaq News*. April 16, 2010. https://nunatsiaq.com/stories/ article/98789_raglans_inuit_royalty_cash_to_go_out_in_coupons/.

Sala-i-Martin, Xavier, and Arvind Subramanian. 2003. "Addressing the Natural Resource Curse: An Illustration from Nigeria." Working Paper 9804. Cambridge, MA: National Bureau of Economic Research.

Schubert, Bernd, and Mayke Huijbregts. 2006. "The Malawi Social Cash Transfer Pilot Scheme, Preliminary Lessons Learned." Paper presented at *Social Protection Initiatives for Children, Women and Families: An Analysis of Recent Experiences, New York, October 30-31, 2006*. New York, NY: UNICEF.

Segal, Paul. 2011. "Resource Rents, Redistribution, and Halving Global Poverty: The Resource Dividend." *World Development* 39, no. 4: 475–489.

–2012a. *Fiscal Policy and Natural Resource Entitlements: Who Benefits from Mexican Oil?* Oxford: Oxford Institute for Energy Studies.

–2012b. "How to Spend It: Resource Wealth and the Distribution of Resource Rents." *Energy Policy* 51: 340–348.

SNC-Lavalin. 2015. *The Raglan Mine Property Beyond 2020 (Phases II and III): Continuation of Mining Operations East of Katinniq. Environmental and Social Impact Assessment: Consultation Report.* https://www.keqc-cqek.ca/wordpress/wp- content/ uploads/2017/03/625472_EI_Vol_2_fr_Partie3.pdf

Soares, Fabio Veras, Rafael Perez Ribas, and Rafael Guerreiro Osório. 2007. "Evaluating the Impact of Brazil's Bolsa Família: Cash Transfer Programmes in Comparative Perspective." *IPC Evaluation Note*. Brasilia: International Poverty Center.

Söderholm, Patrik, and Nanna Svahn. 2014. "Mining, Regional Development and Benefit- Sharing." *Mining and Sustainable Development*. Luleå: Luleå University of Technology.

Standing, Andre. 2014. "Ghana's Extractive Industries and Community Benefit Sharing: The Case for Cash Transfers." *Resources Policy* 40 (June): 74–82.

Standing, Guy. 2008. "How Cash Transfers Boost Work and Economic Security." DESA Working Paper 58. New York, NY: United Nations Department of Economic and Social Affairs.

Tsani, Stella. 2013. "Natural Resources, Governance and Institutional Quality: The Role of Resource Funds." *Resources Policy* 38, no. 2 (June): 181–195.

Wall, Elizabeth, and Remi Pelon. 2011. "Sharing Mining Benefits in
Developing Countries." *Extractive Industries for Development* no. 21.
Washington, DC: World Bank. https://openknowledge.worldbank.org/
handle/10986/18290.

Weinthal, Erika, and Pauline Jones Luong. 2006. "Combating the
Resource Curse: An Alternative Solution to Managing Mineral Wealth."
Perspectives on Politics 4, no. 1 (March): 35–53.

Weiss, Edith B., 1992. "In Fairness to Future Generations and
Sustainable Development." *American University International Law
Review* 8, no. 1: 19–26.

3

Mining Economies, Mining Families: The Impacts of Extractive Industries on Economic and Human Development in the Eastern Subarctic

Stephan Schott, Anteneh Belayneh, Jean-Sébastien Boutet, Thierry Rodon, and Jean-Marc Séguin

Extractive industries in the Arctic have been one of the region's major economic sectors besides public administration. Past mining projects have sometimes left unfavourable changes in landscapes and environmental disasters with an indefinite legacy, for example the massive contamination of arsenic and asbestos at the Giant Mine near Yellowknife, NWT. (Keeling and Sandlos 2015a b) Being in the business of non-renewable resource extraction with long-term environmental consequences seems to contradict the goal of sustainable development. Nevertheless, extractive industries are often one of the few large-scale economic development options for remote Arctic communities. Furthermore, certain minerals and metals, such as lithium, nickel, copper, cobalt, and rare earth metals, are essential inputs for the low-carbon economy of the future. In this context, we need to learn from past experiences and mistakes, and to find a more sustainable path for mining and economic development. This will help ensure that the environmental and social impacts of mining are minimized and acceptable by people that are exposed to them. In addition, in order for extractive resource development to improve the sustainability of northern communities, it is crucial that those local communities and regional self-government institutions benefit from economic gains. These include local job creation and business development, resource rent extraction, and overall human development.

This part of sustainable development is often taken for granted as extractive industries contribute to GDP growth at the national level, while not enough attention is paid at the micro scale to local revenue flows and details of business contracts and employment (Horowitz et al. 2018). Our RESDA project and network discussed the role of the extractive sector for Northern and Arctic development at several meetings and conferences. At the Annual RESDA Workshop in Goose Bay, Labrador, some of the research group members (B. Bradshaw, T. Rodon, A. Keeling, J.-S. Boutet and S. Schott) had in-depth side meetings to discuss the economic impacts, employment conditions, and the fulfillment of Impacts and Benefits Agreements from extractive industries. Then, they decided to apply for ArcticNet funding in order to conduct a thorough comparative analysis of two major mining projects in the Eastern Canadian Subarctic: the Voisey's Bay Mine in Nunatsiavut and the Raglan Mine in Nunavik.[1]

Through close collaboration with the Nunatsiavut Government and Makivik Corporation, funding was secured for the research project *Mining Economies, Mining Families: Extractive Industries and Human Development in the Eastern Subarctic*. For this project, we developed a novel business survey after holding a round of focus groups with Inuit business leaders from both settlement regions. We furthermore combined labour market statistics with employment and training data from both Glencore Corporation and Vale Limited to assess the amount and types of employment and overall development that was created. Here we will discuss the public, business, and individual employment and human development aspects of mining activities in the context of these two Inuit self-governance regions, to get a better understanding of what share of the benefits of resource development stayed in the regions and what practices were most beneficial for people involved with the extractive industry sector.

LITERATURE REVIEW

Generally, economic benefits follow extractive industry development (Rodon and Lévesque 2015). In the context of the Arctic, communities are purported to receive higher wages, transfer payments, and royalties, as well as access to training and education, and growth of entrepreneurial initiatives (Land, Chuhan-Pole, and Aragona 2015). Proponents of resource development often cite higher wage employment as a key benefit that Indigenous communities will receive from participating in resource development. This is based on the assumption that employment is desirable, and that job quality is high. The ability of employment to play a role in regional development, however, is jeopardized when only a few Indigenous workers obtain mining employment and when these workers are segmented in low-skilled positions. The link between extractive industries and economic development goes even further, with studies claiming that those industries promote economic diversification (Dockery 2014). While economic benefits are implicitly accepted as a result of extractive industries, the distribution of these impacts has been called into question. For example, it has been argued that non-renewable resource exploitation can increase income inequality within communities and leave behind a harmful legacy on the land, such as contaminated sites, severe landscape changes, and/or permanent disturbance of ecosystems and wildlife (Sandlos and Keeling 2012).

The extent and scale of local impacts from mining is a contested question. Some studies claim the impacts of mining are felt more on a regional rather than local scale (Ejdemo 2013). However, the impact of mining on local business development is even more under-researched (Kemp 2010). Our project examines the role of extractive industries for employment, business development, and human development in local communities in the Canadian Subarctic. This resource-rich region is home to Indigenous populations, which raises important questions regarding fair participation in local economic development decisions.

Access to mining jobs is often understood as one of the principal benefits of mining for communities. The notion that employment will provide economic benefits to northern Indigenous communities is, however, based on several assumptions. Some of these are that local people will become a large part of the workforce, that the jobs available are high quality and desirable to community members, and that those who gain employment in the mining sector will continue to live in their home communities and not permanently relocate. While several studies have examined the effectiveness of training and hiring programs or the experiences of Indigenous workers at the mine (Voyageur 1997; Industry 2002; Abele 1989), there has been less attention paid to how industrial relations, corporate and union culture, and community agreements affect the quality of mining employment and training opportunities for Inuit (Gibson 2006). Finally, we need to consider how these factors relate to Indigenous peoples' job satisfaction, desire to work, and migration behaviour.

The importance of education, training, and skills development in the Canadian Arctic is often stressed when discussing the economic impacts of mining and other extractive industries (Zhang and Swanson 2014). In the context of small communities, a significant consideration is whether there is enough capacity to take advantage of the economic opportunities. Scholars have highlighted the importance of capacity building, which is the purposeful implementation of measures to address problems and to increase the stock of skills within a community or a region, in order to capitalize on available economic opportunities. That is crucial in the Canadian North (Swanson and Zhang 2015; Zhang and Swanson 2014).

Assessing the economic impacts of mining in the Canadian Subarctic raises additional questions regarding the participation of Indigenous peoples in economic development and whether a fair share of these economic benefits accrue to these communities. Economic development can ensure that residents in the region receive a share of the wealth extracted from their lands, and that they enjoy opportunities which are in line with their counterparts in the rest of Canada (Swanson and Zhang 2015). While there are unanswered questions about the impact of extractive industries on local economic development through employment and training, the effects on business development are even less understood. Our project focuses largely on the effects of major mining projects on local and regional benefits in resource rent sharing, local business development, local employment, and overall human development. The focus on Inuit businesses was firstly due to the limited amount of research that had been conducted on local small business development that is linked to the mining industry, and secondly because of the interest of regional Inuit governments to explore this issue in more detail at both the community and regional level. Another motivation was to assess and compare to what extent these economic benefits trickled down to local businesses in two neighbouring Inuit regions of the Canadian Subarctic.

METHODS

We present a comparative review of the mining sector in each region, followed by an analysis of its impact on the public, business, and employment sectors. To determine the effects of mining on the economic and business development of communities in Nunavik and Nunatsiavut, respectively, we first approached business owners in both regions and conducted four focus groups. The focus group sessions informed the design of a comprehensive business survey that we then administered. The collaboration of the Department of Education and Economic Development of the Nunatsiavut Government (NG) with Mining Development at Makivik Corporation was instrumental in the design and identification of participants for both the focus groups and surveys. The focus groups preceded the survey design process and were conducted with selected businesses from Nunatsiavut, Nunavik, and Happy Valley-Goose Bay. Following a suggestion by the Economic Development division of the NG, we decided to conduct all the focus groups in Nunavik to also allow for networking of businesses from both regions. Three of the focus group sessions were held in Kuujjuaq, and one was held with businesses in Kangiqsujuaq and Salluit in Quebec, in October 2015 (see Belayneh et al. 2018).

Most of the ninety-seven surveys in Nunatsiavut and Happy Valley-Goose Bay were conducted with businesses in person, with the help of local research assistants. A small number were conducted over the phone. Inuit researchers from Nunatsiavut were identified, hired, and trained to administer the survey. These researchers were instrumental in contacting participants and spending approximately ninety minutes in person with them. The questions were loaded onto an iPad through the iSurvey program, which allowed research assistants to conduct the survey remotely and to upload their results when they regained an internet connection.

The survey was designed to identify the types of enterprises in each region, and to differentiate the impacts from mining on business activity over the different stages of mine development. Furthermore, questions referred to several topics: first we asked about the extent to which businesses depended on mining related activities for their revenue; then we inquired about the formation of partnerships within and outside of their region; and last we queried the barriers to business development. An important consideration was local satisfaction with Inuit Impact and Benefit Agreements, and their effects on entrepreneurs and communities. As well, we asked about employment at Inuit businesses, competition for labour, competition with other companies within and outside the region, and training and support programs.

Initially, the survey was piloted on a small scale in Nunatsiavut. Business owners were encouraged that their voices and concerns were being considered through the study and they were generally quite pleased to engage with the research team. Owners who also participated in the initial focus group sessions were happy that the project was advancing, and many agreed to participate in the final survey. There was a general sense that businesses were under surveyed and not consulted enough on their experiences within the extractive industry sector.

The business survey was conducted between May 2016 and April 2017 and was complemented with additional interviews in Salluit in September 2017. On the employment and human development side of the project, we analyzed employment, revenue, income data, and training data from mining companies and Statistics Canada. The project specified several recommendations and policy insights that aim to improve the involvement of Inuit-owned businesses with the mining industry, allowing for more meaningful and substantial participation of Inuit workers in employment, skills acquisition, and training in the major mining companies and in Inuit businesses in Nunatsiavut and Nunavik, where the same methods were used in the comparative study.

COMPARATIVE OVERVIEW OF EXTRACTIVE INDUSTRIES IN THE EASTERN SUBARCTIC

The Eastern Subarctic has two major mining operations in two neighbouring Inuit self-governance regions (Nunavik and Nunatsiavut) that are comparable in size and that extract similar types of minerals. The experience with the mining operations has, however, been quite different in terms of employment, Inuit business development, human development, and training.

Nunavik Region

The Nunavik region has two operating mines. The Raglan Mine and the Nunavik Nickel Mine are both closely located within the Ungava Trough, and both hold several nickel-copper deposits and occurrences which are embedded in extensive magmatic ultramafic geological features (Lamothe 1994).

RAGLAN MINE
The Société Minière Raglan Québec signed an Impact and Benefit Agreement (IBA) with both the communities of Salluit and Kangiqsujuaq, and the Makivik Corporation (Makivik) in 1995.

It was one of the first modern agreements of its kind in Northern Canada, known as the Raglan Agreement (to which we will return below). Since the start of its operations in 1997, and despite highly fluctuating commodity cycles, the Raglan Mine has operated consistently year after year since its beginning. The processing plant increased its annual production from 800,000 metric tons of ore per year in 2007 to approximately 1.5 million tons (Mt) currently (2019) (SNC-Lavalin 2017). Processing of ore material generates nickel and copper concentrates and a series of mineral by-products, such as cobalt, gold, silver, platinum, and palladium. The latter products are increasing in value and importance with the rising demand for these minerals, due to the push for energy transition and technological innovation. Recently, the Raglan Mine received a new certificate of authorization to increase its production output to 1.5 Mt per year, to be implemented during Phase II (starting in 2020) and Phase III (between 2023 and 2031) of underground operations. The extraction period for the new underground operations is expected to last until 2038 (SNC-Lavalin 2017).

On December 31, 2014, the mineral inventory of the Raglan Mine indicated over 7 Mt of proven and probable nickel, copper, and cobalt ore reserves. The resources (measured, indicated and inferred) totalled more than 35 million tons. On an annual basis, the metal sales from the Raglan Mine are estimated to generate $600 million (M). This sales figure is, of course, subject to the constant fluctuations of metal prices. Glencore invested a total amount of $2.1 billion (B) at the Raglan Mine between 1998 and 2016, including all site costs (mine and mill) and ore concentrate transportation.

The 1995 Raglan Agreement provides financial distribution to the signatories, Makivik and the communities of Salluit and Kangiqsujuaq, in two different ways: (1) through guaranteed allocations with varying amounts, depending on the years of operations; (2) with profit-sharing allocations based on a fixed percentage (4.5 per cent) applied on the annual operating net cash flow of the mine. All money transfers are paid to a trust that distributes the allocations to the different beneficiaries described in the trust (the landholding companies of Salluit and Kangiqsujuaq and the Makivik Corporation). Since the start of its operations, the Raglan Mine has distributed more than $140M to the signatories, thus providing significant funds to the region. (For a discussion of the allocation of these revenues, see chapter 2 in this volume.)

Canadian Royalties – Nunavik Nickel Mine

In 2008, the Nunavik Nickel Agreement was signed between the mining promoter Canadian Royalties Inc. and Makivik, along with the landholding corporations of Salluit and Kangiqsujuaq, and the northern village of Puvirnituq. Nunavik's second IBA includes various provisions for the development of the Nunavik Nickel Mine, located near the Raglan Mine area. The objectives of the agreement were to facilitate Inuit participation in the operations, to provide adequate environmental and social mitigation measures, and to specify direct and indirect social and economic benefits to Inuit.

The initial certificate of authorization allows the promoter to develop several open pits to extract nickel-copper deposits identified on the Canadian Royalties property, on a fifteen year time horizon and at a production rate of 3,700 tons per day of ore, starting in 2008. Ore processing was increased to 4,500 tons per day in 2011. Canadian Royalties Inc. started commercial production in 2014 and generated an annual income of $260M in 2015, and $220M in 2016. The total cumulative investment for the Nunavik Nickel Mine until 2013 (before the commercial production) was about $1.5B.

Financial benefits under the Nickel Agreement include fixed annual payments, once the commercial production is executed. Incremental community financial benefits were also paid until the mining project initial investment was recaptured. Once the Nunavik Nickel Project initial investment was recouped, financial allocations have been based on a varying percentage of the gross revenues and the average price of nickel. Since the start of the commercial production in 2014, Nunavik Nickel Mine has generated more than $6M in financial allocations.

Nunatsiavut Region

In the Nunatsiavut region, there is one major industrial mine that is operated by Vale at Voisey's Bay. Operations began there in 2005. Voisey's Bay produces two major types of concentrates, nickel and copper, as well as some cobalt (see table 3.1).

Nickel and copper production increased by 5.7 per cent from 2016 to 2017, while the much smaller cobalt production more than doubled. Unlike operations in Nunavik, where processed ore is shipped for smelting in Sudbury and then for final refining in Nikkelverk, Norway, the nickel concentrate produced at Voisey's Bay is recently being

Table 3.1
Finished production by ore source at Voisey's Bay, 2016-18

Mineral Extraction(Mt)	2016	2017	2018
Nickel	49,000	51,800	38,600
Copper	32,000	34,000	26,000
Cobalt	887	1,829	1,902

Source: Vale 2018

Table 3.2
Ore processing rate and expected mine life comparisons in the Eastern Subarctic mines

	Raglan Mine	Canadian Royalties Mine	Voisey's Bay Mine
Extraction rate	7,000 tons/day	4,500 tons/day	6,000 tons/day
Expected lifespan	2038	2025	2034

Source: MDO 2019

processed solely within the province, at Long Harbour, on the southeastern part of Newfoundland. The processing facility is designed to produce up to 50,000 tons of finished nickel each year once it ramps up to full production. Currently the plant is producing about 35,600 tons per year (Mining Weekly 2018), and Voisey's Bay concentrate will provide 100 per cent of the feed to the plant for the first few years, after which time other sources of concentrate will be brought in.

On 11 June 2018, Vale announced that it will start a $1.7B development of an underground mine and associated facilities, which are expected to extend the Voisey's Bay Mine life to 2034. Vale expects that the underground mine will begin production in 2021 and will ramp up over the following four years, while the current open pit mining operation is expected to continue until 2022.

Vale encourages business capacity building through Indigenous joint ventures designed to meet the supply and service needs of the company's Labrador-based operations. Approximately 80 per cent of the Voisey's Bay Mine and Mill support contracts were handed to Indigenous businesses ("Voisey's Bay," n.d.), although there is evidence that a good proportion of these businesses are not based within the Nunatsiavut settlement region. Generally, mining operations in both Inuit regions are expected to be long lasting (see table 3.2) and have so far generated major revenues and benefits. It is more difficult to assess, however, exactly how individual communities and the Indigenous population have benefitted from these financial flows.

ASSESSING REGIONAL AND LOCAL ECONOMIC
IMPACTS FROM MINING

It is a challenge to comprehensively track economic development trends in Inuit Nunangat. This difficulty arises in part because the Inuit regions are not easily separable from their respective provinces or territories in the gathering of statistics. For example, breaking up Nunavik's gross domestic product (GDP) from the rest of Quebec is a difficult statistical undertaking, as is separating the GDP of Nunatsiavut from the province of Newfoundland and Labrador. While the estimates for economic growth included below are not comprehensive, they do provide an indication of mining activities' impact in Nunavik and Nunatsiavut. The Raglan Mine is a major contributor to the GDP of Nunavik. The mine generates about 40 per cent of the region's GDP, estimated to be close to $1B (Cohen-Fournier 2017). While there is no equivalent estimate available for Nunatsiavut, it is known that mining is important for the GDP of Newfoundland and Labrador, as a whole. In 2016, this sector accounted for 4.7 per cent of the provincial GDP ($1.3B) and 1.5 per cent of employment (Government of Newfoundland and Labrador 2017). The Voisey's Bay Mine (including the Long Harbour Processing Plant) accounts for approximately 42 per cent of the mining jobs currently within the province. Overall, because of the fly-in/fly-out (FIFO) nature of mining operations in Northern Quebec and Labrador, it is difficult for these regions to fully capture the economic benefits derived from the presence of industrial mining. Economic leakages are a central problem of modern mining operations, particularly in remote regions like these (Safali 2015).

The economic benefits from mining in Nunavik and Nunatsiavut can be categorized as either direct, indirect, or induced benefits. These benefits accrue to individuals, businesses, and communities at different stages of mine development. Direct benefits are those that are obtained straight from the mine, such as employment with the mining company and with the contractors. The mines at Raglan and Voisey's Bay have been a source of employment for Inuit in both Nunavik and Nunatsiavut. However, the level of Inuit employment at these mines differs. At Raglan, about 20 per cent of the workforce is Inuit—about 205 Inuit employees (Glencore 2018), while about 44 per cent of the workforce at Voisey's Bay is Inuit—for roughly 252 employees (Vale 2019). These strikingly different proportions of Inuit employees between the two comparable mining operations will require more detailed examination. Although Voisey's Bay has more than double the percentage of Inuit employees, they only have about 20 per cent more Inuit employees in total. Furthermore, it also makes a difference to regional governments and to Inuit in the regions whether these employees reside in the self-governance region or not. We will have a closer look at the number and types of jobs created later. First, we will evaluate the public benefits from mining in terms of the sustainability of rents directly extracted by regional governments and communities from the mining industry.

1. Public Benefits from Mining: The Distribution and Use of Mining Royalties

The share of resource royalties distributed to the communities is calculated based on profits, the value of production, or production rates. There can also be fixed, initial payments that will allow communities to cover expenses associated with IBA administration (Gibson and O'Faircheallaigh 2010) and to establish resource-based trust funds. The money coming from profit-sharing royalties and lump-sum payments are generally paid to the organization and/or community that has signed the benefits agreement with the company. In turn, this organization or community is responsible for distributing the sums paid through mechanisms that can vary across regions. The Nunatsiavut Government (NG) receives direct royalties from Voisey's Bay under the 2005 Inuit Impacts and Benefits Agreement with Vale and invests a portion of these revenues in a trust fund. The NG establishes in their annual budget and decides how to invest or distribute mining revenues and royalties each year. In the most recent budget, the largest expenditure was for improving housing (NG 2019). In Nunavik, the IBA assigns 4.5 per cent of the mine's profits to the communities of Salluit (which receives 45 per cent of this portion of profits) and Kangiqsujuaq (30 per cent), and to Makivik Corporation (25 per cent) (Kativik Regional Government 2007). From 1997 to 2011, Raglan has paid out just over $100M in royalties to Nunavik (George 2012). The use of mining royalties is decided by a vote in Salluit; while the community of Kangiqsujuaq manages the profit-sharing payment through the local landholding corporation and the municipality (see Rodon et al., chapter 2 for more details).

Choices about how mining revenues are used have an impact on the sustainability of mining development. Non-renewable resources are a depletable asset and need to be replaced with other assets such as human capital, physical capital, or new forms of natural capital (e.g., renewable energy or new protected areas). Any exploitation diminishes its availability to the next generations unless the returns,[2] or rents, obtained from the extraction process are reinvested in other forms of assets that can, in turn, benefit these future generations; nevertheless, this process refers only to weak sustainability principles (Hartwick 1977, 1978). This is a minimal prerequisite for a sustainable path from mining. Further criteria related to sustaining crucial and vital ecosystems and to maintaining wildlife in the immediate vicinity of traditional communities meet stronger sustainability criteria (De Groot et al. 2003).

2. *Local Business Development Stemming from Mining*

In addition to local jobs, mining is purported to lead to the growth of entrepreneurial initiatives and business development. However, these indirect and induced benefits are notoriously difficult to measure. In this case, they were partly assessed with the use of a comprehensive business survey administered in Nunavik and Nunatsiavut from 2016-17. This survey revealed that 70 per cent of businesses, in both regions, received less than 10 per cent of their gross revenues from mining and exploration activities (Belayneh et al. 2018). For the Inuit businesses with larger shares of revenues from mining, 64 per cent in Nunavik had either a partnership or joint venture, and 83 per cent in Nunatsiavut. Further examination of the businesses with large shares of mining revenue shows that they are not large employers of Inuit within their respective regions. Many of the major venture Inuit partners are either joint ventures between regional landholding companies, as in Nunavik, or between Nunatsiavut government-owned companies and private enterprises, as in Nunatsiavut. This prevents many sole proprietorships or incorporated companies from bidding on larger, longer-term contracts. One possible solution seemed to be partnerships, however one of the biggest challenges that Inuit-owned businesses identified was to find a trustworthy partner company with which to bid on contracts. In both Nunavik and Nunatsiavut, 20 per cent of business owners declared that the respective IBA provisions were the primary reason for seeking partnerships with non-Inuit companies (Belayneh et al. 2018). According to Inuit business owners that we surveyed, the IBA provides preferential treatment and the opportunity for their company to enter partnerships and increase their chances of procuring contracts within the mining sector. Not all business owners in Nunavik (32 per cent) and Nunatsiavut (41 per cent) are familiar with the contents of the different IBAs. However, the general perception is that an IBA provides good opportunities and preferential treatment to Inuit companies and is generally helpful in the procurement process. The overall satisfaction level with the IBA is similar in both regions, where business owners rated their satisfaction as roughly three on a five-point scale (Belayneh et al. 2019).

Other reasons for partnering up with other businesses varied between the regions. The main reason in Nunatsiavut is to develop business and market access, while in Nunavik, the main motivations for partnerships and joint venture are funding and gaining knowledge and expertise.

3. Employment, Training, and Residency

Employment benefit to Inuit residents in each of the two self-governance regions is an important criterion for justifying mining operations in each region. In this section, we compare policies, targets, and actual trends at both mining sites, in terms of employment ratios of Inuit employees as a proportion of the total labour force, the training and promotion of Inuit employees, and the residency (inside or outside of the region) of beneficiaries.

VOISEY'S BAY
Voisey's Bay Nickel Company's construction of the Voisey's Bay mine began in 2002. The following year saw a substantial increase in infrastructure development; over 1,500 workers were employed on all facets of the project during that year (VBNC 2003b). Employment peaked in 2004, with more than 3,200 employees, including close to 2,500 located in Labrador (VBNC 2004b). Figure 3.1 summarizes the number of employees and the location of work during the construction period.

As shown, the workforce for the Voisey's Bay construction project was at its highest in 2004. During the construction period, Inuit workers represented between 12 and 16 per cent of the total workforce at Voisey's Bay (see figure 3.2). Female participation rate in the workforce

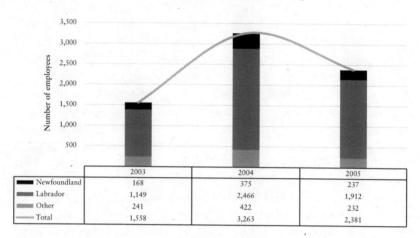

	2003	2004	2005
■ Newfoundland	168	375	237
■ Labrador	1,149	2,466	1,912
■ Other	241	422	232
━ Total	1,558	3,263	2,381

Figure 3.1 Employment by location, Voisey's Bay, 2003–05

was higher for Innu (up to 54 per cent of the total Innu workforce in 2005) and for Inuit (as high as 25 per cent of the total Inuit workforce in 2005) than for the non-Indigenous workforce. Nevertheless, Inuit women, for example, only represented between 2 and 3 per cent of the total workforce during construction.

Overall, about fourteen training programs were offered to Inuit in 2003 at the Voisey's Bay construction site and in various communities. Most of these programs were short term (one to sixteen weeks) and sought to address very specific labour needs at the construction site. One long-term program in industrial instrumentation was also given, but it had only one registered Inuit trainee. In total, approximately 105 Inuit trainees registered in the different programs, and of those that were completed by the end of the year, only three dropouts were recorded. In 2004, training continued at the construction site and in communities throughout Labrador (VBNC 2004b). The company offered information on accommodations, maintenance training, gender sensitivity, and Innu and Inuit cultural awareness. It also had millwright apprentice training and pre-employment programs, such as an introduction to mill processing, which was delivered in Happy Valley-Goose Bay, and one on heavy equipment operation, which was delivered in Natuashish (VBNC 2004b).

During operations, Inuit occupied roughly 40 per cent of all jobs at Voisey's Bay over the period 2007-14. Inuit women typically occupied one-fifth of all positions held by Inuit at the mine, although that

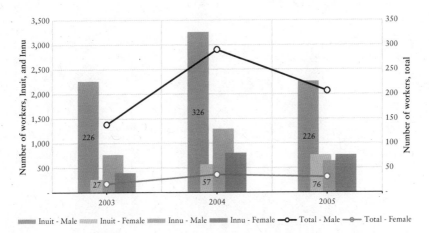

Figure 3.2 Annual employment by Indigenous group (including gender), Voisey's Bay, 2003–05

proportion decreased in recent years. In 2019, for instance, women occupied only forty-one of the 255 jobs held by Inuit (16 per cent). This proportion resembled the overall proportion of all women employed at Voisey's Bay. Figure 3.3 shows that between 2007 and 2014, the proportion of Inuit employed at Voisey's Bay consistently hovered above 40 per cent, while also showing the proportion of Inuit and Innu workers who were female. Notice that there is a much smaller number of Innu employees than Inuit employees, but that the proportion of males and females among the Innu is very equal, unlike the Inuit gender composition.

During the same period, the proportion of Inuit employees who lived within the Nunatsiavut Settlement Region, however, oscillated between only 27 and 35 per cent. Overall, this category of employees – Inuit who live in Nunatsiavut – represented a maximum of 15 per cent of the total workforce at the mine, as a large majority of these employees resided in the Happy Valley-Goose Bay area or beyond (see table 3.3).

The positive trend observed in labour statistics during the operations period, notably the increased labour participation rate and improved employment rate, was largely caused by a rapid growth in the public administration sector, which likely stemmed from the creation of the Nunatsiavut Government in 2005 (see figure 3.4). While government work made up 19 per cent of all employment in Nunatsiavut in 1996, by 2011, that number had jumped to

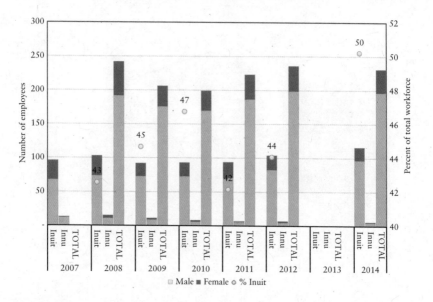

Figure 3.3 Number of employees, Voisey's Bay, December 2007 to December 2014

Table 3.3
Number of Inuit employees by community of residence, Voisey's Bay, 2007 to 2019

Residence	2007	2008	2009	2010	2011	2012	2014	2019
Nain	36	32	26	28	31	37	31	43
Hopedale	7	8	6	6	6	6	13	13
Makkovik	8	9	10	10	10	13	14	9
Postville	4	5	4	4	4	5	5	9
Rigolet	6	5	5	4	7	7	8	9
Nunatsiavut	61	59	51	52	58	68	71	83
Happy Valley-Goose Bay	120	107	104	106	107	99	97	106
North West River	17	14	13	12	15	14	16	17
Sheshatshiu	0	0	0	0	0	0	0	0
Mud Lake	0	1	1	1	1	1	1	1
St. John's	6	5	2	2	2	2	2	6
Other	20	14	15	15	13	13	16	39
Total Inuit	224	200	186	188	196	197	203	252
Total workforce		499	452	450	477	484	469	574
Nunatsiavut: Total Inuit (%)	27	30	27	28	30	35	35	33
Nunatsiavut: Total workforce (%)		12	11	12	12	14	15	14.5

Source: Vale 2007-2014, 2019 (Data not available for 2013.)

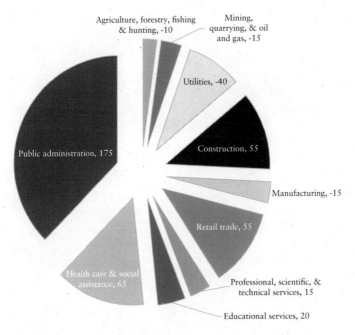

Figure 3.4 Change in number of people employed by principal industry sectors, Nunatsiavut, 1996–2011

29 per cent, with up to 175 public administration jobs being created during that period (Statistics Canada 1996, 2011). We note that, as an aside, the underground expansion of Voisey's Bay is expected to create up to 650 jobs during the peak of the construction phase, and up to 400 new full-time, permanent jobs will be created to sustain this new phase of operations, that will gradually replace the open-pit activities.

THE RAGLAN MINE

At the Raglan Mine, the total number of Inuit employees is increasing significantly, and the monthly surveys show that 20.6 per cent of employees are Inuit, of which 28.4 per cent reside outside of Nunavik (see figure 3.5).

The distribution of Inuit employees by community varied, with Inukjuak having the highest percentage of Inuit employees (14.4 per cent of total Inuit employees), followed by Kuujjuaq with 12 per cent. Salluit and Kangiqsujuaq, which are the two communities closest to the mine, together accounted for 17.8 per cent of all Inuit employees at Raglan (see figure 3.6).

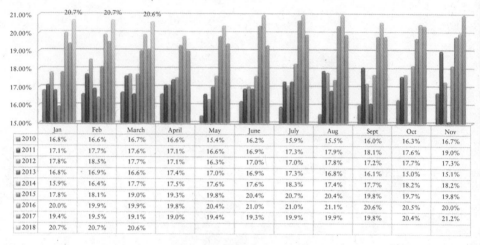

	Jan	Feb	March	April	May	June	July	Aug	Sept	Oct	Nov
2010	16.8%	16.6%	16.7%	16.6%	15.4%	16.2%	15.9%	15.5%	16.0%	16.3%	16.7%
2011	17.1%	17.7%	17.6%	17.1%	16.6%	16.9%	17.3%	17.9%	18.1%	17.6%	19.0%
2012	17.8%	18.5%	17.7%	17.1%	16.3%	17.0%	17.0%	17.8%	17.2%	17.7%	17.3%
2013	16.8%	16.9%	16.6%	17.4%	17.0%	16.9%	17.3%	16.8%	16.1%	15.0%	15.1%
2014	15.9%	16.4%	17.7%	17.5%	17.6%	17.6%	18.3%	17.4%	17.7%	18.2%	18.2%
2015	17.8%	18.1%	19.0%	19.3%	19.8%	20.4%	20.7%	20.4%	19.8%	19.7%	19.8%
2016	20.0%	19.9%	19.9%	19.8%	20.4%	21.0%	21.0%	21.1%	20.6%	20.5%	20.0%
2017	19.4%	19.5%	19.1%	19.0%	19.4%	19.3%	19.9%	19.9%	19.8%	20.4%	21.2%
2018	20.7%	20.7%	20.6%								

Figure 3.5 Inuit employment at Raglan Mine, 2010–18

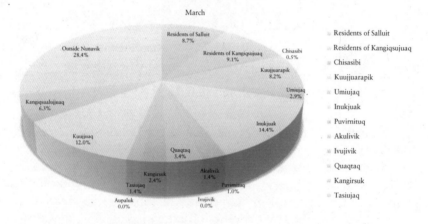

Figure 3.6 Inuit employment at Raglan Mine by residence(Glencore 2018)

The issue of Inuit employment raises several points that are worth discussing. The percentage of Inuit employment is definitely important, but we must also consider the types of jobs that the Inuit workforce occupies, as well as the duration of their employment. By describing how other aspects of Inuit employment and associated statistics have changed over time, we can better grasp the complexity of the issues that are perhaps hidden by the overall increase in the number of Inuit working at the mine. Inuit are mainly employed in relatively low-skilled entry-level positions or medium-skilled positions that are normally available after taking a vocational training course. Accessing local training opportunities and achieving

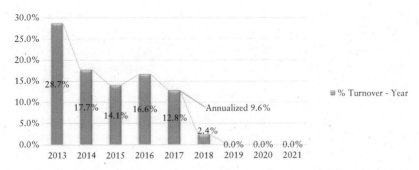

Figure 3.7 Voluntary resignation at Raglan Mine(Glencore 2018)

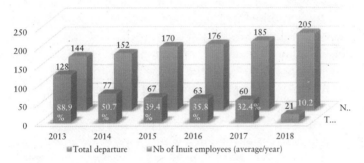

Figure 3.8 Total turnover rate including termination(Glencore 2018)

an education level that is required to move up into different positions are a constant challenge that must be overcome. Turnover rates at the Raglan Mine have, however, steadily improved (see figure 3.7). Unfortunately, the turnover rates for Voisey's Bay were not available. In 2017, for example, about a third of Inuit employees at the Raglan Mine either left or were terminated by the company, down from 89 per cent just four years earlier (see figure 3.8). Glencore is actively investing into new training programs and recruitment in communities, which is helping to improve the turnover rates, and which might have positive long-term impacts on the quality of available jobs and the skill level of the Inuit workforce at the mine.

CONCLUSION

Northern Canada has traditionally depended on the extraction of natural resources for its growth and economic development. This reliance is not likely to change very soon. The Eastern Canadian

Subarctic region has several important deposits of nickel, copper, cobalt, and rare earth metals that could play an important role in the transition to a low-carbon economy, both in Canada and worldwide. Important changes are necessary for Canadians and especially for Indigenous peoples if they are to realize more of the benefits, since they must also live with the immediate consequences of extractive industries. Training problematically lures many people away from smaller communities towards bigger regional centers, with a noticeable strain on the social fabric. Returning to home communities after training can be challenging due to the limited housing options available. Thus, mining training could induce permanent migration to larger hubs in Labrador, or to places outside the region, like Montreal or St. John's. Inuit employment in major mines developed in Inuit territories and in Inuit businesses associated with these mines needs to be improved. Employment numbers are going up, but are still relatively low in Nunavik; however, turnover rates are declining at Raglan, and new employment and training programs by Glencore look promising and could be further enhanced. In Nunatsiavut, Inuit employment is at a relatively high level of around 43 per cent, but many of these employees reside outside of the region. Private Inuit businesses, in addition, are not receiving a large share of the revenues generated from mining. Local business capacity needs to be increased so that Inuit businesses can bid more effectively on larger mining contracts, employ more Inuit, and receive a larger share of mining revenues without the need to always partner up with southern-based companies.

Investments should be made so that training programs for employees can be more accessible in smaller communities to encourage future employees to remain in their home community, if they prefer. Mining companies need to show flexibility in the work schedules to accommodate Inuit who live in a mixed economy of land-based activities and employment in the wage sector. A move to more seasonal employment, flexibility around hunting season and holidays, and job-sharing models could enhance Inuit employment and workplace experiences, and therefore reduce turnover rates. This requires a better understanding of the social consequences of mining, for example, the impact on the land-based economy or the impact of fly-in/fly-out (FIFO) work schedules on workers and their partners and families who stay at home (see Saxinger, chapter 6 for more details). The employment opportunities for Inuit women in the mines should also be better studied and promoted, and gender analysis should be applied to better understand the impact of mining on community well-being and environmental impacts (as further examined by Mills and Simmons, chapter 7). The latter is encouraged, for example, by certain employment programs already introduced in Nunavik.

In 2013-14, the Nunavik regional organizations organized a tour of communities to identify a comprehensive vision of development according to Inuit priorities. This consultation process was referred to as Parnasimautik[3]. In Parnasimautik workshops, Nunavimmiut voiced the urgency for Nunavik to adopt a mining policy that set out the conditions for mineral resource development under existing land, that established wildlife and environmental agreements and legislation, and that encouraged the mining industry to participate in sustainable and equitable social, environmental, and economic development. By explicitly showing what Nunavik Inuit values are and what exactly people want to protect, the Nunavik Inuit Mining Policy, launched in November 2014, guides mining activities in the region, as well as promotes partnerships with industry. A more specific mining policy could address some of the identified concerns and could provide more guidance on how to build Inuit business capacity and training possibilities for current and prospective employees.

In addition, the definition of what constitutes an Inuit business (e.g., ownership share of a business or partnerships, number of Inuit employees, being actively involved in operation of the business, physical location in the region) differs in all four Inuit settlement regions and could be defined more clearly. Doing so could allow Inuit businesses and employees actively involved with extractive industries to receive a larger share of the benefits from extractive industries. In the spring of 2018, Nunavik established a Chamber of Commerce that has the mandate to advocate for local and Inuit business opportunities and that will promote Inuit businesses and inform them about development opportunities. This is something Nunatsiavut might also consider. It could also foster closer collaboration and the sharing of experiences with the mining sector in both jurisdictions.

The question of mining sustainability also needs to be addressed by regional and local decision makers. For example, intra- and intergenerational equity should be considered in the use of royalties and, more generally, a vision should be encouraged where mining development is not necessarily the objective , but rather constitutes an element of a more comprehensive regional development strategy based on economic diversification, skill development, and improvements in the number and quality of local Inuit jobs. In Nunavik, Parnasimautik is a first step in the construction of a Nunavik vision of development, but it can be pursued further. In Nunatsiavut, no comparable process has been undertaken yet. As Vale and Glencore are significantly expanding their business activities, there is an urgent need for the Nunatsiavut Government and governments and institutions in Nunavik to think about economic diversification, investment in housing and infrastructure in the region, and how to avoid migration outside the respective regions. It boils down to the critical question of how the benefits from mining can more directly improve the living conditions in the region. A more thorough and regular examination of local data and the local indicators that measure the circular flow of revenues and jobs in the self-governance regions is a crucial task that will inform us if real progress is going to be realized.

NOTES

1   A. Keeling later withdrew from the application due to conflicts with other applications and projects. However, we would like to thank him for valuable input at the start of the project and feedback throughout the project. We later collaborated on another related publication and he came up with the title of the ArcticNet project.
2   The rent of a unit of extraction is the difference between the price per unit and the marginal cost of extraction
3   An Inuktitut term meaning "what you need to be prepared" (Rogers 2014)

REFERENCES

Abele, Frances. 1989. *Gathering Strength: Native Employment Training in the Northwest Territories*. Calgary: Arctic Institute of North America,
Belayneh, Anteneh, Thierry Rodon, and Stephan Schott. 2018. "Mining Economies: Inuit Business Development and Employment in the Eastern Subarctic. *The Northern Review* 47: 59-78.
– "A Comparative Analysis of Inuit Business Experiences with Extractive Industries in the Canadian Eastern Subarctic." Working Paper.
Cohen-Fournier, Nathan. 2017. "Walking on Thin Ice: Entrepreneurship in Nunavik." *Northern Public Affairs* 5, no. 2.
Cornell, Stephen, Miriam Jorgensen, Stephanie Carroll Rainie, Ian Record, Ryan Seelau, and Rachel R. Starks. 2007. "Per Capita Distributions of

American Indian Tribal Revenues: A Preliminary Discussion of Policy Considerations." *Joint Occasional Papers on Native Affairs (JOPNA)* 2008-02. Tucson and Cambridge: Udall Center for Studies in Public Policy and Harvard Project on American Indian Economic Development.

Cornell, Stephen, and Joseph P. Kalt. 2007. "Two Approaches to the Development of Native Nations: One Works, the Other Doesn't." In *Rebuilding Native Nations: Strategies for Governance and Development*, edited by Miriam Jorgensen, 3–32. Tucson: University of Arizona Press.

De Groot, Rudolf, Johan van der Perk, Anna Chiesura, and Arnold van Vliet. 2003. "Importance and Threat as Determining Factors for Criticality of Natural Capital." *Ecological Economics* 44, no. 2: 187–204.

Dockery, Alfred M. 2014. "The Mining Boom and Indigenous Labour Market Outcomes." In *Resource Curse or Cure? CSR, Sustainability, Ethics & Governance*, edited by Martin Brueckner, Angela Durey, Robin Mayes, and Christof Pforr. Berlin, Heidelberg: Springer.

Ejdemo, Thomas. 2013. "Mineral Development and Regional Employment Effects in Northern Sweden: A Scenario-Based Assessment." *Mineral Economics* 25, no. 2: 55–63. https://doi.org/10.1007/s13563-012-0023-z.

George, Jane. 2012. "Xstrata Mining Dividends Cut Welfare Rolls in Nunavik Communities." *Nunatsiaq News*, September 11.

Gibson, Ginger and Ciaran O'Faircheallaigh. 2010. *IBA Community Toolkit: Negotiation and Implementation of Impact and Benefit Agreements*. Summer 2015 ed. Toronto: The Walter & Duncan Gordon Foundation.

Gibson, Robert B. 2006. "Sustainability Assessment and Conflict Resolution: Reaching Agreement to Proceed with the Voisey's Bay Nickel Mine." *Journal of Cleaner Production* 14, no. 3: 334–348.

Glencore, 2018. "Human Resources Employment – Tamatumani." October.

–2018. "Human Resources Employment – Tamatumani." March 31, 2018.

Gupta, Sanjeev, Alex Segura-Ubiergo, and Enrique Flores. 2014. "Direct Distribution of Resource Revenues: Worth Considering?" *IMF Staff Discussion Note* 14/05. Washington, DC: International Monetary Fund.

Hartwick, John. 1977. "Intergenerational Equity and the Investment of Rents from Exhaustible Resources in a Two Sector Mode and Notes on the Economics of Forestry Exploitation." Working Paper 281. Kingston, ON: Economics Department, Queen's University.

–1978. "Investing Returns from Depleting Renewable Resource Stocks and Intergenerational Equity." Working Paper 294. Kingston, ON: Economics Department, Queen's University.

Horowitz Leah, Arn Keeling, Francis Lévesque, Thierry Rodon, Stephan Schott, and Sophie Thériault. 2018. "Indigenous Peoples' Relationships to Large-Scale Mining in Post/colonial Contexts: Toward Multidisciplinary Comparative Perspectives." *The Extractive Industries and Society* 5, no. 3 (July): 404-414.

Keeling, Arn and John Sandlos,. 2015a. *Guardians of Eternity.* http://www.guardiansofeternity.ca/.

Keeling, Arn and John Sandlos. eds. 2015b. *Mining and Communities in Northern Canada: History, Politics and Memory.* Calgary, Alberta: University of Calgary Press.

Kemp, Deanna. 2010. "Mining and Community Development: Problems and Possibilities of Local-Level Practice." *Community Development Journal* 45, no. 2 (April): 198–218. https://doi.org/10.1093/cdj/bsp006.

Lamothe, D. 1994. Géologie de la Fosse de l'Ungava, Nouveau-Québec. *Géologie du Québec, Les Publications du Québec.* Bibliothèque nationale du Québec, p. 67-73.

Land, Brian C., Punam Chuhan-Pole, and Fernando M. Aragona. 2015. "The Local Economic Impacts of Resource Abundance: What Have We Learned?" Policy Research Working Paper 7263. The World Bank. http://documents.worldbank.org/curated/en/446761467991987706/The-local-economic-impacts-of-resource-abundance-what-have-we-learned.

MDO (Mining Data Online). 2019. https://miningdataonline.com/#top.

NG. 2019. "Budget 2019-20 Focuses on Program Enhancement and Implementation of Key Strategic Initiatives." https://www.nunatsiavut.com/article/budget-2019-20-focuses-on-program-enhancement-and-implementation-of-key-strategic-initiatives/.

O'Faircheallaigh, Ciaran. 2016. *Negotiations in the Indigenous World: Aboriginal Peoples and the Extractive Industry in Australia and Canada.* New York: Routledge.

Rodon, Thierry and Francis Lévesque. 2015. "Understanding the Social and Economic Impacts of Mining Development in Inuit Communities: Experiences with Past and Present Mines in Inuit Nunangat." *The Northern Review* 41: 13-39.

Rodon, Thierry and Stephan Schott, 2013. "Towards a Sustainable Future for Nunavik." *Polar Record* 1-17, March.

Rogers, Sarah. 2014. "Language, Culture are Top Nunavimmiut Concerns, Parnasimautik Consultation Finds." *Nunatsiaq News* March 6, 2014.

Safali, Deogratias. 2015. "Étude des retombées économiques d'une mine: cas de Voisey's Bay et du Nunatsiavut." Master's thesis, Université Laval, Québec.

Sandlos, John, and Arn Keeling. 2012. "Claiming the New North: Development and Colonialism at the Pine Point Mine, Northwest Territories, Canada." *Environment and History* 18, no. 1 (February): 5–34. https://doi.org/10.3197/096734012X13225062753543.

SNC-Lavalin. 2017. *The Raglan Mine Property beyond 2020 (Phases II and III): Continuation of Mining Operations East of Katinniq. Environmental and Social Impact Assessment.*

Statistics Canada. 1996. "Census of Population." Accessed January 28, 2019. http://www12.statcan.gc.ca/census-recensement/index-eng.cfm.

–2011. "National Household Survey." Accessed January 28, 2019. https://www12.statcan.gc.ca/nhs-enm/2011/dp-pd/prof/index.cfm?Lang=E.

Swanson, Lee A. and David D. Zhang. 2015. "The Base Requirements, Community, and Regional Levels of Northern Development." *The Northern Review* 38.

Vale 2018. *Annual Report.* http://www.vale.com/EN/investors/information-market/annual- reports/2of/2oFdocs/Vale_20-F%20FY2018%20-%20final_i.pdf.

–2019. Employment Data, acquired from Matthew Pike, February 14, 2019.

Voisey's Bay Nickel Company (VBNC). 2003. *Inuit Impact and Benefit Agreement Report.*

–2003-05. *Social Responsibility Report.*

Webb, Mariann. 2018. "Vale to Proceed with Voisey's Bay Expansion." *Mining Weekly,* June 11.

Zhang, David D. and Lee A. Swanson. 2014. "Toward Sustainable Development in the North: Exploring Models of Success in Community-Based Entrepreneurship." *The Northern Review* 38.

4

Local Benefits of Education, Training, and Employment with Resource Industries

Andrew Hodgkins

Vocational Education and Training (VET) and related employment of local northern Indigenous people in Canada are an important part of public-private partnerships negotiated between Indigenous governments and non-renewable resource developers. Partnerships receive the support from both federal and territorial governments, they and are often associated with Impact and Benefit Agreements that are negotiated during the mining life cycle. Hence, while the "impacts" of resource development include environmental degradation of traditional lands, potential "benefits" include gaining much needed training and employment. Employment opportunities are particularly attractive for local communities considering their isolation from mainstream economic development. The degree to which resource development can promote sustainable employment opportunities – that is, employment that is long-term, skilled, and associated with the trades – warrants further attention. This is particularly the case since the skilled trades are well paid, relatively secure, and include expertise that is transferable to other employers once the mining cycle has run its course. The question remains: to what degree can resource development benefit northern Indigenous communities in terms of achieving sustainable employment?

This chapter shares findings from two case studies that were conducted in northern Canada. The first was an oil sands-sponsored training program that operated in Northern Alberta's Regional Municipality of Wood Buffalo. The second was a RESDA-sponsored project in Nunavut that examined the experiences of Inuit employees at the Baffinland Iron Ore's Mary River Mine operating in North

Baffin Island. In both cases, qualitative research methods were used and VET participants, workers, and stakeholders were interviewed over a period of a year. While sharing the same objective of training and employing a local Indigenous labour force, the cases vary considerably in terms of region, ethnicity, stages of the mining life cycle in which the studies occurred, as well as outcomes achieved at the time the research was conducted. These different cases aim to highlight significant challenges and opportunities experienced by local northern communities and industry players alike in achieving training and employment. I begin by providing an historical overview of northern resource development VET partnership programs, followed by a presentation of both cases, including historical labour force participation in mining, a description of the cases, methods used, and some key findings. In summary, I present a series of recommendations that various stakeholders may want to consider when developing similar programs.

BACKGROUND

Training and employment of northern Indigenous peoples gained considerable attention during the post-World War II era—the apotheosis of which came in 1957, with Prime Minister Diefenbaker's "Roads to Resources" northern vision, which literally paved the way for resource development. Government policy shifted towards encouraging Indigenous people to participate in the wage economy, which, in northern Canada, meant primarily the extraction of non-renewable resources. Programs sought to provide vocational training for skilled labour positions (Dickerson 1992).

The provision of adult education began in the 1960s, although it usually led to menial entry-level jobs, with few staffing positions made available. (McLean 1997). Adult vocational training centres were established in Fort McMurray in 1965 and in Fort Smith in 1968. By 1974, there were permanent positions for adult educators in twenty-six of the sixty settlements, located both in the Northwest Territories and Nunavut (Lidster 1978). A decade later, Arctic College was created, including campuses in Iqaluit and Fort Smith. In 1995, Arctic College was divided into Aurora College, in present day NWT, and Nunavut Arctic College. Aurora College has twenty-three learning centres across the territory, and Nunavut has five campuses in twenty-five communities.

Besides these projects that were run at a federal level, other programs linked to the oil and gas industry were introduced and coordinated between the colleges and private employers (Abele 1989; Hobart
and Kupfer 1974). Men had also been sent outside the NWT for apprenticeship training prior to the development of college programs (Hill
2008; Jenness 1964).

One of the most significant events shaping northern resource
development VET policy was the discovery of oil in the Beaufort Delta
of the Northwest Territories in 1970. In response to the discovery, the
Canadian government proposed guidelines for the construction and
operation of oil and gas pipelines in the North. That led to public
hearings known as the Berger Inquiry (1974–77), named after Justice
Thomas Berger, who led hearings into a proposal to develop a pipeline
along the Mackenzie Valley. The Inquiry gathered testimony from
close to 1,000 witnesses in thirty-five communities (Berger 2004). The
general sentiment expressed was for development to proceed only
after land claims had been negotiated. While industry and government
proponents raised the promise of local employment, Aboriginal oppon-
ents wanted to ensure their lands would be protected first. The standoff
resulted in the landmark decision to put a moratorium on development
until land claims were settled.

Since then, the politics of protest characterizing the Inquiry has been
transformed into a politics of partnerships, which characterized the
public hearings that began in 2006 for this megaproject when it was
revived as the Mackenzie Gas Project. With land claims settled, newly
formed Indigenous governments and their industry partners resur-
rected the hopes for the project. Public-private training-to-employment
programs with a federal price tag of $12.7 million began in 2003. It
was intended that Indigenous northerners would benefit from employment in building the pipeline, with estimates of eventually offering
approximately 100 to 130 full time annual jobs (JRP 2009). However,
the project never resulted in the timelines identified, and employment
program graduates were forced to seek opportunities elsewhere (AFS
2008). The training society reported that there were 650 training
projects delivered, costing approximately $10 million, and that the
1,366 program participants attained 1,090 jobs at over 150 employers
(AFS 2008). However, of these, only nine indentured apprenticeships
with industry partners were subsequently formed, requiring apprentices to receive both their technical training and on-the-job training
in Alberta, where industry partners were operating (AFS 2008).

The partnership program developed to train northerners for a pipeline, that remains a "pipe dream," illustrates some of the challenges associated with training a northern labour force. VET partnerships involve varying levels of support and coordination from federal, territorial, and provincial governments, Indigenous governments, and industry partners. In addition to providing in-kind support, industry partners provide on-the-job training, including apprenticeships. Governments generally provide student funding and training delivery agents—usually through an established post-secondary vocational college system. And so, while Indigenous groups have greater control of how a proposed development can proceed, including the types of potential benefits, the federal government still retains significant control of how funding tied to training will be allocated; in regions where resource development occurs, federal government funding agreements for training are tied to the ability of local communities to partner with resource developers (ASEP, n.d.; GOC 2000, 2009, 2013). This means that in order to receive federal funding, Indigenous groups must first partner with resource developers operating in the region. That can then lead to tenuous agreements, due to changes to the global economy and the ability of partner groups to develop just-in-time training delivery, for projects that may never come together, as in the case of the Mackenzie Gas Project.

CASE STUDY 1: WOOD BUFFALO

The Regional Municipality of Wood Buffalo is located in northeastern Alberta, bordering both Saskatchewan and the Northwest Territories. The region contains the largest known reservoir of crude bitumen in the world. The main employer is the oil sands mining industry.

Wood Buffalo contains five First Nations of Cree and Chipewyan descent, as well as significant populations of Métis. At the time of research, of the 51,496 residents of Wood Buffalo, approximately 5,365 were Indigenous, with 2,530 identifying as Métis, and 2,225 registered as status Indians (Statistics Canada 2008). Approximately 10 per cent of the oil sands workforce, or 1,700 workers, identify as Indigenous (Government of Alberta, n.d.). According to local labour market studies, Indigenous people in the municipality want full-time work in the oil sands industry, primarily in the skilled trades (ATC 2005, 2007).

Historically, the industry has made provision to train and hire an Indigenous labour force, beginning with the construction of Suncor in 1964. As part of its licence to operate, and in conjunction with the Department of Indian and Northern Affairs, Suncor sponsored and developed on-the-job training programs. A second mine, Syncrude, created a Native Personnel Specialist role, responsible for designing and implementing specialized training programs. Canadian Bechtel (the company that constructed Syncrude) also worked in conjunction with both the Construction and General Workers Union and Keyano College to develop a five-week pre-employment course (Littlejohn and Powell 1981; Krahn 1983). As well, mine operators were encouraged to award contracts to Indigenous businesses (Littlejohn and Powell 1981). Ultimately, though, most Indigenous employees occupied low-skilled and low-paying positions.

This specific case examines a training program sponsored by a prominent oil sands mine in the province of Alberta. The mining industry is important to the Canadian economy, but it has also garnered significant media attention over environmental degradation, which has been linked to the adverse social and health impacts on Indigenous communities located downstream from its operations (Kelly et al. 2010). The Regional Municipality of Wood Buffalo was chosen as the case site for several reasons: (1) it is the municipality where most oil sands mining activity occurs; (2) it has well-established training-to-employment programs; and (3) Indigenous groups in the region have a complex, contested relationship with the mining industry.

The mine donated $5 million to a local college, and of that, about $600,000 was dedicated annually to a training program over a three year period. This program began in 2009, with a cohort of forty students, and ended in 2012 with the last cohort of forty students. The other half of the program funding was provided by the provincial government and local Indigenous groups. The twenty-nine-week program was designed to upgrade students' knowledge and skills to allow them to pass various trades entrance exams. A one-month work placement at the mine gave them practical on-the-job training and experience. Participants had to be eighteen years or older, have a minimum of Grade 9 education, be drug and alcohol free, and indicate an interest in working in the mining industry. The program aimed to develop the skills and competencies needed to proceed onto a trade's apprenticeship path with the mine. Recruitment of candidates involved three selection protocols performed by different partners in the following order: Indigenous employment coordinators (program eligibility), the college (academic eligibility), and the mine (employment eligibility). Program admission was based upon a scored selection criteria.

· The training program was offered at four community learning centres and two campuses. The in-class portion of the program lasted twenty-nine weeks, with students attending a minimum of thirty hours of instruction per week. Program materials included seven modules of "Connecting to College and Careers" (704 hours). The final month of the program was spent at the mine site, where students performed a variety of jobs related to a chosen trade under the guidance of a mentor. . Upon successful completion of the program, students were offered an apprenticeship with the mine.

A cohort of students were interviewed while they were attending classes (fall 2010), and again, six months later, during the apprenticeship phase at the mine site (spring 2011). Throughout the inquiry, I was able to observe classes, attend induction and graduation ceremonies, and have ongoing conversations with students and program stakeholders. Semi-structured interviews with student participants (n =10) and program stakeholders (n =10) occurred during both the classroom and work phases of the program. In-person student interviews were held in the fall at a regional college in the city of Fort McMurray and involved two focus groups (FG-1:3 M; FG-2:3 M:2F) and two individual interviews (2F). Individual follow-up interviews occurred individually, either in-person or over the phone. Stakeholders came from organized labour (four), schools (two), a community college (one), Indigenous organizations (two First Nations and one Métis), and the provincial government (one).

With the assistance of college personnel, students were selected based on gender (6M:4F), ethnicity (seven First Nations, three Métis), and home community (four local communities and one Alberta city). Student ages ranged from eighteen to twenty-seven years, with an average age of twenty-two. Interview questions initially related to recruitment, reasons for taking the program, career choice, educational and family background, role models, and challenges associated with acquiring an education and training. Follow-up interviews focused on their experiences apprenticing with the mine, perceptions of the training program, and future career paths. Individuals who were no longer in the program were asked why they chose to drop out. Stakeholder interviews were less structured, and were used to gain further insight about the program, including the history of training partnerships in the municipality, and associated challenges and successes.

FINDINGS

The findings and results presented here relate to VET participant demographics, including levels of education and program participation, followed by perceptions shared during the learning-to- work transition. For purposes of providing a coding designation, "I" designates interview (e.g., I-1 refers to Interview 1), "FG" designates focus group, and "M" and "F" designate gender.

All VET interview participants had a fragmented K-12 schooling experience. None of the students had graduated from high school with an advanced diploma, and most had dropped out by Grade 11. The levels of schooling reported by students included Grade 12 equivalency (2M; 1F); Grade 11 (4M; 2F); and Grade 10 (F). Only one student reporting a Grade 12 equivalency graduated from a local high school. Four students had upgraded at a community college campus. Four were labourers and two were unemployed before entering the training program. By the time the college portion of the program had ended, three males and one female had dropped out without gaining pre-trade certifications; the other six students—three male and three female—graduated and went on to a successful apprenticeship with the mine. These varied backgrounds by which participants entered the training program illustrate the fragmented learning-to-work transitions frequently experienced by vulnerable populations.

Given the available data, figure 4.1 illustrates participation and attrition of VET participants in the three cohorts. According to a college manager, the increase in number of applicants in the second year was attributed to the program's popularity, which was communicated through word of mouth, as well as in stories in the local newspaper. By attracting a larger sample of applicants in the second year, the selection of suitable candidates most likely improved. Evidence of improved candidate selection can be seen in a decline in attrition rates occurring during benchmark junctures of the program. For instance, the number of participants offered an apprenticeship at the time of graduation and who were still apprenticing a year later remained the same in the second cohort. By offering the program over three years, facilitators were also able to learn from previous mistakes and to improve program delivery.

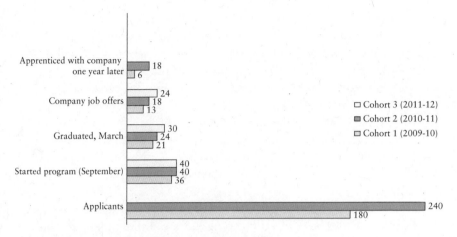

Figure 4.1 Program participant recruitment and retention

As the program screened applicants based upon ethnicity, credentials, and attitudes towards work at the mine, responses relating to career choice had already been somewhat predetermined. For example, credentials and attitudes were scored on a points system. However, other factors mentioned during interviews relating to reasons for working at the mine included wages, family background, and other personal associations with the industry. Some respondents also felt that by relocating to Fort McMurray they would have more options available to them in terms of lifestyle and education, and therefore be able to enjoy a higher quality of life.

In addition to the mine's reputation of workplace safety, participants also described the lucrative wages they can earn there. Positive attitudes about wage earnings appeared to remain throughout the course of the program. At the time of follow-up interviews, several apprentices had already undertaken overtime shift work. A typical twelve-hour shift rotation cycle involves three-day and three-night shifts, followed by six days off. Overtime allows workers to earn time and a half for the first twelve hours of work on scheduled days off and double time thereafter. As one apprentice reasoned, by doing overtime and working statutory holidays like Christmas, he can earn an extra $10,000 dollars, which he can then put towards a truck (1–16). Another apprentice cited the signing bonus with the mine which amounts to $80,000 towards a home mortgage for ten years of service (1–19).

The mine sponsor also offers a fly-in/fly-out (FIFO) program for employees who want to return to their home community on their days off. At the time of follow-up interviews, two of the three students interviewed from the same community were currently returning home on the FIFO program; however, these students indicated that they may later relocate to Fort McMurray as it is tiring to make the regular commute to and from their home community. From these accounts, it can be surmised that, as mobility patterns shift to accommodate both a mine employee lifestyle and earning potential increases, new identities and values will begin to form.

Many participants described family members as labourers who possessed trades related skills. Some of these family members had also worked at the sponsoring mine or were indirectly associated with it through Indigenous organizations (such as employment coordinators). The following responses illustrate the intersection of these relationships on participants' career aspirations:

Well my uncle is a welder and that's kind of why I want to get into it. He's a good welder. And I want my family to be proud of me. (FG-2, M)

I remember when I was younger, my dad – he built our home...he put up the whole system for us; he gave us lights and everything. I was right along-side him helping him and so that's kind of why I picked that trade. (FG-1, F)

My dad worked there [the mine sponsor] for 30 years and he retired so why not keep it going? (FG-2, F)

Indigenous business and political leaders and related organizations provided critical networks of support. In addition to administering and distributing student financial aid, Indigenous organizations also sponsored various oil sands career development programs. For instance, the Athabasca Tribal Council in Fort McMurray sponsored summer "career camps" for youth. Guests included Indigenous leaders and Elders, employment personnel, as well as industry representatives who were present at induction and graduation ceremonies in Fort McMurray. Honoured guests offered words of encouragement to students to urge them to complete their education and fulfil their career goals with the sponsoring mine. Similar messages were reinforced in inspirational posters displayed at the regional college during the in-class portion of training.

CASE STUDY 2: NUNAVUT

Inuit participation and retention rates in the mining sector have always been low. The first modern mine in Nunavut opened near Rankin Inlet in 1957; the five-year underground nickel mine employed eighty Inuit full time and around twenty others part time by 1961(Rodon and Lévesque 2015). The Nanisivik lead–zinc mine, which began construction in 1974 and operated near the Baffin Island community of Arctic Bay from 1976 to 2002, also employed relatively few Inuit, with most working no more than one or two months despite reporting that they liked their work and the money they earned (Hobart 1982). The proportion of workers quitting before completing a six-week work period in Nanisivik generally increased from 11 per cent in 1975, to 22 per cent the following year, 30 per cent in 1977,, and 23 per cent in 1978 (Hobart 1982). Of those Inuit employed, only 5 per cent were apprentices, 23 per cent were semi-skilled workers, and 8 per cent were heavy equipment operators. The remainder were labourers (43 per cent) and helpers (16 per cent) (Hobart 1982).

Hobart (1982) noted three distinctive characteristics of the Inuit workforce at the Nanisivik Mine: 1) the very wide area from which they had been recruited (twenty-three home communities across the Northwest Territories); 2) their youthfulness (median age of twenty-eight); and 3) the surprisingly high proportion who were single (49 per cent). These factors, he suggested, were indicative of "the relative disinterest of many Inuit for the existing conditions of mine employment."

Reasons for low participation and retention rates usually involve a combination of factors that have been consistently reported by Inuit workers relating to the nature of shift work separating workers from their families. Hobart (1982) noted that while respondents indicated that they liked working at the mine, the six-week on and two-week off rotation cycle was cited as interfering with traditional family ties and roles such as hunting. More recently, similar observations have been made by Rodon and Lévesque (2015) regarding the nature of shift work impacting home life and interfering with hunting (Bell 2017; Rodon & Lévesque 2015). Rodon and Lévesque (2015) also speculate as to how the evolution of mining has impacted employment in recent years. In the 1950s, it was not an easy undertaking to fly-in/fly-out the miners needed. Now, however, this practice has become the norm for northern mines, and local workers are no longer so necessary.

In addition, the mining profession has changed a lot over the years. Mines have gone from primarily manual jobs that required much physical ability and little education or training, to skilled trades that require higher education and proficiency in complex technologies (Lévesque 2015). The authors also note that Inuit have not always been interested in the jobs made available to them, and that when work is made available, there is usually little opportunity to gain promotions.

The case examines outcomes of training and employment provisions of The Mary River Inuit Impact Benefit Agreement (MRIIBA 2013). This agreement was negotiated in 2013 between the Qikiqtaaluk Inuit Association (QIA), representing the Inuit of Baffin Island (Qikiqtaaluk) in Nunavut, and Baffinland Iron Mines Corporation (BIMC), an international iron ore company. The Mary River Project is located on Inuit-owned lands in the North Baffin region of Nunavut. The project is a multibillion-dollar, open-pit, high-grade iron ore mine, with a life expectancy of twenty-one years and opportunities for expansion (Dalseg and Abele 2015). Over this time frame, the project is expected to generate thousands of jobs and to triple the growth rate of the territory's annual gross domestic product. It may provide nearly $5 billion in tax revenue and royalties (Ritsema et al. 2015).

Based upon Article 26 of the Nunavut Land Claims Agreement (NLCA), project developers must negotiate an Inuit Impact and Benefit Agreement (IIBA) with regional Inuit organizations (RIAs), and offer compensation, royalties, local employment and training, and business contracts. The MRIIBA is supposed to ensure that the five signatory communities in the Qikiqtaaluk Region specifically benefit from the mine through royalties, priority-hiring arrangements, and education and training programs including trades apprenticeships. The goal of a minimum of 25 per cent Inuit employment for Baffinland staff and for all new contracts awarded since 2016 has also been identified. So far, only half that rate has been achieved (Bell 2017; Brown 2017). To achieve the goal, Article 8.1.6 of the agreement requires that various stakeholders, including the company, the Qikiqtani Inuit Association, territorial training institutions, and the different North Baffin communities "use their best efforts … to promote communication among education and training participants" (MRIIBA 2013).

Mittimatilik (Pond Inlet; 72.7°N, 77.9°W) was chosen as the Nunavut case site as it is both the most populous North Baffin community (pop. 1,617; 90 per cent Inuit), and is located nearest to the project (approximately 160 km southwest). Politically, it is classified as an incorporated hamlet. Like most communities, Mittimatilik's formal economy is service based, with its labour force dominated by public administration, retail trade, educational services, construction, health care, and social services sectors (Statistics Canada 2013, quoted in Ritsema et al. 2015). Subsistence hunting, fishing, and gathering, as well as the practice of cultural arts in tourism and sales in arts and crafts, are also important parts of the overall economy (Ritsema et al. 2015). Mittimatilik has an elementary/high school, as well as a community learning centre that offers programs and courses for adults.

Field work took place over two summers (July 21-August 8, 2015; July 4-21, 2016). Unlike Wood Buffalo, where there was no VET program, workers and people who had been previously employed at the mine in Nunavut were interviewed instead. A total of forty-eight interviews and one focus group were conducted with workers using semi-structured and open-ended interviews. Of the initial twenty-two workers interviewed in 2015, fourteen were subsequently re-interviewed the following year; information concerning the status of five workers was also provided through word of mouth, and no information was gained about the status of three individuals. An additional worker was also interviewed once in 2016. Eleven interviews and two focus groups were conducted with stakeholders as well, during both phases of the field work. Stakeholders included community members involved in education and health care, residents who had a longstanding knowledge of the community, and representatives from various levels of government, including municipal, regional, territorial, federal, and Inuit organizations. While several conversations occurred with mine representatives, no formal interview was granted. In addition to stakeholder interviews, other informal conversations and observations, including those made during community and regional socio-economic meetings, occurred throughout the inquiry.

Recruitment of worker interview participants initially occurred with the assistance of hamlet council employees, as well as the use of a community Facebook site and local radio announcements; thereafter participants were recruited through word of mouth. Most interviews occurred at the hamlet office, although some took place in other public places (n=4) and residences (n=3). Initial interviews lasted for approximately thirty to forty minutes, with questions focusing on work experience, levels of training and education, lifestyle, and career aspirations. Follow-up interviews were subsequently tailored to different individual circumstances and lasted between twenty and fifty minutes. Questions concerned the identifying of education and career goals, the degree to which workers were successful in fulfilling those, and their level of success in gaining employment, receiving training, or getting job promotions.

Most interviews were audio recorded, while eight relied solely on note-taking. Several interviews (n=5) required the assistance of an interpreter. Prior to being interviewed, participants in both case sites were explained the purpose of the research. A copy of the letter explaining the research and ethical considerations was also provided and signed by participants. All interview notes and recordings were transcribed and sent to interview participants to verify the accuracy of the statements made. Interviews were subsequently thematically coded.

FINDINGS

The findings and results presented here are organized into two sections: 1) worker demographics relating to employment with the project; and 2) experiences of workers at the mine. Stakeholder interviews are incorporated as a means of providing additional context to the perceptions shared by workers. For purposes of providing a coding designation, "w" designates worker (e.g., w-1 refers to Worker 1).

Worker Demographics

Most workers (n=23) were men (n=19) whose ages ranged from twenty-one to sixty-three; however, the majority were in their twenties. In terms of their employment history with the project, workers were organized into four cohorts: Cohort 1 (n=5; 4 men, 1 woman) were continuously employed at the time of interviews; Cohort 2 (n=4; 3 men, 1 woman) were employed and subsequently laid off between interviews; Cohort 3 (n=12; 10 men, 2 women) had been laid off for over a year; and Cohort 4 (n=2) had applied, but never gained employment. Whereas Cohort 1 workers are characterized as having relatively high job security, the other cohorts experienced job insecurity. By combining the last three cohorts together, job insecurity reflects the majority of interview participants (n=18).

According to the National Occupation Classification (NOC) matrix, occupations are organized into five categories and are based on skill level: level A occupations require university education; level B occupations usually require college education, specialized training, or apprenticeship training; level C occupations usually require secondary school and/or occupation-specific training; and for level D occupations, on-the-job training is usually provided (https://noc.esdc.gc.ca/Structure/Matrix). Most respondents can be categorized into level D occupations, with the majority either indicating employment as general labourers (pouring concrete and laying rebar), or in service support (dishwashers, assistant cooks, housekeeping). Level C occupations reported included heavy equipment operators (HEOS) and community liaison officers (CLOS). The role of a CLO is to act as community contact and to assist in communication between workers and human resource personnel.

In Cohort 1, all men were employed as heavy equipment operators (HEOS). Duration of employment ranged from two to three years, and one individual (W-17) had been employed with the project since 2002. One woman (W-13; aged 46), who had been a housekeeper at the mine for the past three years, reported that she had been on medical leave and was returning to her job the following week.

In Cohort 2, duration of employment ranged from four months to six years. All workers had been employed by contractors, although one employee (W-2) had also worked for the mine and was a general labourer for one and a half years; W-6 had worked as a dishwasher and general labourer for six years; and W-21 had been an HEO for two years. The female worker (W-23) had been employed in service support as a baker's assistant, housekeeper, and dishwasher for four months.

Most Cohort 3 workers had worked for a contractor (n=9) and were employed on a temporary seasonal basis in service support (dishwashers, assistant cooks, janitors, housekeepers). The term of employment for contract employees ranged from one month (two rotations) to six years, with an average duration of one and a half years. Three workers had worked for the project in the following positions: community liaison officer (four years), administrative assistant (three months), and HEO (six weeks). Duration of these jobs ranged from one month to four years. Cohort 3 workers had been employed from 2008 to 2014. However, most indicated that they were last employed at the project in 2013 and 2014 (n=5).

In Cohort 4, two people had applied, but had never worked for the project. One man (W-19) had applied for a managerial position, but was incarcerated at the Baffin Correctional Centre between

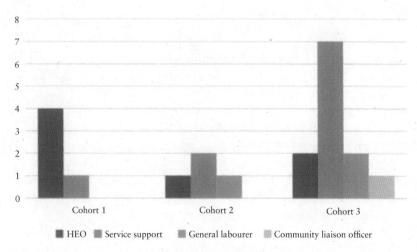

■ HEO ■ Service support ■ General labourer ▨ Community liaison officer

Figure 4.2 Project occupations reported by workers

interviews, while the other man (W-10) did not respond to my request for a second interview. In this case, it is assumed he is not working at the project. Figure 4.2 summarizes the occupations reported by Cohorts 1 to 3.

Education and Training

Education grade levels are based upon the Alberta curriculum, which Nunavut uses portions of in its own programming. The highest education grade levels achieved fell into the following groups: less than a Grade 10 (n=7); Grade 10 (n=8); Grade 11 (n=3); and Grade 12 (n=3). However, several community educators indicated the grade levels reported would not be commensurate with provincial levels of achievement. Reasons for this discrepancy included a policy of passing students along with their peers ("social pass"). Also, since few students go onto high school in these communities, it is not realistic to staff the schools with high school subject specialists. Limited course offerings also impacted learning outcomes, with English 30 being the only academic Grade 12 course offered at the local high school.

Some interviewees mentioned changes to the way skills are assessed, which affects access to mine-related employment. Whereas people were once screened on demonstrable skills, educational credentials now preclude workers from accessing the same jobs. This means that older workers are able to access better jobs (and pay) because of their work experience; younger workers are left out because they do not have the educational credentials or the work experience. These observations appear to contradict Article 7.3.2 of the MRIIBA (2013), which states "Where appropriate, the Company will consider ability, skills and experience as an equivalent to formal qualifications, and Inuit applicants with experience equivalencies will be treated equally with all applicants with formal training."

When asked about supports to get further education or training, most workers were unclear about where to get assistance, or the career paths that they needed to follow in order to achieve their stated career aspirations. Younger males in Cohort 1 were best able to articulate clear linkages between education, training, and employment. This most likely relates to their better facility with the English language, which would also allow better access to sources of information, either through the internet or other educational supports available to them through the local school or community learning centre.

Where would you apply for that [training]?
Guess through Arctic College, or I don't know where I would apply, just pretty lost right now in Pond. (W-10, 2015)

So, you don't know what the qualifications were, or how to get them if you want to work?
I'm not sure how to get those qualifications.
How about going to Arctic College—is that part of it?
I've been meaning to go back to school. (W-16, 2015)

As the above two passages suggest, the relationship between the college and the project had not significantly developed by the time of the interviews. Nunavut Arctic College offers programs that are tied to funding, however, owing to a combination of Nunavut's policy of decentralizing government administration and services to regional communities and the territory's enormous size and sparse population, resources are spread thin. With respect to the delivery of post-secondary education, this means that some courses may not always be offered at one of Nunavut's five campuses or the community learning centres located in each of its twenty-five communities. As one stakeholder explained, courses that are offered are the ones that are cheaper and less complicated, given available personnel and housing. Several community stakeholders also noted that some courses are readily offered (e.g., Office Admin, the Teacher Education Program, Adult Basic Education); whereas programs like the Environmental Technology

Program (funded by the company during the second year) were considered too expensive to run a second time, despite reportedly graduating eleven students (compared with two from the Teacher Education Program). A sixteen-week pre-trades program was offered the previous year, yet only two of the nine students who participated passed the Trades Entrance Exam. The success rate was so low because the exam takers lacked basic English language comprehension, math, and scientific literacy. Some stakeholders also felt that a greater emphasis on trades was needed in the school in order to generate further interest from students.

On-the-job training offered by the company includes the following: 1) operations procedures (e.g., site orientation); 2) health and safety (e.g., first aid); 3) spill response (cleanup and reporting); 4) supervisor training for those with experience as managers; 5) other areas such as HEO assessment and training (Brubacher 2015b). With respect to on-the-job training, some workers were able to articulate opportunities. However, rather than providing a clear sense of career pathways, recollections were oftentimes vague or noncommittal, meaning that respondents generally did not express a definite course of action.

And do you see yourself continuing as a dishwasher and as a housekeeper, or do you see yourself getting other types of work there?
To be honest, I don't know. Keep working as a dishwasher and a housekeeper until I can see what the money to go to a bakery course.
To take a bakery course? And would [your employer] be able to support you in doing that, in taking that course? Would they be able to train you or get you sent out for training?
I was talking to my supervisor about that but they were very busy, and didn't have much time so I didn't really get my answers. (W-2, 2015)

So you're a heavy equipment operator?
Yeah.
And when you say they trained you up, what does that mean?
I was in the classroom and the paperwork, simulators—they have simulators for the equipment.
Oh right, they have the classroom component.
At main camp site.
How long was that for?

Good month training back-and-forth in-and-out of the field, and in the class.

Did you get a certificate out of that?

Just based on test, test score at the end, yeah....

Would you be interested in getting more training?

Yeah, I'm always wanting to learn more.

And what's the opportunities for more training? Or did they talk to you about that?

Yeah, there's um, site supervisor, yard supervisor, the crew that's out in the oilfield, there's the office department, exploration, geology, other departments where they need your diplomas.

Are you looking to enter into a trade?

They offer all. There was talk of offering apprenticeship grants in other departments, mechanics electrician stuff like that if the mine grows. (W-12, 2015)

A year later, W-12 had taken a three-month unpaid leave to sort out family tensions and to help raise his children. He also informed me that his circumstances required that he quit his job, and he was subsequently not eligible to apply for Employment Insurance.

With respect to apprenticeships and trades, Brubacher (2015) reports that in 2014, one third of all work performed at Mary River fell into the construction and maintenance trades areas, yet only 2 per cent of this work was performed by Nunavummiut (12). The one worker (W-14) who was able to discuss his hopes of entering into a trade described his work situation this way:

I'm a fuel truck driver waiting on being apprentice mechanic.

You're the first person that I've talked to who is talking about apprenticeships ... you're also the first person I talked to that's gotten their Grade 12 education. So, is there a relationship between getting your Grade 12 and apprenticing?

Ah, yeah, sort of. I would like to get into that kind of stuff like apprentice.

And where did you hear about the apprenticing? Did they approach you at [the project]?

Yeah, they did when they were interviewing, and talking about the mine opening again. They were saying there's going to be apprentice jobs and training at the work site, so I went ahead with fuel truck driver since that's part of maintenance which is with the mechanics, welders, and, so got to work alongside mechanics and welders, and people like that.

Did they approach you about apprenticing, or did you approach them?

Well, I asked them about it, and then after my work probation they asked me if I wanted to get into trades, and I went with it. (W-14, 2015)

A year later, W-14 informed me that there was a "barrier in getting my apprenticeship" that had to do with writing the Trades Entrance Exam, an exam that he had once written, but which had since "expired." He had challenges in passing the mathematics part of the exam, and he felt the local college could assist him with that. However, making time to study was challenging: "there's also family time; it gets very stressful … I was already signed up for overtime, and so I did not work on the course work."

This worker's unfortunate set of circumstances illustrates the precarious learning-to-work transitions accompanying shift work. It also underscores the importance of establishing support systems for prospective apprentices. A component of on-the-job training should therefore include appropriate time to upgrade, because, as the above passage indicates, such training cannot be realistically completed during time off, when workers need to rest and attend to other responsibilities.

DISCUSSION

In both case sites a combination of on-the-job and classroom training was used to teach participants. These programs received support from partner organizations. Various levels of government, with their administrative centres located in regional hubs (Iqaluit and Fort McMurray), helped to coordinate programs. They also provided financial and administrative support. In addition, Indigenous governments, acting on behalf of the locally affected communities, negotiated agreements and monitored program outcomes. They then used the results to pressure other partners to improve outcomes.

As noted in the interviews with VET participants and workers, both groups clearly hoped to gain employment with the mine. Despite the remoteness of operations, both groups were also willing to participate in FIFO shift work. Participants in Wood Buffalo were given the option of relocating to Fort McMurray, which was more attractive, partly because of its proximity to the mine. However, low educational achievement of VET participants and workers alike indicates that the ability to pass trades entrance exams continues to represent a significant barrier to entering apprenticeship programs.

Aside from these similarities, differences between the cases significantly impacted outcomes. Unlike VET participants in Wood Buffalo, miscommunication between workers and employers in Nunavut can be attributed to language barriers, as many workers did not speak English as their primary language. It is also reasonable to assume that cultural barriers pertaining to both wage labour and FIFO employment also affected understanding and expectations concerning roles and responsibilities and ultimately, conflict resolution. Miscommunication contributes to misunderstandings relating to recruitment, training, and termination of employment. Consequently, there is an urgent need to support the role of community liaison officers, especially considering the remoteness and isolation of communities, given that the mine's head office in located in Ontario. Relatedly, considering the high turnover rates and termination of workers, work ready programs" should also have a section outlining worker rights in addition to their responsibilities, including a list of who they can contact when problems arise. These measures are especially important given the region's geographic remoteness and the absence of a union or other community liaison that can actively intervene and mediate disputes.

However, miscommunication existing between workers and employers is symptomatic of larger overarching issues occurring at an institutional level. As Abele (2006) notes, training programs intended for megaproject developments are episodic, uncoordinated, disjointed, and suffering from "program amnesia" (i.e., repeated introduction of new programs to address persistent problems). Whereas the Regional Municipality of Wood Buffalo has had the same major mining companies operating in the region since the 1970s, the mine in Nunavut is new, having only recently entered an operation phase of development. The stage of the mining life cycle impacts the labour force: while construction phases require a large workforce for a relatively short period of time, the operational phase requires a smaller, but longer-term, stable workforce. During this latter phase of development, stable linkages can begin to form between different partners, including improved communication as to what is realistically possible in achieving training and employment.

Stable linkages can be seen in the Wood Buffalo case as noted in the (relatively) clear communication between the different partner organizations. In turn, VET participants received consistent support from the different partner organizations as already mentioned in induction and graduation ceremonies. They also received support from family members and friends who had worked at the mine, and who were able to convey to them realistic expectations about working there. Stable linkages can also be seen in the program itself. By offering three cohorts, lessons learned could be applied to the following year. Evidence of reducing "program amnesia" can be seen in the increased recruitment, retention, and apprenticeship completion rates over the three years the program was offered.

Conversely, analysis of the weaker linkages in the Nunavut case helps provide insight into some of the communication gaps existing, and their resulting impact on training and employment pathways for workers. As Ritsema et al. (2015) note with respect to Nunavut's multilayered system of governance, the "convoluted system of competing interests, perspectives, and goals" has resulted in confusion among local communities. The resulting complexity makes it easy to lose track of who is responsible for what service or policy. This confusion is extended to the ad hoc provision for VET partnerships. Owing to Nunavut's policy to decentralizing services, Nunavut Arctic College has three regional campuses and twenty-five community learning centres. A trades training centre is located in only one region—at Rankin Inlet's Kivalliq campus. This stands to reason given Nunavut's sparse population. However, the MRIIBA requires that local labour be preferentially hired. This means that other than funding short-term programs at community learning centres, resource developers are more likely to provide on-site training rather than develop third-party agreements with the college. Residents who do not live in the Kivalliq Region are therefore constrained in terms of accessing VET opportunities.

At the time of the research, no formal agreements had formed between the college and the mine. Given its present operational phase of production, it seems reasonable that a formal agreement will inevitably occur. That could result in the formation of a new trades centre. However, if a new trades centre is to be created in the region, then questions remain unanswered. For instance, how many local candidates will there be to fill programs being offered? Will there be provision for residential housing to accommodate both instructors and students from neighbouring communities? What will be the *long-term* funding commitments from the various partner organizations for such a centre? How much of the funding will be devoted to pre-apprenticeship programs, and how much for upgrading? And finally, when the mine closes, what will become of the centre?

SUMMARY

Given the relatively recent time frame in which development has occurred, northern regions share similar experiences with resource extractive industries in other areas. As the chapter has shown, they also share similar challenges, including those associated with training and retaining a local labour force. If sustainable employment is that which is long-term, skilled, and associated with the trades—i.e., jobs that are well paid, relatively secure, and transferable once the mining cycle has run its course—then various factors must coalesce to support such a worthy goal.

By sharing findings from two very different case studies, this chapter has attempted to shed some light on both the enormous challenges as well as the possibilities that exist in achieving sustainable employment with resource extractive industries. Compared to the newly developed programs and initiatives examined in Nunavut, Wood Buffalo illustrates similarities, differences, and features that help make programs and partnerships successful in achieving the intended outcomes. Many participants in Wood Buffalo transitioned into trades-related employment, and several factors contributed to their success. These include: (1) a committed employer who has been in the region for several decades; (2) tight linkages existing between education, training, and employment; (3) role models, particularly family members who have been or are currently employed at the mines; (4) rigorous selection and screening processes involving numerous stakeholders; (5) a large pool of candidates to draw from; (6) multiple program cohorts which allow stakeholders to improve delivery; and (7) lucrative salaries and career opportunities.

REFERENCES

Abele, Frances. 1989. *Gathering Strength: Training Programs for Native Employment in the Northwest Territories.* Calgary, AB: The Arctic Institute of North America.

–2006. "Education, Training, Employment, and Procurement." *Submission to the Joint Review Panel for the Mackenzie Gas Project.* Yellowknife, NT: Alternatives North.

AFS (Aboriginal Futures Society). 2008. *Annual Report 2007–2008.*

ASEP (Aboriginal Skills and Employment Partnerships). (n.d.). Retrieved from http://Aboriginalskills.ca/_downloads/ASEP_1pager.pdf.

ATC (Athabasca Tribal Council). 2005. *2005 – 2012 Labour Market Analysis.* Prepared for the Sustainable Employment Committee, APCA. Fort McMurray, AB.

–2007. *2006 Labour Market Analysis.* Prepared for the Sustainable Employment Committee, APCA. Fort McMurray, AB.

Bell, Jim. 2017. "QIA to Push Hard for Inuit Employment at Mary River." *Nunatsiaq News Online.* May 9, 2017. http://www.nunatsiaqonline.ca/stories/article/65674qia_to_push_hard_on_inuit_employment_at_mary_river/.

Berger, Thomas. (2004). *Northern Frontier, Northern Homeland: The Report of the Mackenzie Valley Pipeline Inquiry* [electronic resource]: Ottawa, ON: Department of Indian Affairs and Northern Development, Government of Canada.

Brown, Beth. 2017. "Inuit Org Hopes to Boost Job Numbers at Mary River." *Nunatsiaq News Online.* October 10, 2017. http://www.nunatsiaqonline.ca/stories/article/65674inuit_org_hopes_to_boost_inuit_job_numbers_at_mary_river.

Brubacher, Doug. 2015. *Baffinland Iron Mines Corporation Mary River Project: 2014 Socio-economic Monitoring Report, March 2015.* Ottawa: Brubacher Development Strategies, Inc.

Dalseg, Sheena and Frances Abele. 2015. "Language, Distance, Democracy: Development Decision Making and Northern Communications." *The Northern Review* 41: 207–240. https://doi.org/10.22584/nr41.2015.009.

Dickerson, Mark. 1992. *Whose North? Political Change, Political Development, and Self-Government in the Northwest Territories.* Vancouver: UBC Press.

Government of Alberta. (n.d.). "Alberta's Oil Sands." http://oilsands.alberta.ca/economicinvestment.html.

GOC (Government of Canada). 2000. *Gathering Strength: Canada's Aboriginal Action Plan*. Ottawa: Minister of Indian Affairs and Northern Development.

–2009. *Federal Framework for Aboriginal Economic Development*. Ottawa: Minister of Indian Affairs and Northern Development.

–2013. *Budget 2013:Private Sector Participation in Aboriginal Training*. https://www.budget.gc.ca/2013/doc/plan/chap3-1-eng. html#a23-Job-Opportunities-for- Aboriginal-Peoples.

Hawthorn, Harry B., ed. 1967. *A Survey of the Contemporary Indians of Canada: Economic, Political, Educational Needs and Policies*, Volume II. Ottawa: Indian Affairs Branch.

Hill, Dick. 2008. *Inuvik a History, 1958–2008: The Planning, Construction and Growth of an Arctic Community*. Victoria, BC: Trafford Publishing.

Hobart, Charles W. 1982. "Inuit Employment at the Nanisivik Mine on Baffin Island." *Études/Inuit/Studies* 6, no. 1: 53–74.

Hobart, Charles W. and George Kupfer. 1974. *Inuit Employment by Gulf Oil Canada: Assessment and Impact on Coppermine, 1972-73*. Edmonton, AB: Westrede Institute.

Jenness, Diamond. 1964. *Eskimo Administration, Vol. II: Canada*. Montreal, QC: Arctic Institute of North America.

JRP (Joint Review Panel). 2009. *Foundation for a Sustainable Northern Future: Report of the Joint Review Panel for the Mackenzie Gas Project*, Volume I. https://aeic-iaac.gc.ca/155701CE- docs/Mackenzie_Gas_Panel_Report_Vol1-eng.pdf.

Kelly, Erin, David Schindle, Peter V. Hodson, Jeffrey Short, Roseanna Radmanovich, and Charlene Nielsen. 2010. "Oil Sands Development Contributes Elements Toxic at Low Concentrations to the Athabasca River and its Tributaries." *Proceedings of the National Academy of Sciences of the United States of America*, September 14, 2010. 1-6.

Krahn, Harvey J. 1983. "Labour Market Segmentation in Fort McMurray, Alberta." PH.D diss., University of Alberta, Edmonton, AB.

Lidster, Echo. 1978. *Some Aspects of Community Adult Education in the Northwest Territories of Canada, 1967–1974*. Government of the Northwest Territories.

Littlejohn, Catherine and R. Powell. 1981. *A study of Native Integration into the Fort McMurray Labor Force*. Edmonton, AB: Alberta Oil Sands Environmental Research Program.

Mclean, Scott. 1997. "Objectifying and Naturalizing Individuality:
A Study of Adult Education in the Canadian Arctic." *The Canadian Journal of Sociology* 22, no. 1 (January): 1–29.

MRIIBA. 2013. *The Mary River Project Inuit Impact and Benefit Agreement.* http://qia.ca/wp-content/uploads/2017/05/Complete_Mary_River_IIBA_from_QIA.pdf.

Ritsema, Roger, Jackie Dawson, Miriam Jorgensen, and Brenda Macdougall. 2015. ""Steering Our Own Ship?" An Assessment of Self-Determination and Self-Governance for Community Development in Nunavut." *The Northern Review* 41: 157-180. https://doi.org/10.22584/nr41.2015.007.

Rodon, Thierry and Francis Lévesque. 2015. "Understanding the Social and Economic Impacts of Mining Development in Inuit Communities: Experiences with Past and Present Mines in Inuit Nunangat." *The Northern Review* 41: 13-39. https://doi.org/10.22584/nr41.2015.002.

Statistics Canada. 2008. *Aboriginal People in Canada in 2006: Inuit, Métis, and First Nations, 2006 Census.* www.statcan.gc.ca.

5

Finding Space for Subsistence and Extractive Resource Development in Northern Canada

Rebecca Rooke, Harvey Lemelin, and David Natcher

Indigenous peoples of Northern Canada have for centuries maintained lasting connections with the environment through hunting, fishing, and gathering of wildlife resources. Today, as in the past, Indigenous peoples continue to harvest, process, distribute, and consume considerable volumes of wild foods annually. Collectively, these activities are known as "subsistence" and together comprise an essential component of northern Indigenous cultures (Gartler 2018). Since the late 1800s, mining has also assumed a significant presence in Northern Canada, and in the lives of Indigenous peoples. With substantial investment, mining industries have, for more than a century, looked northward in pursuit of resource wealth. The Canadian government has been equally committed to northern industrial development in their efforts to "modernize" the northern economy (Natcher 2019).

Despite more than a century of northern resource development, subsistence economies continue to demonstrate considerable resilience and remain integral to the health and well-being of northern Indigenous communities. This resilience has, in some cases, been achieved by incorporating mining into the contemporary mixed economies of Indigenous communities.

Rather than displacing subsistence economies, the wages earned through mining provide an economic basis for the continuance of wildlife harvesting, thereby invigorating social institutions and perpetuating traditional land-based values (Natcher 2009, 2019).

By creating employment opportunities and diversifying local skills, mining has contributed to the social and economic development of the North (Southcott and Natcher 2018).

Notwithstanding its purported economic benefits, mining has also resulted in significant environmental degradation. Beginning with exploration and continuing through mine operation to ultimate closure, mining has an impact on the land, air, and water of northern landscapes. The extent of practices such as clear-cutting and other habitat modification required for northern mining has also adversely affected wildlife populations (Bowman 2011; Johnson et al. 2005; Weir et al. 2007). The fragmentation of critical habitats has been particularly challenging to migratory species like birds and caribou (Boulanger et al. 2012; Klein 2000; Male and Nol 2005; Weir et al. 2007; Parlee et al. 2018). The environmental impacts from mining have, by extension, created vulnerabilities within Indigenous communities by putting their subsistence economies at risk. The inevitable tension resulting from mining and subsistence harvesting remains a matter of considerable contention between the industry and Indigenous communities in the North (Walton et al. 2001).

In recent years the evolution of the mining industry has been quite apparent. The industry has demonstrated an ability to incorporate a much broader range of values into development planning and management, in addition to those already required by legislative and regulatory regimes. Northern communities have also been empowered through the settlements of comprehensive land claims and other legislation. That, in turn, has effectively influenced the terms by which development can occur. Now more than ever, mining companies and Indigenous communities appear to be making considerable strides in ensuring that their respective economic interests can coexist, and that one form of production does not unnecessarily have to displace the other.

In this chapter, we examine the complex relationship between Indigenous subsistence economies and the mining industry in Northern Canada. The focal point of this study is on the mitigation strategies employed by mining companies to reduce the adverse effects of their activities on subsistence harvesting. This chapter is based on research that examined strategies used by the mining industry to mitigate potential impacts on wildlife and the environment, thereby supporting the subsistence economies of northern Indigenous communities. By way of conclusion, we offer recommendations for improving efforts that can contribute to the sustainable development of Northern Canada.

BACKGROUND

Following World War II, Indigenous communities in Northern Canada experienced a period of considerable change. This change was due in large part to the collapse of the fur trade, which left many northern Indigenous peoples challenged to provide income for themselves and their families (McMillan 1988; Bowman 2011). In some cases, government interventions included the forced resettlements of northern Indigenous communities in order to deliver social services (e.g., health care, food rations) and vocational training (Berkes and Jolly 2002; Watson et al. 2003) (see Hodgkins in this volume). It was at this time that the mining industry gained considerable government support to help transition Indigenous peoples into the industrial economy (Natcher 2019), and into more "socially acceptable lifestyles" (Hipwell et al. 2002; Tester and Blangy 2013; Tester et al. 2013). From that point forward, mining began to assume a significant presence across the North.

One of the first major mining operations was the Keno Hill Mining District, along with the Faro and Watson Lake projects in Yukon. Beginning in the mid-1940s these Yukon mines were among the world's largest producers of silver, lead, and zinc. In the 1950s, the Rankin Inlet Nickel Mine was established along the west coast of Hudson Bay (Green 2013). Other major mining projects were initiated throughout the 1970s. These continued to prioritize the employment of Indigenous peoples. Indigenous employment continued to be encouraged by government based on the belief it would lessen the need for health and social service funding (Silke 2009; Green 2013; Tester et al. 2013). However, the conditions at these mine sites proved often to be inadequate, with limited, poorly constructed housing, unhealthy sanitation, and alarmingly high rates of workplace mortality (Tester et al. 2013). For these and other reasons it was quite common for Indigenous employees to abandon mining employment and return to their subsistence lifestyles.

By the 1970s, social and environmental recognition of Indigenous rights was advancing (McMillan 1988). With the Berger Inquiry of the Mackenzie Valley Pipeline in 1974-77, the environmental and social impacts of northern industrial projects began to receive critical attention (Walton 2001). Other political advances included the settlement of comprehensive land claims. These agreements provided Indigenous governments with the authority to dictate the terms by which industrial development would occur within their territories (Walton 2001). Today, there are roughly a dozen large-scale mining operations in Northern Canada. These include the Minto Mine in Yukon, the Meadowbank Gold Mine in Nunavut, the Mary River Mine on North Baffin Island, and the Vale Mine at Voisey's Bay in Nunatsiavut. In addition to these operating mines, there are several other projects in exploration and development phases.

Research conducted on the impacts of mining on Indigenous communities has often focused on the effects to wildlife populations and, specifically, the physical alterations made to wildlife habitat (Bowman 2011; Couch 2002; Johnson et al. 2005; Weir et al. 2007). For example, industrial disturbances to critical habitat or migration routes can alter distribution, increase predation risk, and contribute to population declines (Boulanger et al. 2012; Bowman 2011; Couch 2002; Johnson et al. 2005; Klein 2000; Weir et al. 2007). Necessary infrastructure, such as runways, ports, tailing ponds, processing sites, living quarters, and transportation corridors, have all contributed to the loss of critical habitat (Bowman 2011; Couch 2002; Johnson et al. 2005; Weir et al. 2007). Some developments leave legacies long after mine closure. The cumulative effects of these development sites have shown to have spatial and temporal effects on wildlife (Johnson et al. 2005). For example, Boulanger et al. (2012) determined that the zone of influences (ZOI) of the Ekati Diamond Mine extends fourteen kilometres beyond the actual mine site. Within this ZOI, the probability of encountering caribou during their annual migration is greatly reduced due to human activity.

Mining in Northern Canada has also contributed to the increased incidences of human-wildlife conflicts (Bowman 2011; Johnson et al. 2005; Weir et al. 2007). Mining projects typically bring large numbers of transient labourers to remote northern regions (Boulanger et al. 2012; Johnson et al. 2005; Klein 2000). The increased anthropogenic pressures can result in wildlife mortality through vehicle collisions, occurrences of self-defence, and increased competition between hunters for fish and wildlife (Bowman 2011; Johnson et al. 2005; Weir et al. 2007). In addition, the scent of food and waste at these industrial sites can attract animals, often resulting in the extermination of "nuisance" animals (Bowman 2011; Johnson et al. 2005; Weir et al. 2007).

The impacts of mining to the natural environment also affects Indigenous communities. Many northern Indigenous communities continue to be dependent upon wildlife species for subsistence purposes, including the provisioning of nutritional country foods, and the maintenance of Indigenous identity through the traditional sharing and harvesting practices. Declining wildlife populations due to lost habitat caused by mining have had an adverse effect on traditional subsistence economies, as well as on community health and well-being (Bernauer 2011; Cameron 2012; Manley-Casimir 2011). Indigenous communities situated near mining projects have long expressed concerns regarding interrupted migratory patterns of wildlife, the lack of protection of critical breeding grounds, and contaminated food and water sources (Bernauer 2011; Ellis 2005). These impacts have significant consequences, especially when coupled with the broader impacts associated with climate change (Cameron 2012; Johnston et al. 2012).

Additionally, the boom-bust nature of mining operations poses additional risk to Indigenous communities who often fall victim to cycles of development (Tester and Blangy 2013) (see Schott et al. in this volume). As mining operations typically last for twenty years, once the resources have been extracted, many of the financial and infrastructure services cease operation, and the companies and economic development opportunities soon leave these remote regions. In these instances, the resident Indigenous communities are forced to live with the environmental impacts left in the wake of mining developments, typically with little to no compensation for damages (Manley-Casimir 2011) (see Dance et al. in this volume). For this reason, northern communities have historically borne the disproportionate burden of northern development (Cameron 2012; Manley-Casimir 2011; Tester and Blangy 2013).

Notwithstanding the environmental impacts, mining does create employment opportunities, even if these are often temporary, in regions where there are few other wage-earning opportunities. The lack of local employment in many Arctic regions has been attributed to patterns of out-migration, where residents of remote communities are relocating to more populated regional centres in search of employment and other services unavailable locally. With few employment opportunities in communities, mining is often seen as a way of keeping residents, particularly the youth, at home. For this reason, many community leaders look favourably upon mining projects that provide jobs that would otherwise not exist (Bowman 2011).

Yet, in communities across the North, competing interests over subsistence activities and mining operations often results in local tensions, with public conflicts emerging among residents over how to respond to large-scale mining proposals (Peterson 2012). Communities often feel they must choose between the protection of the environment and the continuance of their traditional subsistence economies, and employment and the potential revenues generated from mining agreements (see Rodon et al. in this volume). The duality of this proposition too often leaves communities in development dilemmas (Simon 2003), where middle ground accommodation is difficult to find. It is in this context that this research was conducted. Specifically, we set out to identify the regulatory and collaborative tools used by industry and northern communities to create space for the continuance of both mining and subsistence economies in the North.

METHODS

This study was informed by previous research conducted by the Resources and Sustainable Development in the Arctic (RESDA) research network and its predecessor, the Social Economy Research Network of Northern Canada (SERNNOCa). The research conducted by these investigate networks provided critical insight on resource development and traditional economic and social activities of northern Indigenous communities. For example, a key objective of SERNNOCA was to gain a better understanding of the dynamic relationship between subsistence harvesting and wage-earning, and how this mixed economy contributes to the social cohesion of northern communities (Natcher et al. 2012; Natcher et al. 2015; Natcher et al. 2016; Natcher et al. 2017; Jean et al. 2018). Continuing with this area of research, RESDA brought together researchers from a broad range of disciplines and organizations representing communities, government, the private sector, and non-profit organizations. All these researchers are committed to ensuring a larger share of the benefits of resource development stay in the region with fewer costs to communities (Southcott and Natcher 2018; Southcott et al. 2018; Southcott et al. 2019; Natcher 2019). These two research programs greatly helped to influence the design of this research.

Informed by the outputs of SERNNOCA and RESDA, we conducted a series of key interviews with industry representatives, Indigenous community leaders, and government officials, who all shared valuable experience and informed views on northern resource development, albeit from very different points of view (Rooke 2014). The interviews explored: the benefits and limitations of northern regulatory regimes;

the evolution of environmental and socio-economic impact assessments; how best to secure benefits for both industry and communities; and the challenges of engaging Indigenous communities in decision-making. Furthermore, key informants (listed here as Participant 1 to Participant 8) offered recommendations based on their own direct experience on how best to protect the subsistence-based interests of communities, without sacrificing the economic needs of industry.

In the following sections, interview responses are summarized. The main research themes regarding mitigation, regulation, and relationship building are discussed. Key informants offer several examples of effective mitigation strategies. Important tools and frameworks for improving wildlife management protocols and ensuring the sustainability of traditional harvesting practices are reviewed.

RESULTS

Management Planning and Mitigation Strategies

In order to receive permission to operate in Canada's northern territories, mining companies must conform to legislated requirements to conduct Environmental Impact Assessments, including Socio-economic Assessments and Strategic Environmental Assessments (Couch 2002). These guide the development and implementation of effective plans for managing, minimizing, and mitigating the effects of development upon the natural environment and nearby communities. Consistent with these requirements, wildlife management plans are created to identify and prioritize the different species of interest, the expected effects of operations on these populations, and the preventative measures and mitigation strategies intended to alleviate these pressures to wildlife. These plans also include various monitoring and reporting strategies to assess efficacy, such as the Terrestrial Ecosystem Management Plan, created for the Meadowbank Gold Mine in Nunavut (Cumberland Resources Ltd. 1996).

As mining operations commence, the monitoring and reporting protocols established by the wildlife management plans are documented and submitted to impact review boards, so that the latter can monitor and assess the effectiveness of these wildlife management plans. Mechanisms used to reduce human-wildlife interactions at the mine site are an important component of wildlife management plans (Klein 2000). These include prevention of vehicle collisions with wildlife along haul roads through the development of Right of Way policies. Waste management plans seek to manage food waste that attract wildlife to mine sites. This would prevent wildlife from entering the potentially dangerous mining operation areas. Hunting and fishing activities by mine personnel are commonly not permitted on site, nor is the possession of firearms and harvesting equipment allowed.

Companies also devise community engagement strategies, typically submitted to Impact Review Boards, to uphold certain licenses and approvals for operation. One such example is the Community Engagement Report created by De Beers Snap Lake Diamond Mine (De Beers Canada Inc. 2013). This consists of plans for consultation, information sharing, community and site visits, and socio-economic initiatives. The strategies for incorporating communities into the planning and management of mines are typically consistent with negotiated Impact and Benefit Agreements between companies and the local communities, as well as socio-economic agreements with government (Caine and Krogman 2010; Sosa and Keenan 2001). Requirements mandated by these agreements are intended to provide an overall mechanism for minimizing, as well as improving, interactions between parties, and ensuring adequate benefits to all those involved.

Mining companies also work with communities to conduct research and monitoring projects in order to better understand how their activities affect wildlife populations. For instance, the diamond mines in the Northwest Territories, along with government agencies and local communities, have conducted population studies on grizzly bears and wolverines through DNA sampling of hair tufts found around these large mine sites. These projects often incorporate a relatively large sample area that extends beyond the immediate areas of mining sites. This provides a better understanding of how these species have responded to resource development (Dominion Diamond Ekati Corporation 2014). Another example is the Bruce Head Narwhal Monitoring Program, established by Baffinland with the Mary River Project. In this case, the company has partnered with the North Baffin communities to conduct an extensive monitoring initiative of the company's shipping activities through Milne Inlet and Eclipse Sound, and the potential effects from these activities on the narwhal population. This project was largely inspired by community concerns for this culturally and ecologically significant species, and has incorporated local residents in the design and conduct of the program. Results from this study will be of great interest to the global marine biology community (Participant 3).

Wildlife management plans also include baseline information and population surveys that draw on both Western science and Traditional Ecological Knowledge. Wildlife management plans outline the status of certain wildlife species, anticipate wildlife responses to anthropogenic stressors, create monitoring timelines, and implement, where possible, preventative measures. "Species of interest" are identified in collaboration with local communities, and are typically those most sensitive to operations, and that are of great ecological importance and cultural significance. Also included in wildlife management plans are monitoring and reporting protocols for impacts to wildlife populations. These activities create transparency, improve mitigation efforts, and tend to be more cost-effective in the long term. In these ways, "the monitoring and reporting and the enforcing become extremely important" (Participant 4).

Interviewees also discussed the mitigation strategies used to reduce incidental impacts on wildlife. Most often these mitigation strategies were designed to reduce certain anthropogenic stressors to wildlife populations by reducing human-wildlife interactions due to the mining operations. Occurrences such as vehicle collisions on haul roads, self-defence, off-trail machine use, and recreational hunting and fishing around the mine, have all been found to impact wildlife populations and, in some instances, reduce the availability of wildlife for traditional harvesting. These impacts can typically be managed with company policies and employee training, like providing the right of way for wildlife (Participants 4, 7, 8), restricting hunting and fishing near the mining camp or during work hours (Participant 4), providing regular site surveillance (Participant 4), appropriately disposing of waste and attractants (Participants 4, 7), and educating workers (Participant 7).

The level of daily disturbance to wildlife habitat through mining operations can also be managed by determining the zone of influence. This can help ensure that disturbances to surrounding habitats are minimized and managed (Participants 4, 7). Impacts such as dust fall and excessive noise from the operation of the mine can be monitored daily and controlled if necessary. Roadways, bridges, and wildlife corridors can all be constructed or modified if they are recognized to have an impact on wildlife range and distribution, migratory routes, or critical breeding habitat. For example, several participants discussed the partnership in the Northwest Territories between the diamond mining companies, the territorial government, Hunters and Trappers Organizations, and northern communities, to investigate the decline of the Bathurst caribou herd. Although the mining industry is "working to develop a strategy on caribou as part of a larger group," and "focusing on what can be done to ensure that caribou don't continue to decline, and what the potential impacts are" (Participant 7), the decline of this specific caribou herd is of significant concern to several Indigenous groups, and to the biodiversity of the region. Information of this kind will be used in future policy, regulation, and mining wildlife management plan updates and for further monitoring.

Notwithstanding the efforts being made, there are still calls for more reflexive and adaptive corporate policies to reduce their impacts on wildlife populations (Meek 2012). Instability of continual growth of the mining industry in the Arctic creates an environment of unpredictability and increasing cumulative impacts, although there have been several instances in which individual wildlife management plans have been mindful and successful in dealing with the risk they pose to wildlife. One example is the Ekati Diamond Mine, which has specially designed road edges for use by caribou and wolves. However, there remains a fair amount of variance across the region (Klein 2000; Kofinas 2005). The creation of one overarching management plan encompassing strategies designed for all species of wildlife, landscape characteristics, and the unique needs of individual communities is unlikely (Meek 2012). Each socio-ecological system is unique and dynamic, and therefore must be managed as such. This necessarily involves the equitable inclusion of Traditional Ecological Knowledge (TEK) to help understand the long-term effects that mining development may have on the landscape, how best to mitigate these impacts, and how to potentially adapt in the future (Cameron 2012; Ellis 2005; Parlee et al. 2012).

Environmental and Social Impact Assessments

The environmental assessment legislation implements a "hugely consultative process" (Participant 2), ensuring that mining developments take the best care of the environment and formally incorporate social and economic components with the local communities for the life of the project (Participants 2, 3, 8). Environmental assessments work to reduce impacts by anticipating potential risks and planning for their management by requiring scoping studies, baseline data collection, and a complete plan for the management of the mine throughout its lifespan. Requirements for adequate closure and remediation plans, including sufficient security-bonding, ensure the mining plan will be responsible and financially supported before, during, and after closure (Participants 2, 3).

Community engagement and consultation is another component required by the environmental assessment legislation. This requires the consideration of specific questions and concerns of Indigenous communities regarding project impacts to traditional territories and cultural practices, including the collection and incorporation of Traditional Knowledge (TK). By encouraging early engagement, the environmental assessment process can lead to the establishment of mutual understandings and agreements, resulting in more productive relationships and stronger protection and mitigation for environmental and socio-economic resources (Participant 4).

Demonstrating such knowledge and understanding of the project and the associated lands and resources is critical in the review of the Environmental Impact Statement. The environmental assessment process "tests your assumptions and it tests your proposed actions...the whole intent of that is to create a project that has no significant adverse environmental effects" (Participant 2). The due diligence required in preparing the Environmental Impact Statement ensures new projects will correspond with official land-use plans for the region, and will receive substantial input from local communities and stakeholders, thereby invoking greater support for development (Participant 2).

Land-Use Planning, Resource Management, and Regulatory Boards

In addition to Environmental Impact Assessments, mining companies must obtain a myriad of licenses and permits to operate (Participants 1, 2, 3, 7, 8). Specific requirements for licenses and permits, as well as the administrative body regulating such documentation, can vary by territory and region. Many of these management and regulatory bodies are administered by government agencies at federal, territorial, and local levels. Typically, it is necessary for all mining projects to gain approvals and licenses from land-use boards, water boards, wildlife boards, and community governments, all of which may have their own regulatory processes, ensuring for the safety and sustainable use of resources. These boards actively encourage and require companies to consult local communities and Traditional Knowledge holders of the region (Participants 2, 3). These requirements, explains Participant 2, generally lead to more effective regulation and enforcement of development projects, thus reducing overall impacts. Impact Review Boards and other institutions of public government (IPGs), which were created by land claim agreements, were identified as critical institutions for managing the impacts of mining projects. Impact Review Boards and IPGs incorporate multiple stakeholder groups and perspectives, including the Indigenous communities, in the decision-making process.

Interview participants did acknowledge that inconsistent leadership on the part of territorial governments is sometimes a challenge. Respondents noted that the high rates of turnover, and a general lack of northern experience among some government administrators, can unnecessarily place industry and communities in opposition. This lack of capacity results in difficulties in enforcing regulations and can end up slowing down the decision-making process (Participant 3). Others noted that the inexperience of regulatory agency personnel often results in environmental legislation going unenforced (Participants 7, 8).

Another level of oversight improving the regulation of industry has been the relatively recent creation of independent monitoring agencies. These agencies work to ensure mining operations conduct impact mitigation and community engagement in accordance with regulatory standards and established agreements. These agencies are made up of members from industry, government, and local communities, and are largely funded by the mining companies. They strive to provide additional reviews of operations, management, monitoring, and reporting, thereby ensuring regulations are followed and local considerations are incorporated (Participants 3, 7, 8).

Strategies and Agreements for Socio-Economic Mitigation and Advances

Negotiated agreements between mining companies, Indigenous communities, and territorial governments were identified as the main mechanisms available for mitigating socio-economic impacts on communities and harvesters. Impact and Benefit Agreements were regarded as the most effective tool for communities to secure important benefits from mining operations on their traditional territories. Participants discussed how these agreements are typically negotiated early in the development process, and usually outline how local concerns will be addressed and protected. Participants discussed the importance of

IBAs for social and cultural well-being, economic opportunities, and the importance of the financial compensation to be received in return for accepting certain levels of environmental impacts and the loss of resources.

For example, Hunter Support Programs (HSPs) are sometimes included in the terms of IBAs. These HSPs are often negotiated to compensate communities for the loss of access to traditional lands due to non-compatible land developments. One recent example occurred in the James Bay region, where a hydroelectric development led to increased levels of mercury in fish, an important part of the Cree diet. To help mitigate the loss of fisheries and to restore and strengthen the Cree fisheries in ways which respond to Cree aspirations, a Mercury Agreement was signed in 2001 between the Grand Council of the Crees, the Cree Regional Authority, and Hydro-Québec. Specific initiatives funded by the agreement include: funding for fish harvesting, processing, storage, and transportation infrastructure; feasibility assessments of commercial and sport fisheries, including fish marketing; facilitating Cree access to fishing sites; the enhancement and construction of fishing camps; the introduction of training programs designed to perpetuate Traditional Knowledge of the fishery and related harvesting activities; provisions for the purchase and maintenance of fishing equipment; the development and maintenance of trails to fishing sites; and fish restocking programs. These and other program initiatives were designed to ensure the sustainability of the fishery for the long-term benefit of the Cree.

Socio-economic agreements are another effective strategy for mitigating impacts to communities and traditional practices. Likened to IBAs, these agreements are a way for companies to support local economic and community development. One main feature of these agreements is the hiring targets negotiated between companies and communities. These targets outline employment and training opportunities for northern and local residents, and provide a measure for economic development, ideally beyond the life of the mine (Participant 8). In addition, provisions are also made to support traditional activities, such as providing air support to distant harvesting locations.

Other important measures identified were those related to the social and cultural well-being of the communities. These included initiatives such as the annual fish tasting and berry picking events that De Beers holds at the Snap Lake Diamond Mine for local communities. These types of initiatives bring local people to the site to involve them in environmental monitoring and to illustrate how Traditional Knowledge and local values can be incorporated into mining operations (Participant 7). Companies will also conduct additional programs in the communities to support social and cultural well-being. These can include a variety of activities such as developing Traditional Knowledge databases, supporting charity events, and creating scholarship funds for Indigenous students. Participant 8 stated that companies have, in fact, been quite "proactive in providing resources for folks to build life skills" and have been "reasonably effective for providing career paths for trades in particular."

In summary, the key points expressed by interview participants regarding the success of existing mitigation strategies and the critical characteristics working to foster this success requires combined public and private resources, and the co-operation of government, industry, and communities. The strategies outlined in this section demonstrate the ways in which critical attention to the state of wildlife populations and community concerns can protect wildlife and ensure that subsistence uses are sustained.

CONCLUSION

There have been major advances since mining projects first began in the Canadian Arctic. However, the legacy of past mining activities, along with unrealized promises, continue to foster a significant amount of fear and distrust in the North when it comes to mine development (Participants 3, 7). These past experiences continue to influence relationships between communities and the mining industry to this day. Yet, a belief shared by all participants, is that to establish truly meaningful relations, capable of safeguarding the environment, subsistence economies, and the economic interests of the mining industry, will require the formation of sincere and meaningful relationships: "the nature of the relationship is most important; even more so than any specific strategy or regulation" (Participants 1, 4, 6). Key characteristics of healthy relationships include not only honouring regulatory requirements, but also building respect: "there is already an obligation to consult and involve communities, but then you actually have to listen to what they say." For many of the participants in this study, this is best achieved by engaging communities as partners, rather than consulting them simply to meet a regulatory obligation. As noted by one senior mining official: "communities are hungry to engage and want to have partnerships and relationships. The most successful mines with respect to building relationships have been the ones working within the community directly, having a presence in the community as much as possible."

The ambitious nature and large-scale operations characterizing northern mines has great potential for environmental, social, and economic impacts, and these are expected to continue with advances in mining technologies. The complexities of the cumulative effects from industrial development, coupled with other climate-related stressors, will continue to rise and become more challenging if they are not carefully managed (Johnson et al. 2005; Nassichuk 1987). It is therefore imperative that the impacts of mining on Indigenous subsistence economies are accurately assessed and mitigated (Male and Nol 2005). This will best be achieved through meaningful collaboration rather than protracted litigation.

REFERENCES

Agnico Eagle Mines Limited. 2013. *Meadowbank Gold Project 2013 Annual Report*. Agnico Eagle.
–2014b. *Final Environmental Impact Statement – Meliadine Gold Project, Nunavut*. Golder Associates.
–2014a. *Meadowbank Mine 2013 Wildlife Monitoring Summary Report*. Nunavut Environmental Consulting Ltd.
Armitage, Derek, Fikret Berkes, Aaron Dale, Erik Kocho-Schellenberg, and Eva Patton. 2011. "Co-Management and the Co-Production of Knowledge: Learning to Adapt in Canada's Arctic." *Global Environmental Change* 21, no. 3: 995-1004.
Baffinland Iron Mines Corporation. 2012. *Mary River Project Final Environmental Impact Statement*. Baffinland.
–2014. *Mary River Project 2013 Annual Report to the NIRB*. Baffinland.
Berkes, Fikret and Dyanna Jolly. 2002. "Adapting to Climate Change: Social-Ecological Resilience in a Canadian Western Arctic Community. *Conservation Ecology* 5, no. 2 (January): 18-32.
Bernauer, Warren. 2011. *Mining and the Social Economy in Baker Lake, Nunavut*. Centre for the Study of Co-operatives, University of Saskatchewan.

Bohnet, Darryl. 2013. *Report of Environmental Impact Review and Reasons for Decision – Gahcho Kué Diamond Mine Project.* Mackenzie Valley Environmental Impact Review Board.

Boulanger, John, Kim G. Poole, Anne Gunn, and Jack Wierzchowski. 2012. "Estimating the Zone of Influence of Industrial Developments on Wildlife: A Migratory Caribou Rangifer Tarandus Groenlandicus and Diamond Mine Case Study." *Wildlife Biology* 18, no. 2: 164-179.

Bowman, Laura. 2011. "Sealing the Deal: Environmental and Indigenous Justice and Mining in Nunavut." *Review of European Community & International Environmental Law* 20, no. 1 (April): 19-28.

Caine, Ken J. and Naomi Krogman. 2010. "Powerful or Just Plain Power-Full? A Power Analysis of Impact and Benefit Agreements in Canada's North." *Organization & Environment* 23, no. 1 (March): 76-98.

Cameron, Marc. 2012. *Diavik Wildlife Management Policy.* Diavik Diamond Mine.

Corbin, Juliet and Janice M. Morse. 2003. "The Unstructured Interactive Interview: Issues of Reciprocity and Risks when Dealing with Sensitive Topics." *Qualitative Inquiry* 9, no. 3 (June): 335-354.

Couch, William J. 2002. "Strategic Resolution of Policy, Environmental and Socio-Economic Impacts in Canadian Arctic Diamond Mining: BHP's NWT Diamond Project." *Impact Assessment and Project Appraisal* 20, no. 4 (December): 265-278.

Cumberland Resources Ltd. 2006. "Meadowbank Gold Project: Terrestrial Ecosystem Management Plan."

Dominion Diamond Corporation. 2013. *Ekati Diamond Mine 2013 Wildlife Effects Monitoring Program.* Dominion Diamond Ekati Corporation.

Dowsley, Martha. 2009. "Community Clusters in Wildlife and Environmental Management: Using TEK and Community Involvement to Improve Co-Management in an Era of Rapid Environmental Change." *Polar Research* 28, no. 1 (January): 43-59.

Ellis, Stephen C. 2005. "Meaningful Consideration? A Review of Traditional Knowledge in Environmental Decision Making." *Arctic* 58, no. 1 (March): 66-77.

Fereday, Jennifer and Eimear Muir-Cochrane. 2006. "Demonstrating Rigor Using Thematic Analysis: A Hybrid Approach of Inductive and Deductive Coding and Theme Development." *International Journal of Qualitative Methods* 5, no. 1 (March): 80-92.

Galbraith, Lindsay, Ben Bradshaw, and Murray B. Rutherford. 2007. "Towards a New Supraregulatory Approach to Environmental

Assessment in Northern Canada." *Impact Assessment and Project Appraisal* 25, no. 1 (March): 27-41.

Gartler, Susanna. 2018. "One Word, Many Worlds: The Multivocality of Subsistence." *Alaska Journal of Anthropology* 16, no. 2 (January): 49-63.

Green, Heather. 2013. "State, Company, and Community Relations at the Polaris Mine (Nunavut)." *Études/Inuit/Studies* 37, no. 2: 37-57.

Gunn, Jill H. and Bram F. Noble. 2009. "Integrating Cumulative Effects in Regional Strategic Environmental Assessment Frameworks: Lessons from Practice." *Journal of Environmental Assessment Policy and Management* 11, no. 3 (September): 267-290.

Haley, Sharman, Matthew Klick, Nick Szymoniak, and Andrew Crow. 2011. "Observing Trends and Assessing Data for Arctic Mining." *Polar Geography* 34, no. 1-2 (March): 37-61.

Hipwell, William, Katy Mamen, Viviane Weitzner, and Gail Whiteman. 2002. "Aboriginal Peoples and Mining in Canada: Consultation, Participation and Prospects for Change." Working Paper 10. Ottawa: North-South Institute.

Klein, David R. 2000. "Arctic Grazing Systems and Industrial Development: Can We Minimize Conflicts?" *Polar Research* 19, no. 1 (February): 91-98.

Kofinas, Gary P. 2005. "Caribou Hunters and Researchers at the Co-management Interface: Emergent Dilemmas and the Dynamics of Legitimacy in Power Sharing." *Anthropologica* 47, no. 2 (January): 179-196.

Maracle, Tobi J., Glenna Tetlichi, Norma Kassi, and David C. Natcher. 2018. "Caribou and the Politics of Sharing." In *When the Caribou Do Not Come: The Social-Ecological Complexity of Community-Caribou Relations in Canada's Western Arctic*, edited by Brenda Parlee and Ken Caine, 136-147. Vancouver: University of British Columbia Press.

McMillan, Alan. D. 1988. *Native Peoples and Cultures of Canada: An Anthropological Overview.* Vancouver: Douglas & McIntyre.

Meek, Chanda L. 2013. "Forms of Collaboration and Social Fit in Wildlife Management: A Comparison of Policy Networks in Alaska." *Global Environmental Change* 23, no. 1 (February): 217-228.

Megannety, Michèle. 2011. "A Review of the Planned Shipping Activity for the Baffinland Mary River Project: Assessing the Hazards to Marine Mammals and Migratory Birds, and Identifying Gaps in Proposed Mitigation Measures." Master's thesis, Dalhousie University. Retrieved from DalSpace.

Metcalfe, Leroy. 2013. *Helping Inuit Adapt to Climate Change*. Sikumiut Environmental Management Ltd.

Nassichuk, Walter. 1987. "Forty Years of Northern Non-renewable Natural Resource Development. *Arctic* 40, no. 4 (December): 274-284.

Natcher, David C. 2009. "Subsistence and the Social Economy of Canada's Aboriginal North." *The Northern Review* 30, 69-84.

–2019. "Normalizing Aboriginal Subsistence Economies in the Canadian North." In *Resources and Sustainable Development in the Arctic,* edited by Chris Southcott, Frances Abele, Dave Natcher, and Brenda Parlee, 219-233. New York: Routledge.

Natcher, David C., Damian Castro, and Lawrence Felt. 2015. "Hunter Support Programs and the Northern Social Economy." In *Northern Communities Working Together: The Social Economy of Canada's North,* edited by Chris Southcott, 189-203. Toronto: University of Toronto Press.

Natcher, David C., Lawrence Felt, and Andrea Procter, eds. 2012. *Settlement, Subsistence and Change Among the Labrador Inuit: The Nunatsiavummiut Experience*. Winnipeg: University of Manitoba Press.

Natcher, David C., Tobi Maracle, Glenna Tetlichi, and Norma Kassi. 2017. "Maintaining Indigenous Traditions in Border Regions of Northern Canada." In *Indigenous Peoples and Resource Development in Canada,* edited by Robert Bone and Robert Anderson, 262-280. Ontario: Captus Press.

Natcher, David, Shea Shirley, Thierry Rodon, and Chris Southcott. 2016. "Constraints to Wildlife Harvesting Among Aboriginal Communities in Alaska and Canada." *Food Security* 8, no. 6: 1153-1167.Natural Resources Canada. 2014. *Interactive Map of Aboriginal Mining Agreements*. http://www2.nrcan.gc.ca/mms/map-carte/MiningProjects_cartovista-eng.html?utm_campaign=MMS_Home.

Nunavut Impact Review Board. 2011. *Public Information Meetings Summary Report: Created as Part of the NIRB's Review of Baffinland Iron Mines Corporation's Mary River Project*. NIRB.

–2012. *Public Information Meetings Summary Report, May 22-May 30, 2012, for the NIRB's Review of AREVA Resources Canada Inc.'s Kiggavik Project*. NIRB.

Parlee, Brenda, Karen Geertsema, and Allen Willier. 2012. "Social-Ecological Thresholds in a Changing Boreal Landscape: Insights from Cree Knowledge of the Lesser Slave Lake Region of Alberta, Canada." *Ecology and Society* 17, no. 2 (June): 20.

Parlee, Brenda, John Sandlos, and Dave Natcher. 2018. "Undermining
Subsistence: Barren-ground Caribou in a "Tragedy of Open Access.""
Science Advances 4, no. 2 (February). e1701611.

Simeone, Tonina. 2008. *The Arctic: Northern Aboriginal Peoples.* Library
of Parliament Social Affairs Division.

Simon, David. 2003. "Dilemmas of Development and the Environment in
a Globalizing World: Theory, Policy and Praxis." *Progress in
Development Studies* 3, no. 1: 5–41. https://doi.
org/10.1191/1464993403ps048ra.

Sosa, Irine, Karyn Keenan, and Canadian Environmental Law Association.
2001. *Impact Benefit Agreements Between Aboriginal Communities and
Mining Companies: Their Use in Canada.* Ottawa: Canadian
Environmental Law Association.

Southcott, Chris, Frances Abele, Dave Natcher, and Brenda Parlee. 2018.
"Beyond the Berger Inquiry: Can Extractive Resource Development
Help the Sustainability of Canada's Arctic Communities?" *Arctic* 71,
no. 4 (December): 393-406.

Southcott, Chris, Frances Abele, David C. Natcher, and Brenda Parlee, eds.
2019. *Resources and Sustainable Development in the Arctic.* New York:
Routledge.

Southcott, Chris and Dave Natcher. 2018. "Extractive Industries and
Indigenous Subsistence Economies: A Complex and Unresolved
Relationship. *Canadian Journal of Development Studies / Revue canadienne d'études du développement* 39, no. 1 (January): 137-154. https://
doi.org/10.1080/02255189.2017.1400955.

Tester, Frank J. and Sylvie Blangy. 2013. "Introduction: Industrial
Development and Mining Impacts." *Études/Inuit/Studies* 37, no. 2
(January): 5-14.

Tester, Frank J, Drummond E. J. Lambert, and Tee W. Lim. 2013. "Wistful
Thinking: Making Inuit Labour and the Nanisivik Mine Near Ikpiarjuk
(Arctic Bay), Northern Baffin Island." *Études/Inuit/Studies* 37, no. 2
(January): 15-36.

Therivel, Riki and Bill Ross. 2007. "Cumulative Effects Assessment: Does
Scale Matter?" *Environmental Impact Assessment Review* 27, no. 5:
365-385.

Walton, Lyle R., H. Dean Cluff, Paul C. Paquet, and Malcolm A. Ramsay.
2001. "Movement Patterns of Barren-Ground Wolves in the Central
Canadian Arctic." *Journal of Mammalogy* 82, no. 3: 867-876.

Watson, Alan, Lilian Alessa, and Brian Glaspell. 2003. "The Relationship
Between Traditional Ecological Knowledge, Evolving Cultures, and

Wilderness Protection in the Circumpolar North." *Conservation Ecology* (11955449) 8, no. 1 (December): 2.

Weir, Jackie N., Shane P. Mahoney, Brian McLaren, and Steven H. Ferguson. 2007. "Effects of Mine Development on Woodland Caribou Rangifer Tarandus Distribution." *Wildlife* Biology 13, no. 1 (March): 66-74.

6

The FIFO Social Overlap: Success and Pitfalls of Long-Distance Commuting in the Mining Sector

Gertrude Saxinger

Commuting between homes and mining camps, and living a multi-locality life with extremely different spheres of social interaction, demands a committed workforce and considerable support from partners and families of commuting workers. The individual social, cultural, economic, and psychological well-being of mobile workers and their families is key to maintaining healthy communities.

The study of the impacts of mining and mineral development on local communities has been a concern in research for quite some time (Davison and Howe 2011). Inevitably, in the contemporary global mineral extraction sector, and in Canada in particular, so-called long-distance commuting (LDC) or fly-in/fly-out (FIFO) operations are increasingly the norm in remote, sparsely populated regions (Storey 2018; Eilmsteiner-Saxinger 2011; Saxinger 2015; Taylor et al. 2016) such as Canada's northern provinces and territories. Single-industry company towns are usually no longer built or even considered due to the financial burdens for the corporations to maintain a fully-fledged infrastructure for workers and their families. As well, there are risks for the state when enterprises go bankrupt or simply shut down in bust times and leave behind a whole mining town without jobs and a virtually worthless real estate market in collapse (cf. Storey and Hall 2018; Saxinger and Petrov et al. 2016). In regions with sufficient demographic potential, LDC/FIFO is sometimes preferred by the companies who are not willing to train locals sufficiently for employment, but rather prefer to bring in already skilled personnel (Taylor and

Carson 2014). Interregional LDC sees workers from all parts of the country, or even from abroad, commute to the mine, while intraregional LDC exists where mining sites are so remote from local towns and villages that a daily trip to and from work is not possible. This is the case in the example of the Canadian territory of Yukon, where a mix of inter- and intraregional LDC prevails (cf. Jones 2013; Jones and Southcott 2015). LDC/FIFO operations are on the rise but have not been closely studied. They require special scholarly attention.

Overall, this volume looks at the structural conditions for increased benefits from resource extraction for local and Indigenous communities while minimizing negative impacts. It also asks whether the extractive industrial activities enable sustainable development. However, in this chapter on rotational shift work or LDC/FIFO operations in the Yukon, my focus is *not* on structural conditions and impacts of incoming FIFO workers to remote regions, or the economic and social consequences for sending or receiving communities— these topics are already reflected in most of the contemporary FIFO publications (McKenzie 2014; Harwood 2012; Storey 2001; Storey 2010; Saxinger 2011; Saxinger and Öfner et al. 2016; Saxinger and Nuykina et al. 2016; Storey and Hall 2018; Vodden and Hall 2016). Instead, this chapter concentrates on the personal practices of rotational shift workers (cf. Sibbel 2010; Barclay et al. 2013; Barclay et al. 2016; Bissell and Gorman-Murray 2019; Saxinger 2021) and offers some general insights into spatio- and socio-structural conditions, and how people might better cope with hypermobility and multi-locality (cf. Hilti 2009; Saxinger 2016a; Saxinger 2016b; Saxinger 2021; Verne 2012; Weichhart 2015).

The challenge is being away from home for extended periods of time while on a remote site, confined with peers in a camp setting, and doing hard work every day in harsh environments. I argue that good LDC/FIFO conditions facilitate the well-being of the mobile workforce (as well as their partners and children) and reduce drop-out rates, thus contributing to sustainable benefits for local communities in terms of stable local employment. This chapter draws attention to both the pitfalls and the successful coping strategies among workers in relationships and with families, as well as for Indigenous and female employees.

While much of the literature and public discourses focus on the negative sides of LDC/FIFO for receiving communities (Carrington et al. 2011) and families (Bissell and Gorman-Murray 2019), I attempt to draw in my work (cf. Saxinger 2016a; Saxinger 2016b; Saxinger 2015; Saxinger and Nuykina et al. 2016; Saxinger and Öfner et al. 2016) a more nuanced picture. Based on extensive research of LDC/FIFO workers in Russia and Canada, the study portrays successful coping strategies by individual workers or families that pursue an alternative model to the sedentism of mainstream society. As stated above, LDC/FIFO operations are increasing worldwide, and mining professionals are, to a large extent, globe trotters. This is especially the case during the frequent busts and downturns in mining activity. Workers follow the work, which can be far away from their place of residence. Having a mixed set of skills allows workers to be flexible, so that when a mine shuts down suddenly, workers can find other jobs. This was the case with several operations in the Yukon over the past decades; when those mines closed, workers went to others in British Columbia or Alberta instead.

For the purpose of this chapter, I focus primarily on workers with families and the specifics of Indigenous and female workers. Racial and gendered aspects are particularly understudied in research on LDC/FIFO operations. I argue in the following that socio-spatial spheres of LDCS/FIFOS – home and being on site (in the camp) – are overlapping, but at the same time separated. I draw on the development of my concept of the FIFO social overlap on my long-term LDC research in the Russian Arctic oil and gas industry (Saxinger 2016a; Saxinger 2016b). The FIFO social overlap denotes not only the home and camp division, but also – in the case of the Indigenous workforce – the dichotomy of living off the land and working in the mines, and the pressure of switching back and forth. In the case of women working in mining, the dichotomy that comes with LDC/FIFO operations is the constant shift between the everyday female reality in a mixed environment when off shift, and the hyper male environment while on site. I will argue in my conclusion that the successful transgression back and forth between these spheres, and thus a successful coping with LDC/FIFO operations, is closely tied to the personal capacity of simultaneous separation and integration of the socio-spatial spheres (Saxinger 2016a; Saxinger 2016b). Of course, it is also tied to structural conditions in the society at large. Successful coping rests especially on the company that facilitates the conditions for healthy LDC/FIFO operations and the well-being of the staff and their impacted families. The company's responsibility rests on its role as the one who is defining the working conditions, its shift roster setup, the design of and amenities in camp, as well as its facilitation of the conditions for journeying back and forth. The role of labour unions in lobbying for better LDC/FIFO work facilitation should not be underestimated, while elaboration on this is outside the scope of this chapter.

METHOD

This chapter highlights pitfalls based on the experiences of both male and female workers involved in different capacities in the mining industry in the Yukon. They come from different communities of the Yukon and further afield. In in-depth interviews, the Indigenous and non-Indigenous workers were asked to give advice to newcomers in the sector and to those thinking about a mining career in the future. The present book chapter ethnographically recapitulates these inter-locutors' insights into the LDC/FIFO operation system, their manifold problems, and the personal attitudes and support structures that help make the best out of a situation that is both particular to newcomers and very unusual compared to the societal average life. The findings in this chapter are based on an anthropological in-depth and long-term study of LDC/FIFO operations from 2014 to 2019 in the Yukon within the framework of the ReSDA project LACE – Labour Mobility and Community Participation in the Extractive Industries (see Acknowledgements). It is based on over one hundred narrative inter-views on a variety of research topics within the project's scope. Interviews and conversations were conducted with mobile workers, their family members, representatives of First Nation self-governments and territorial administration, experts for employment, labour mobil-ity, and labour safety in the Yukon, company representatives, HR experts, and other people with relevant insights into the project's theme. The LACE research team (myself and PhD candidate Susanna Gartler) was based primarily in Mayo, Yukon. There, we worked together with the First Nation of Na-Cho Nyäk Dun, on whose tra-ditional territory the Keno Hill Silver District includes an abandoned mine site with on and off extraction operations in recent years. Also located nearby are the newly established Eagle Gold open pit mining project, as well as placer mining and exploration activities.

I undertook visits to mining camps to carry out interviews, not only with local intraregional workers from all over the Yukon, but also with interregional workers from other provinces in Canada and abroad. The interviews, participant observation, and informal talks were analyzed according to the principles of grounded theory (Corbin and Strauss 2008; Charmaz 2014; Charmaz and Mitchell 2001). We engaged in community-based participatory research with the First Nation of Na-Cho Nyäk Dun, publishing *The Mobile Workers Guide* (Saxinger and Gartler 2017) as a source of information for soon-to-be employees from the mining sector community when the large Eagle Gold Project was still only on the horizon, in 2017 (Saxinger and First Nation of Na-Cho Nyäk Dun 2018). The issue of LDC/FIFO operations and its impacts on communities and families was also raised in a 2017 video created by the LACE team (Saxinger and Gebauer et al. 2017), and which was produced by the First Nation of Na-Cho Nyäk Dun. This varied set of information and sources forms the basis of this chapter.

LDC/FIFO LIFE WITH A PARTNER AND CHILDREN

There are no comprehensive statistics available on divorce rates among LDC/FIFO workers to support the widely held assumption that LDC/FIFO operations would hurt families and relationships. My own research among LDC/FIFO workers in the Russian oil and gas sector (Saxinger 2015; Saxinger 2016a; Saxinger 2016b), as well as a survey by Nichols Applied Management (2007) in Alberta's oil sands region revealed that the divorce rates among the mobile workers are no different from the national average. This finding is borne out in the present research in the Yukon. However, this does not detract from the fact that long-distance commuting places special burdens on relationships and families connected to the rotational shift work leading to the absence of one parent/partner. Then, when that partner returns home, there are unusually high social and emotional expectations of one another. It is a matter of partners and families getting used to the specific circumstances of the mobile lifestyle. When divorces and breakups do occur, many times the shift work is not the sole reason. In fact, many mobile workers and partners of LDC/FIFO workers report enjoying the freedom and independence it gives both partners, while at the same time not downplaying the hardships.

There are a variety of challenges when balancing family life with being away from home for two or more weeks during a work rotation. Some couples quarrel about things such as loneliness and the division of labour in the household. Other families manage these on and off conditions very well. Since mining is male dominated, it is usually the father who is absent from home, but there are also many examples of commuting mothers. For many families, this lifestyle becomes an established routine and there are numerous examples of people who work long-term, even to retirement, as LDCS/FIFOS.

In many cases, the spouse being off shift devotes more time to the family. Jason, a mobile worker, describes his situation, "You know, when I am at home, I can take out the kids and relieve the burden from my wife's shoulders. I love to cook for them. My wife also works during the week, so I can help her with running the house." Jessica, a spouse of a miner, states, "We celebrate the time together when he is here, like cooking his favorite meal or hanging out at the camp-fire together and maybe inviting friends." She adds, "We also go out for a nice dinner just him and me and value our time together or travel somewhere."

A big downside of LDC/FIFO operations is the fact that mobile workers often cannot be at home for important family events such as birthdays, religious holidays, children's sports games, graduations, and other special events – or in emergency situations, such as when a family member gets sick. Therefore, it is important to be mentally and emotionally prepared for missing important events. In severe cases, however, companies allow for special leaves, such as when there are substantial health issues or a death in the family.

Not seeing a parent for a prolonged time is especially hard for children. On the upside, however, they can spend more time with the mobile parent while they are at home and off work. Some take their children on holiday or, like many Indigenous families, to fish camp or on a hunt. Many grown up kids cherish these memories from their childhood, while at the same time remember missing the parent who was not there. Being seriously committed to one's partner and children is really important during time off work: "If you are not careful, the years pass by and you have not seen your kids growing up," Jason states. Many workers raise the issue of feeling guilty about not being at home, but also highlight that it is important to cope emotionally. They try to remember that the whole family is benefitting from this job and the associated social mobility.

Staying connected during the mobile work rotation is essential. Fighting over the phone or criticizing a spouse's parental choices and parenting style can be a source of fundamental conflict. A social worker told me, "I have a couple of clients who come to me with troubles in the family. Most of them are young parents and not very far along yet in their relationship. So they do not have too much experience together." It is a gradual process to figure out how to manage one's private life in the best way as a rotational shift worker. As people get older and become more experienced, being multilocal becomes a routine (cf. Saxinger 2016a; Saxinger 2016b).

The days before a shift ends might seem longer and the longing for loved ones or friends back home grows stronger. The expectations of a "perfect" time off shift grow as well. For many mobile workers, the first days after coming home mean sleeping and relaxing, being by themselves and getting relief from the structured industrial environment of the mine site. Others enjoy a lot of social interaction with their friends and family at home when they come off shift, before they relax and "zone out" for some time. For others, it is a time to go on holiday to the beach or travel abroad. Sometimes this transition period may also lead to frictions with the expectations of the partner or children, who have built up in their heads what this "perfect" time is going to be like. At home, the worker who returns is confronted by different family routines. Sometimes this is tricky when, for example, different styles of parenting conflict.

FIFO families are also "Skype families." Regular (phone and video) communication from the mine site to home helps family members stay connected and allows workers to keep track of what is going on at home. On the one hand, it makes sense to communicate frequently when there are major household decisions to be made. Staying connected is necessary for emotional well-being. On the other hand, solving troubles at home over the phone might be less feasible, and one must accept that one cannot be involved in all matters at home. Barry explains, "I do not call home every day. I do not want to get too much involved in the little things that occur at home. The time in the camp is when I concentrate on my job and on earning money."

Seemingly, the older that people become, the more they get used to a mobile relationship, making breakups less likely to occur (cf. Saxinger 2016a; Saxinger 2016b). Sometimes jealousy is involved, either on the part of the mobile worker, who does not trust the spouse back home, or by the person at home, who might think that their partner might be cheating on them with someone in the camp. Freddie, an electrician, argues, "Having a trustful relationship and staying away from paranoia and jealousy is key for making a good FIFO relationship happen. Most of what drives you crazy is only in your head anyway. You cannot concentrate on your job and you end up in silly fights for nothing. I trust and love my wife and she does the same. You can never predict things, but we are both committed to our relationship." He adds, "Sometimes it is just a myth that hook-ups happen on site in the camp. The company for sure does not like to see that because it just creates troubles. I think you should just work together and stay out of trouble, that's what I do." The interviews and personal observations reveal that most breakups occur among those who are not yet used to this mobile lifestyle, but other aspects come into play, too. Mistrust, domestic violence, or simply not caring about things going on at home, as well as not being engaged in the relationship while off shift, can undermine personal ties.

DRUGS, ALCOHOL, AND BINGE SPENDING

Substance abuse is often mentioned when asking workers or their family members about LDC/FIFO related troubles that they experience. It is, at the same time, a prejudice among other circles that mobile workers are heavy drinkers and drug users. But most mobile workers have no such problems. Nevertheless, substance abuse including alcohol binging, massive gambling, and spending substantial amounts of money, due to high wages and large disposable incomes, are indeed troublesome phenomena that are sometimes triggered by and certainly facilitated by this lifestyle.

Most LDC/FIFO companies have strict anti-alcohol and drug policies. Urine tests to identify drug usage must be passed when being hired. "If they suspect something they can also do tests randomly. This is the companies' right, you signed that in the contract. They can check you anytime," a heavy vehicle driver explains. In the event of a collision or other accident, there is an immediate drug and alcohol screening of all parties involved. Furthermore, negative behaviour, such as not showing up at the pick-up point for getting on site, or showing up hungover or otherwise impaired, can lead to dismissal. Not being able to work is one thing, but the most severe consequences are accidents: mine sites are dangerous workplaces. Workers that are under the influence of drugs or alcohol put themselves and their colleagues at risk. Tanya, responsible for HR, says: "If we find marijuana or alcohol, they are done. We immediately charter a plane and send them home. We have strict policies and a clear procedure for that."

In Canada, mining industry dry camps, with a constant and complete prohibition of alcohol on the premises, are common today. Not least it is a matter of liability and labour safety, but it also promotes safety in the camp by helping to prevent fights or other drug or alcohol associated negative behaviour. Many interviewed workers would like to unwind with a drink after work; however, they are used to the camp prohibition during their weeks on shift. Moreover, many workers welcome this break and see it as part of a healthy lifestyle. There are different aspects to be considered: while some workers continue their non-drinking habit into their time off shift, others binge drink during their off-shift period. Those who have alcohol problems benefit from dry rules. Brenda, a first aid responder, told me: "Sometimes I see these guys drunk and not knowing what to do with themselves back home in the community. It is great when I see the same people sober in camp; they are proud of what they can do and are so strong!"

Overspending money during off-shift periods is a temptation for everyone and it requires good skills in financial management to balance immediate "treats," necessary expenses, and long-term savings. There is no doubt that many jobs on LDC/FIFO mine sites are very lucrative. The high wages associated with mining jobs are one of the main reasons people take up a life on the move. Some people enter the mining sector for a limited time in order to make a lot of money quickly for special investments, such as to buy real estate or to start a family. Others who pursue a permanent mining career become used to a high income and an expensive lifestyle. These people are especially vulnerable to the boom and bust cycles in mining, which make the sector so unpredictable over long periods of time. Ben, an electrician, puts money aside, however "... I was actually saving up for a house, but now could not really work this year because of the mining downturn, so I spent it all already. Yeah when I first started off in the mining industry I didn´t care about money. I needed a truck? I went and bought a truck. I wasn't thinking ahead." Many workers save their money for further professional training (and from higher wages), for times when there is no work available, or when health issues do not allow them to work in the mines anymore.

CAMP LIFE

Mining camps in remote regions are enclosed facilities that house and feed the workforce and, in most cases, provide recreational facilities for workers where they can unwind after a difficult workday, such as a gym or TV lounge. These camps vary in size, from just a couple of people in exploration camps to a few hundred in larger mining sites. In the Alberta oil sands region, where the camps are not necessarily remote, populations can number over a thousand. Camp accommodations may consist of single rooms or shared bedrooms. Often, the bathrooms are shared, but there are TV sets in every room. Sometimes the style of accommodation in camps can be "hot-bedding," which means that two people share the same bed is one being on shift while the other rests in the room. However, "hot-bedding" is currently not the case in the mines in the Yukon. Being enclosed in a controlled setting can be mentally challenging. When signing the work contract, a person also agrees to obey the camp rules and regulations. For safety reasons, most of the larger camps have surveillance infrastructure. Leaving the camp is only possible with a permit and prior notification of where the person plans to go. Due to the wildlife in the surrounding area, as it is the case in the Yukon, safety – such as carrying bear protection spray – is especially relevant when people walk in the bush for recreation after work.

Most of the interviewed camp residents report that being so close to colleagues can lead to strong social bonds, as one interlocutor put it: "Our time is split 50/50 between camp and home, so it's almost your second home. You need to be open minded here, so many different folks from all walks of life." Francis, an environmental technician, explains, "Yelling is a no go. People who swear or yell a lot do not last long in the camp. The managers have to be careful about the way they talk too. Otherwise it might end up causing trouble among the crew." Another interlocutor says, "Shit flows downhill. If the boss yells it affects all others in the lower ranks." Sometimes outright conflicts can occur in camps. However, in such cases, the management is called to calm matters and to establish a safe and emotionally peaceful environment.

In particular, the last couple of days of a work shift can be extra stressful for mobile workers because they might still have a lot of work to complete in a short space of time, or they may be exhausted after a long shift stretch and want nothing more than to go home. During these times, it is particularly important to remain calm and civil with fellow workers. It is generally agreed that it is important to mutually support each other by doing things like listening to your fellow workers if they are having trouble. At the same time, it is also a good idea not to engage in too much social activity with colleagues, and to reserve some "me time." Rather than socializing after work, many workers prefer instead to spend their time in their bedroom watching TV, chatting to family and friends over Skype, reading, or playing video games.

It is necessary to balance the connection and close companionship with fellow workers alongside the need for recreation and alone time. Alan, a safety manager, explains, "You must be careful not to get too involved in other people's private lives and you don´t need to listen to everything people tell you. Sometimes it is better not to say what you think or voice your political opinions. It just makes the life out here easier." This shows how important it is to satisfy one's own need for privacy and to be aware of how one engages with colleagues. The interlocutors stress that relationships in camp should be professional and based on respect and kindness. Many report that they have left previous employment settings because of tensions and arguing on site.

The social amenities differ from camp to camp. In mid-sized mining camps, there may be TV rooms, pool tables, a small library, badminton courts, or a gym for working out. In larger camps, there may be a store that provides essentials, or a café. Leisure activities after a long work-day are essential to relax and unwind, and to help get a good night's sleep, ensuring workers have energy for the next day. Although mining camps are mostly dry, alcohol-free celebrations can occur from time to time, for example on New Year's Eve, etc. Private initiatives, like decorating the camp or establishing a hiking or workout group, can create the feeling of still having some autonomy in a highly routinized and monotonous setting. For example, Brenda reported: "We started our own little library here and shared books. We had a shelf and everyone put the books on there. This was pretty fun, you know, we did something together."

Bonnie, a camp services worker, reports, "Especially young folks out here sometimes have a hard time to entertain themselves." It is very important that LDC/FIFO workers have the ability to have a good time being on their own. Entertaining oneself in the camps is a skill that takes practice and some getting used to. Some people, however, like being by themselves out at site without having family duties. The routines and the strict schedule of the industrial mine camps can help workers to unwind from the stresses at home.

Satisfaction with food in the canteen is essential for morale among the crew, which, in turn, increases the satisfaction with FIFO operations in general. The chef and kitchen staff play a vital role, as John, a fore-man, explains: "Before the new management came here it was just convenience food, like pre-packaged food that was just warmed up. It was the number one reason for complaints and people were grumpy. Ever since they cook fresh food and have a steak night now and then or a special pizza day people are looking forward to their meals and sit together in the canteen with a smile on their face."

WOMEN IN MINING AND GENDERED CAMP SITUATIONS

The mining sector is still dominated by men, but there are increasingly more women working in mining camps. However, as Andrew, a geologist, puts it, "The camp and the workplaces have still a strong 'testosterone culture,' which is not appealing to all people. Some men even find it annoying too." A certain "male" way of talking can make the situation uncomfortable for women. And yet most male workers are not only comfortable with more females in the camp, on a personal level, they also feel that the overall atmosphere in the crew is better with both genders, as for instance, yelling and coarse language are reduced when there are more women employed. "I like that there are more and more women joining us. I also find this masculine stuff often annoying. Men work hard and play hard. Women have a more balanced attitude and this makes the whole atmosphere better," a male worker asserts.

Women are increasingly hired in a variety of professions at mine sites. This is the result of more women training in fields relevant to mining. Today, women are among the ranks of geologists, environmental engineers, heavy vehicle operators, hauler drivers, mechanics, welders, electricians, first aid responders, and safety managers. While all professions are now open to females, there is still a disproportionate number of women who are employed in entry-level, lower paying jobs, such as site maintenance and kitchen staff. Women also work in a variety of administration jobs on site. Compared to the previous generation, access to jobs in mining for women has certainly improved, but some female workers still report that they need to show much more ambition than their male co-workers, and to prove themselves to their male peers. Sydel, a welder in her forties, explains that "especially in the beginning it was really annoying that the males talked behind my back, asserting that 'this is not a job for a woman.' They considered me as less qualified, although I had all my certificates and was even better qualified than many of my male colleagues. It is important not to get stressed out about other people's opinions. It is your job – be confident that you do it well."

When it comes to sexual harassment in the mining world, some women are unsure how to proceed. Some interlocutors report that such situations are not necessarily the rule anymore, but that they still occur, unfortunately. For women, it is important to know that they are protected by laws of the state and, most often, by the rules and regulations of the companies, too. Sexual, racial, and other discrimination at the workplace is against human rights laws. Camp managers and company representatives clearly state, that in case of sexual and other (like racial) harassment, the perpetrator is fired immediately.

Sexual harassment in male-dominated fields like mining appears to be less of a problem today than it was in the past, but women still report of structural discrimination and harassment. However, often unpleasant situations are not blatant harassment. Discrimination is sometimes more subtle and it is not always easy for an individual to judge certain behaviour at first glance. There is a fine line between false comradery and making inappropriate sexual remarks or physical advances. So how to deal with such situations? All women I spoke with asserted that it is necessary to be direct. Marilyn, part of the cleaning staff, advised, "Talk to the guy who is behaving in a weird way. Often there are only a few weird guys who do not know how to behave. The others will support you. Just be upfront. Do not hesitate; do not be ashamed. Remember: nobody has the right to put you in an unpleasant situation." Such assertive behaviour does not come easily for everyone. It can be especially tricky for some women to succeed in a testosterone laden workplace and to maintain a healthy socio-psychological well-being. A lot still needs to be done by the companies to change gender discrimination in a practical and structural sense. The data shows that improvements are being made by company management, but that there is still much work to be done.

MIXED ECONOMIES: INDIGENOUS EMPLOYMENT AND LIVING OFF THE LAND

For First Nation communities and Indigenous employees, it is crucial that mining is done in a manner that does as little harm as possible to the environment, and that does not jeopardize their subsistence or traditional ways of life. The First Nation interview partners stress that mining should not be done only for short-term revenues. Instead, a goal is to preserve some of the wealth in the ground for future generations to profit. It is a tightrope walk between the ideas of environmental protection that are embedded into Indigenous worldviews, on the one hand, and on the other hand, the wish and need for wage work. The mines often offer the only employment opportunities in the region, but they are intrinsically detrimental to nature, despite how responsibly they are operated. For example, open pit mines often remove enormous quantities of topsoil, even whole mountains, regardless of whether the processing of the ore is done in an environmentally friendly way. In the Yukon, industrial hard rock mining began in the first half of the twentieth century and it has left a toxic legacy. Remediation and reclamation of the two major old mines, Faro and Keno, has not yet started, despite the mines having been closed for around twenty and thirty years respectively; currently they are under care and maintenance.

"Living off the land, means being connected to nature or being one part of a holistic environment, and it is key for many First Nations people, including those in the Yukon. Hunting, fishing, trapping, and picking berries or medicinal plants have been part of First Nation cultures since time immemorial. Cultural identities and survival for First Nation peoples is based on sharing food harvested from the bush and the rivers with their friends, Elders, and relatives in the community. This is important for fostering and maintaining social bonds and mutual help within community groups (cf. Kuokkanen 2011). It is also essential for the families of harvesters to fill their freezers and to store dried fish, moose, or caribou meat for the winter. Mike works a shift roster of two weeks on, two weeks off. He explains, "For me, working in the mine is good. You know, the hunting gear, the fuel, the vehicles and the boat are expensive. A lot has changed compared to the old days. It is hard for others who don't have well-paying jobs to afford to go out on the land."

A critical point is having time off work for traditional activities in the season when animals or fish are more active. Some companies acknowledge these cultural needs for First Nation peoples in Impact and Benefit Agreements (IBAs) negotiated with an Indigenous community. Based on such agreements, First Nation workers are sometimes entitled to take longer leaves during the hunting season in order to harvest for their families and communities during peak animal migration. Tanya, an HR person at a mid-sized mine, explains, "Today companies want to employ locals. Therefore, we have to adjust to the local cultural needs of our workers. If they give us a note in advance, we can manage to find someone else and they can take longer leaves than two weeks."

However, it does not always work out like that: some mining companies are not sensitive towards the cultural needs of their employees. Since it is such an important part of Indigenous cultural identities and livelihoods to harvest subsistence foods, local workers sometimes feel forced to quit their jobs during hunting season if the company is not being sensitive or flexible enough. This, in turn, leads to unemployment, job churn, and other potential troubles. Therefore, facilitating the needs of Indigenous employees is also a commitment to local communities in the region. It is the basis for structural conditions that prevent dropouts and allow people to participate in the labour market, while being on the land at the same time.

Attending community celebrations or family functions such as weddings and funerals has a very high priority for Indigenous people. This often leads to distress and dissatisfaction with work, because days off at short notice are usually possible only in case of a severe emergency at home, or a death in the close family. Not only is a colleague required to suddenly step in at the workplace, but the logistics for transport out of the camp can also make these flexible work arrangements difficult.

Tanya, the HR representative, addresses specific experiences with First Nation employees, saying, "We have great people from the community here. They are an important part of our multicultural environment. We must consider that First Nation people have a very specific culture and attachment to the land. By recognizing certain cultural needs, we try to strengthen the commitment to the mining company. For instance, we plan to regularly celebrate not only Canada Day but also the Annual Aboriginal Day." This acknowledgment of Indigenous events is important to make First Nation employees feel respected and welcome in an industrial mining environment that is still sometimes described as racist and prejudiced.

Not unlike sexual harassment, the corporate regulations regarding racial discrimination and harassment on site are strict, and any reported violation leads to dismissal. However, discrimination is often subtler than outright harassment, and is also symptomatic of a wider structural problem. One factor in the racial division of labour in camps is a structural one—Indigenous people working in entry-level jobs are less able to climb up the career ladder because of a lack of vocational skills. The different professions can lead to isolation between the different peer groups, both during and after work. This, however, is also a matter of general camp atmosphere and the demographic composition in a camp.

Most importantly, the companies are called out by Indigenous representatives to explicitly exercise their anti-discrimination laws and regulations.

CONCLUSION

This volume tries to understand sustainability and regional development in the context of northern resource extraction. In this chapter, I tried to outline how workers can successfully manage long-distance commuting (LDC/FIFO) and rotational shift work, which is a fundamental feature of mining in Canada's northern provinces and territories. There are numerous pitfalls associated with this mobile and multilocal lifestyle, but there is also immense experience to be gained from it that mining workers and their families can share with the younger generation. Furthermore, it is up to the sector to improve the living conditions in camps and the mobility to and from home, as well as to develop mechanisms to support, socio-structurally, institutionally, and practically, the LDC/FIFO families to stay healthy and connected while being separated. Thus, the rotational shift work lifestyle requires specific coping strategies on the part of LDC/FIFO workers, as well as responsible employers and camp management, who should provide a safe and supportive environment.

Even if a camp is laid out and operated in the best possible way, it is still difficult for people to adjust to long times away from family and other social networks, as well as being contained in a remote location and under constant surveillance. Camps are intersectionally structured with different hardships and opportunities for people of different gender, age, professional status, or ethnic identity (Saxinger 2021).

Spouses and children face similar challenges in coping with the routines of a shift-working partner or parent. Besides emotional problems, it is obvious that household management issues differ substantially from those of sedentary families. This also entails alternative gender roles in relationships and in the household. Loneliness in camps, binge drinking, and drug abuse may occur in some workers who are off shift, while others may live a physically and psychologically healthier lifestyle than others.

For Indigenous people in the region, jobs in the mines are reported to be essential since there are not many other employment opportunities. However, professional advancement in the mining industry is often limited due to a lack of qualifications and the high Indigenous dropout rates from school. These factors are embedded in a larger socio-structural context of colonial legacies and ethnic-based discrimination in the wider society. The technology used on site is state of the art, and thus, up-to-date vocational training is crucial. At Yukon University, there is a Centre for Northern Innovation in Mining that offers qualification programs for locals. I did come across successful Indigenous miners and other Indigenous professionals higher up the career ladder, but the reality so far is that Indigenous people are very likely employed in entry-level jobs or in the much lower paid camp service sector. However, the employment on a rotational shift basis provides extended periods at home which can be used for activities like traditional hunting trips, fishing, and taking care of the trap line. These activities help to maintain facets of the traditional Indigenous lifestyle. Moreover, the substantially higher income associated with the mining sector often allows workers to buy sufficient hunting and trapping equipment and machinery for use in those traditional activities.

The structural conditions for women in mining also deserve special attention. The male-dominated mining sector creates an atmosphere on site that is often detrimental to emotional well-being for women. Outright sexual harassment or subtle discrimination and belittling of women's skills in male-dominated professions is still many women's experience. Companies are increasingly hiring women to attenuate the "testosterone culture" – as it was called by one interlocutor – in camp and at the workplace. Coarse language is just one example of issues women face in a male-dominated environment. While the number of women in the mining industry is growing overall, females still face the problem that they are primarily employed in the low paid service sector in the camp.

This chapter highlights how usually, the longer that workers and their families are involved in LDC/FIFO operations, the better they cope with this specific lifestyle, eventually perceiving their mobile and multilocal life as standard and not deviant. New employees are often not supported by the company to adjust to the mining work life and its associated LDC/FIFO realities. Early dropout is a frequent consequence. For local and Indigenous people in particular, this is a severe problem since mines are often the only large employers in the region, forcing them to commute to regions farther away if they want to continue to pursue wage work in mining.

LDC/FIFO workers require strong personalities, strong families, and strong relationships from the start, but as the data show, it becomes emotionally easier with time and routine. Greater flexibility in shift rosters (like an extended period off during the hunting season), creating a safe environment for Indigenous and female employees, providing healthy food and diversified recreation facilities in the camp, and acknowledging Indigenous culture on site are just a few examples of how LDC/FIFO operations can be improved by the companies with little financial investment. Responsible corporate conduct and care is needed for meaningful facilitation of living in camps, for the commute, and for supporting families in staying connected.

The FIFO social overlap phenomenon, as I call it in the introduction to this chapter, is specific to a mobile and multilocal way of life, and it requires, both from the companies and the involved people, meaningful and explicit investments in connecting the different socio-spatial spheres – e.g., at home and on site [3]– involved in LDC/FIFO lives. At the same time, it is important that people can also consciously live a separated but good life at camp. The healthy conditions on site have fundamental repercussions on a healthy life in the home sphere. Thus, for LDC/FIFO workers, the socio-spatial spheres of home and on-site overlap. To manage this successful separation and integration—the FIFO social overlap—requires strong personalities, long-term experience of rotational shift work, and camps designed to support the well-being of workers by the corporations. Not least, such careful facilitation of LDC/FIFO operations reduces dropout rates and thus the Indigenous and local communities in the Yukon adjacent to the mining operations can benefit from the stable employment of their people. In this way, sustainable jobs in a local community are facilitated by supportive conditions for mobility and multilocality, that is, for LDC/FIFO operations.

ACKNOWLEDGEMENTS

This research took place within the framework of an insightful co-operation between the First Nation of Na-Cho Nyäk Dun (FN NND) and the project LACE – Labour Mobility and Community Participation in the Extractive Industries. I gratefully acknowledge that this research was allowed to be conducted on a variety of different First Nation traditional territories across the Yukon. I would like to thank the following people for their support: Chief Simon Mervyn and the Council of FN NND for welcoming LACE in their community; Joella Hogan (then Heritage Manager of FN NND) for her endless support of the LACE research activities; Bobbie-Lee and Herman Melancon and Andrew

Harwood for friendship, conversation, and their deep insights into LDC/FIFO; all the numerous interview partners who patiently narrated their stories and viewpoints; community members who supported LACE in one way or another; Susanna Gartler (PhD student, co-author of *The Mobile Workers Guide* and enthusiastic field work companion); Valoree Walker (LACE coordinator at Yukon University) and Chris Southcott (Lakehead University, RESDA PI/LACE CO-PI) for their inspiring support along the way; Jessica Dutton, Tara Cater, Jenifer Rasell for proofreading and comments; Andrijana Djokic (CEO of the FN NND Development Corporation) and Greg Finnegan (then CEO of the FN NND Development Corporation) for supporting LACE's community outreach activities and publications. I also thank the mining companies' representatives who made the visits at the mining sites possible. The LACE project is funded by the SSHRC (Social Sciences and Humanities Research Council), the research network RESDA (Resources and Sustainable Development in the Arctic), the Yukon Government/ Department for Economic Development, and the University of Vienna in Austria. LACE has been affiliated to the Yukon Research Centre at Yukon University.

NOTES

1 Henceforth "LDC/FIFO" is used as a general term for rotational shift work and all its sub-forms, like bussed-in/bussed-out, drive-in/drive-out, or rail-in/rail-out. LDC/FIFO by bus and car, as well as by airplane, are the most prevailing types in the Yukon.

2 All interview partners are anonymous and pseudonyms used.

3 Elsewhere I argue for considering extended journeys to and from work as one of the involved socio-spatial spheres: HOME – JOURNEY – ON DUTY as a meaningful social triad (Saxinger 2016a; Saxinger 2016b).

REFERENCES

Barclay, Mary A., Jill Harris, Jo-Anne Everingham, Philipp Kirsch, Susan Arend, Shirley Shi, and Julie Kim. 2013. *Factors Linked to the Well-Being of 'Fly-In, Fly-Out' Workers*. Research Report, University of Queensland, Brisbane: CSRM and MISHC, Sustainable Minerals Institute, University of Queensland. https://www.csrm.uq.edu.au/publications/factors-linked-to-the-well-being-of-fly-in-fly-out-fifo-workers.

–2016. "Geologists, FIFO Work Practices and Job Satisfaction." *Applied Earth Science* 125, no. 4 (October): 221-230. Doi:10.1080/03717453.2016.1239036.

Bissell, David and Andrew Gorman-Murray. 2019. "Disoriented Geographies: Undoing Relations, Encountering Limits." *Transactions of the Institute of British Geographers* 44, no. 4 (December): 707-720. Doi:10.1111/tran.12307.

Carrington, Kerry, Russell Hogg, and Alison McIntosh. 2011. "The Resource Boom's Underbelly: Criminological Impacts of Mining Development." *Australian & New Zealand Journal of Criminology* 44, no. 3: 335-354. Doi:10.1177/0004865811419068.

Charmaz, Kathy. 2014. *Constructing Grounded Theory*. London: Sage.

Charmaz, Kathy and Richard G. Mitchell. 2001. "Grounded Theory in Ethnography." In *Handbook of Ethnography*, edited by Paul Atkinson, Amanda Coffey, Sara Delamont, John Lofland, and Lyn Lofland, 160-174. Los Angeles-London-New Dehli-Singapore-Washington DC: Sage.

Corbin, Juliet and Anselm Strauss, eds. 2008. *Basics of Qualitative Research: Techniques and Procedures for Developing Grounded Theory*. London: Sage.

Davison, Colleen M. and Penelope Hawe. 2012. "All That Glitters: Diamond Mining and Tåîchô Youth in Behchokö, Northwest Territories." *Arctic* 65, no. 2: 214-228.

Eilmsteiner-Saxinger, Gertrude. 2011. "'We Feed the Nation': Benefits and Challenges of Simultaneous Use of Resident and Long-distance Commuting Labour in Russia's Northern Hydrocarbon Industry." *Journal of Contemporary Issues in Business & Government* 17, no. 1: 53-67.

Harwood, Sharon. 2012. *Cloncurry Shire Community Plan. Background Report 4: Results from the Long Distance Commuter Survey*. Cairns, QLD: Centre for Tropical Urban and Regional Planning. https://researchonline.jcu.edu.au/34828/.

Hilti, Nicola. 2009. "Here, There, and In-Between: On the Interplay of Multilocal Living, Space, and Inequality." In *Mobilities and Inequality*, edited by Hanja Maksim, Manfred Max Bergman, and Timo Ohnmacht, 145-164. Farnham: Ashgate.

Jones, Christopher. 2013. "Mobile Miners: Work, Home, and Hazards in Yukon's Mining Industry." MA thesis, Department of Sociology, Lakehead University, Thunder Bay, Ontario.

Jones, Christopher and Chris Southcott. 2015. "Mobile Miners: Work, Home, and Hazards in Yukon's Mining Industry." *The Northern Review* 41: 111-137. https://thenorthernreview.ca/index.php/nr/article/view/475.

Kuokkanen, Rauna. 2011. "Indigenous Economies, Theories of Subsistence, and Women. Exploring the Social Economy Model for Indigenous Governance." *American Indian Quarterly* 35, no. 2: 215-240.

McKenzie, Fiona H. 2014. "Fly-In/Fly-Out, Flexibility and the Future: Does Becoming a Regional FIFO Source Community Present Opportunity or Burden?" *Geographical Research* 52, no. 4 (October): 430-441. Doi:10.1111/1745-5871.12080.

Nichols Applied Management. 2007. "Report on Mobile Workers in the Wood Buffalo Region of Alberta – December 2007." Wood Buffalo: Nichols Applied Management, Management and Economic Consultants. Unpublished report.

Saxinger, Gertrude. 2015. ""To you, to us, to oil and gas" – The Symbolic and Socio-economic Attachment of the Workforce to Oil, Gas and its Spaces of Extraction in the Yamal-Nenets and Khanty-Mansi Autonomous Districts in Russia." *Fennia International Journal of Geography* 193, no. 1 (March): 83-98.

–2016a. "Lured by Oil and Gas: Labour Mobility, Multi-Locality and Negotiating Normality & Extreme in the Russian Far North." *The Extractive Industries and Society-an International Journal* 3, no. 1: 50-59. Doi:10.1016/j.exis.2015.12.002.

–2016b. *Unterwegs. Mobiles Leben in der Erdgas- und Erdölindustrie in Russlands Arktis.* With an Extended Summary in Russian and English. Wien: Böhlau. Open Access: https://www.vr-elibrary.de/doi/book/10.7767/9783205201861" https://www.vr-elibrary.de/doi/book/10.7767/9783205201861

–2021. "Rootedness along the Way: Meaningful Sociality in Petroleum and Mining Mobile Worker Camps". *Mobilities* 16, no 2: 194-211. DOI: 10.1080/17450101.2021.1885844

Saxinger, Gertrude and First Nation of Na-Cho Nyäk Dun. 2018. "Community Based Participatory Research as a Long-term Process: Reflections on Becoming Partners in Understanding Social Dimensions of Mining in the Yukon." *The Northern Review* 47: 187-207. https://thenorthernreview.ca/index.php/nr/article/view/758.

Saxinger, Gertrude and Susanna Gartler. 2017. *The Mobile Workers Guide. Fly-in/Fly-out and Rotational Shift Work in Mining. Yukon Experiences.* Whitehorse, Yukon: RESDA, First Nation of Na-Cho Nyäk Dun, Yukon College. https://fifo-guide.jimdo.com/.

Saxinger, Gertrude, Robert Gebauer, Jörg Oschmann, and Susanna Gartler. 2017. "Mining on First Nation Land: The First Nation of Na-Cho

Nyäk Dun in Mayo/Yukon Territory." YouTube video, 13:14. First Nation of Na-Cho Nyäk Dun (Producer). Mayo, Vienna. https://www.youtube.com/watch?v=u4UXywmkoqM [25.09.2019].

Saxinger, Gertrude, Elena Nuykina, and Elisabeth Öfner. 2016. "The Russian North Connected: The Role of Long-Distance Commute Work for Regional Integration." In *Sustaining Russia's Arctic Cities: Resource Politics, Migration, and Climate Change*, edited by Robert W. Orttung, 112-138. New York: Berghahn Books.

Saxinger, Gertrude, Elisabeth Öfner, Elvira Shakirova, Maria Ivanova, Maksim Yakovlev, and Eduard Gareyev. 2016. "Ready to Go! The Next Generation of Mobile Highly Skilled Workforce in the Russian Petroleum Industry." *The Extractive Industries and Society-an International Journal* 3, no. 3 (July): 627-639. Doi:10.1016/j.exis.2016.06.005.

Saxinger, Gertrude, Andrey Petrov, Natalia Krasnoshtanova, Vera Kuklina, and Doris A. Carson. 2016. "Boom Back or Blow Back? Growth Strategies in Mono-Industrial Resource Towns – 'East' and 'West'." In *Settlements at the Edge. Remote Human Settlements in Developed Nations*, edited by Andrew Taylor, Dean B. Carson, Prescott C. Ensign, Lee Huskey, Rasmus Ole Rasmussen, and Gertrude Saxinger, 49-74. Cheltenham: Edward Elgar Publishing.

Sibbel, Anne M. 2010. "Living FIFO: The Experiences and Psychosocial Wellbeing of Western Australian Fly-in/fly-out Employees and Partners." Doctoral thesis, Edith Cowan University, Perth.

Storey, Keith. 2001. "Fly-in/Fly-out and Fly-over: Mining and Regional Development in Western Australia." *Australian Geographer* 32, no. 2: 133-148. Doi:10.1080/00049180120066616.

–2010. "Fly-in/Fly-out: Implications for Community Sustainability." *Sustainability* 2: 1161-1181. Doi:10.3390/su2051161.

–2018. "From 'New Town' to 'No Town' to 'Source', 'Host' and 'Hub' Communities: The Evolution of the Resource Community in an Era of Increased Labour Mobility." *Journal of Rural and Community Development* 13, no. 3: 92-114.

Storey, Keith and Heather Hall. 2018. "Dependence at a Distance: Labour Mobility and the Evolution of the Single Industry Town." *The Canadian Geographer* 62, no. 2 (June): 225-237.

Taylor, Andrew and Dean B. Carson. 2014. "It's Raining Men in Darwin: Gendered Effects from the Construction of Major Oil and Gas Projects." *Journal of Rural and Community Development* 9, no. 1: 24-40.

Taylor, Andrew, Dean B. Carson, Prescott C. Ensign, Lee Huskey, Rasmus Ole Rasmussen, and Gertrude Saxinger, eds. 2016. *Settlements at the Edge. Remote Human Settlements in Developed Nations.* Cheltenham: Edward Elgar Publishing.

Verne, Julia. 2012. *Living Translocality: Space, Culture and Economy in Contemporary Swahili Trade.* Stuttgart: Franz Steiner Verlag.

Vodden, Kelly and Heather Hall. 2016. "Long Distance Commuting in the Mining and Oil and Gas Sectors: Implications for Rural Regions." *The Extractive Industries and Society-an International Journal* 3, no. 3 (July): 577-583. Doi:10.1016/j.exis.2016.07.001.

Weichhart, Peter. 2015. "Residential Multi-Locality: In Search of Theoretical Frameworks." *Tijdschrift voor economische en sociale geografie* 106, no. 4 (September): 378-391.

7

Reframing Benefits: Indigenous Women and Resource Governance in Northern Canada

Suzanne Mills, Deborah Simmons, Leon Andrew,
and Johanna Tuglavina

Increasingly, Indigenous women's voices and actions are emerging as interventions in regional and national discourse about resource extraction and environmental stewardship. Indigenous women's contributions are often overlooked, both as a result of narrow interpretations of their messages by decision-makers, and because women continue to be a minority in environmental governance institutions (Natcher 2013). When Indigenous women have contributed to environmental decision-making processes, their messages have not been confined to reflection on gender issues, but instead have been broad in scope, and called decision-making processes themselves into question (Kennedy Dalseg et al. 2018).

Additionally, several authors, including Brenda Parlee in Chapter 12 of this volume, have argued that the concept of benefits that underpins plans for resource development in Indigenous territories tends to default to a narrow, masculinized scope that may in fact serve as an obstacle to achieving Indigenous visions for well-being and self-determination (Kuokkanen 2019; Cox and Mills 2015). In sum, Indigenous women's contributions to environmental assessments (EAs) as individuals or as organizations have highlighted the need both for women's voices to be included in formal decision-making processes, *and* for environmental governance to be re-centred around community needs and desires and away from a myopic focus on resource extraction. Following this reasoning, a gendered approach to environmental governance would be broad in scope, and would centre on well-being and Indigenous values rather than the extractive activity itself.

This chapter examines how approaches to gendering environmental governance materialize in practice by examining two cases, the Voisey's Bay Nickel Mine in Nunatsiavut, and the prospect of shale oil development in the Sahtú Region of the Northwest Territories. The cases are a study in contrasts since they present "before" and "after" snapshots of major resource extraction initiatives and represent markedly different approaches to resource governance. Whereas Sahtú communities were just beginning to feel the effects of a brief exploration boom and the prospect of future shale oil development during 2012-14, Nunatsiavut reflected the experience of a full cycle of development, including a formal EA and a decade of surface level nickel mining, beginning in 2005. The cases also differ in the degree to which women participated in discussions about development, consciously addressing gender issues. In the Sahtú Region, women were not formally involved in talks about environmental governance, but they did participate as community members. On the other hand, Nunatsiavut women and their organizations provided submissions that addressed gender explicitly throughout the EA process, and they were involved in the negotiation of an Inuit Impact and Benefit Agreement (IIBA).

Recently, women in Nunatsiavut gathered to describe their own experiences of impacts and benefits, as well as those that they observed in their communities. In the two cases, environmental governance processes differed in scope. In the Sahtú Region, discussions were much broader than typical environmental decision-making processes and examined development in the context of wider community goals. Nunatsiavut women's contributions were part of a ten-year review of the Voisey's Bay project initiated by the Nunatsiavut Government. As such, discussions were limited to examining mining development, mirroring the scope of the initial EA process. Thus, while gender analysis in the Sahtú case requires an etic or interpolative approach, "reading between the lines" to gendered spaces of Indigenous agency and aspirations, the Nunatsiavut case allows for a direct emic reading of women's own perspectives on gender impacts of resource extraction.

These two cases are brought together in part because like-minded researchers were engaged in dialogue through the RESDA network and began to learn from the stories arising from our respective community-based research contexts. Applying mixed methods, including focus groups (Nunatsiavut) and participant observation (Sahtú), we consider how a gendered approach to resource decision-making can be transformative by more fully accounting for the scope of Indigenous aspirations. When examined together, the Nunatsiavut and Sahtú cases shed light on the gendered nature of Indigenous environmental governance. This prefigured a path for gender analysis, which is an essential component of Indigenous environmental planning and governance in resource extraction contexts.

GENDER, RESOURCE EXTRACTION, AND ENVIRONMENTAL GOVERNANCE

Much of the previous research about gender and resource development has shown how the negative impacts of resource extraction, as well as the distribution of benefits, are gendered (Archibald and Crynkovich 1999; Nightingale et al. 2017; Pauktuutit et al. 2015). Communities affected by mining and other forms of resource extraction often have higher than average gender income disparities as a result of the gendered nature of employment in resource extraction (Sharma 2010; Lahiri-Dutt 2012; Dorow 2015). Additionally, qualitative studies recount how participation in resource extraction can lead to increased levels of substance abuse, gambling, single parenthood, and family violence in Indigenous communities (Brubacher and Associates 2002; Kuyek 2003; Gibson and Klinck 2005; National Aboriginal Health Organization 2008; Gibson 2008; Government of NWT 2009; Davidson and Hawe 2012). What this research tends to overlook, however, is how resource extraction often dominates community planning discussions, undermining the full consideration of non-extractive forms of development such as tourism, arts production, or an increasing reliance on subsistence activities and sharing networks. This omission has been highlighted by northern Indigenous women, who, in their submissions to EAs, draw attention to how the masculinity embedded in resource extraction affects overall approaches to development and environmental governance in northern communities by overwhelming local processes in scale of funding and resources (Kennedy Dalseg et al. 2018).

Both gender income gaps and negative sociocultural impacts can be partially attributed to the masculinity that permeates all aspects of resource extraction. The gender typing of resource jobs and work cultures that are hostile to femininity and non-hegemonic masculinities excludes women from many jobs in mining, forestry, and oil development, while segregating them into a narrow range of feminized occupations with lower hourly wages than similarly skilled male-typed occupations (Cox and Mills 2015; Mills 2006; Brandth 1995; Brandth and Haugen 2000; Coen et al. 2013; Filteau 2014; Miller 2004). Women across all occupation types face discrimination, harassment, and violence (Lahiri-Dutt 2012). The masculinity of extractive industries also affects social and cultural relations outside of the workplace. The substantial wealth associated with resource extraction amplifies the social status of those working in these industries, magnifying gender inequalities.

The effects of extractive employment on gender relations within families have also been well documented. Several community reports have provided qualitative data suggesting that participation in resource extraction has often brought a rise in alcoholism and other addictions, as well as increases in domestic violence, family breakdown, and psychological stress for families and individuals (Weitzner 2006; Gibson 2008; Czyzewski et al. 2014). Saxinger (Chapter 6 of this volume), presents a more nuanced picture, outlining the benefits of fly-in/fly-out work for individuals and communities, while also describing mining camps as having a "testosterone culture," undesirable for both women and many men. Last, O'Faircheallaigh's (2007) research about how wealth generated from IBAS is distributed in Australia suggests that most models of income distribution do not account for unequal effects that resource extraction has on different sub-populations, such as women. Together these studies have shown women to receive fewer benefits from resource extraction and to experience greater negative impacts. This is notwithstanding environmental governance systems that have been established purportedly to support fair participation in decision-making and fairness in distribution of benefits and mitigation of impacts.

We use the term environmental governance to include pre-project decision-making processes such as EAS, community consultations and Impact and Benefit Agreement (IBA) negotiations, ongoing environmental monitoring and decision-making about the allocation of royalties and cash transfers resulting from IBAS, and the monitoring of IBA employment and environmental monitoring commitments. Unlike the academic literature, which has often focused on gendered effects of extraction, Indigenous women's submissions to EAS have emphasized process. Submissions to EAS by women's groups in several regions called for a gendered lens to all environmental governance and for an understanding of gender "beyond a variable" (Cox and Mills 2015; Kennedy Dalseg et al. 2018; Status of Women Council of the NWT 1999; National Aboriginal Health Organization 2008; Pauktuutit 2012; TIA 1997; Hallett and Baikie 2011; Archibald and Crynkovich 1999). In their submissions, women sought to centre everyday social and cultural life and how social relations would be affected by resource extraction. In several of the above examples, Indigenous women were able to contribute to EAs through conscious and collective organizations of women rather than as individuals. We suggest that the inclusion of women's perspectives in environmental governance has become more critical as decision-making processes about extractive development have grown to become central sites where northern communities chart their future trajectories.

Community-driven planning was not the original impetus for either EAS or IBAS. EAS began as a state-driven process of environmental regulation; IBAS arose both as a way for companies to secure access to resources, and as a means for Indigenous communities to ensure that they secure some benefits from development on their territories (Gibson 2002; Peterson St-Laurent and Le Billon 2015). As a result, neither EAS nor IBA negotiations embody democratic principles common in other governance institutions. Yet, as Kuokkanen (2019) points out, since the magnitude and scale of extractive projects can overwhelm northern economies and communities, environmental planning processes are critical for the articulation of self-determination. Ensuring that governance processes include all perspectives, particularly those of groups who face greater negative consequences from extractive development, is therefore paramount.

Unfortunately, this imperative is not reflected in current environmental governance processes. Research documenting the participation of women in such processes has suggested that women's perspectives are often either discounted or excluded altogether. In some cases, women have made distinct submissions to both EAS and/or IBAS with little measurable success (Cox and Mills 2015). Staples and Natcher (2015) found that women are underrepresented on hunter and trapper organizations and that, as a result, women's traditional subsistence activities are often pushed aside in co-management. More recently, White (2020) showed that 81 per cent of appointed members on eight selected co-management boards were men. In other situations, however, women have not had a formal role in environmental planning but have been well represented on negotiating committees and as participants or leaders (O'Faircheallaigh 2011). Consequences of not adopting a gendered analysis are that models of wealth distribution do not account for the increased social risks experienced by women (O'Faircheallaigh 2007), and narrowly scoped environmental governance processes are unable to capture the breadth of concerns and desires of Indigenous women (Kennedy Dalseg et al. 2018).

Among environmental governance processes, EAS have often provided the greatest space for the participation of women. In their submissions, women have called for a shift in the approach to environmental governance in various ways. First, they have broadened the scope of issues relevant to the cost-benefit assessment of resource extraction projects. As individuals and as groups, women sought to ensure that environmental governance processes centered on social issues that affect all members of families and communities, such as income disparities, food security, and domestic violence, as well as subsistence economies (Brockman and Argue 1995; Hallett and Baikie 2011). Moreover, women argued that governance processes as a whole need to account for gender from the outset by adopting gender analysis (Brockman and Argue 1995). Kennedy Dalseg, Kuokkanen, Mills, and Simmons (2018) contend that repositioning of subsistence economies and social concerns to the centre of environmental planning is the foundation of a gendered analysis since this repositioning is critical to providing more balanced gender relations. This critique of EAS echoes the work of some Indigenous feminists who have connected the social relations of the body, the family, and the community to larger governance goals of self-determination and Indigenous resurgence (Kuokkanen 2019; Moreton-Robinson 2000; Green 2017). These authors have argued gender and social relations at the scale of the body and family cannot be separated from larger-scaled self-governance processes.

In this chapter we contribute to these discussions by examining two cases of decision-making in relation to resource extraction. This may help us to better understand what gender justice might look like in an environmental governance context. In the Sahtú case, we use an etic gender lens to examine three community environmental governance initiatives (or what we might retrospectively call "ethical space" processes, borrowing the concept more recently developed by the

Indigenous Circle of Experts [2018]). These include a combination of knowledge exchanges, research, and planning exercises in response to the short-lived Canol shale oil exploration boom during 2012-14. The Sahtú case offers an opportunity to apply gender analysis to an environmental decision-making process outside of the confines of EAS and IBA negotiations. In the Nunatsiavut case, we draw on previous research projects as well as two focus groups conducted with the Nunatsiavut Government and AnânauKatiget Tumingit with sixteen Inuit women from each of the five coastal communities (Nain, Hopedale, Postville, Makkovik, and Rigolet). Focus groups discussed the relative merits and gendered implications of the Voisey's Bay Mine ten years following its opening, reflecting on the outcomes of EA and IBA decision-making.

RESTRUCTURING ENVIRONMENTAL GOVERNANCE IN THE SAHTÚ

The sudden pressures for shale oil exploration and development in the Sahtú Region during 2012-14 gave rise to intense and unprecedented conflict. There was polarization between those strongly advocating for the socio-economic benefits and jobs that resource extraction could bring, and those concerned about negative impacts on Dene and Métis ways of life, social well-being, and the environment. To mediate the conflict and support informed decision-making, environmental planning processes were initiated by communities that were broader than a formal EA, including a variety of regional and local activities to elicit broad community participation. These initiatives did not include women as a distinct constituency; however, they did reframe the discussion of environmental planning to decentre resource extraction and allow for a more balanced consideration of community aspirations for well-being and land-based ways of life—or traditional economies—in accordance with the various impacts and benefits of resource extraction.

While the level of internal conflict and innovative community planning processes that emerged from shale oil play were unprecedented, the issues were by no means new to the Sahtú Region. The key difference between past and present development scenarios is the shift in the political economy of the region since the signing of the Sahtú Dene and Métis Comprehensive Land Claim Agreement (SDMCLCA) in 1993. The "discovery" (by non-Indigenous explorers) of oil at Tłegǫ́hłį

(Norman Wells) and subsequently radium/uranium on the east side of Sahtú (Great Bear Lake) in the early twentieth century rendered the region an international focal point for resource extraction a century ago. The oil discovery precipitated negotiation of Treaty 11 in 1921, the centenary of which is being commemorated as we go to press. In the early 1970s, the prospect that a gas pipeline might traverse the region (extensively discussed through the Berger Inquiry and deferred pending land claim settlements) eventually led to the signing of the 1993 land claim agreement. The SDMCLCA provided certainty in land ownership, an integrated regional collaborative (co-management) framework for ensuring Dene and Métis participation in decision-making (partially enabled by the 1998 Mackenzie Valley Resources Management Act [MVRMA]), and a process for negotiating self-government agreements.

The settlement region defined by the SDMCLCA encompasses five communities and an ecologically diverse area of 280,238 km². Though there is no explicit mention of women or gender in the Agreement, one of the overarching objectives is "to recognize and encourage the way of life of the Sahtú Dene and Métis which is based on the cultural and economic relationship between them and the land." This Chapter 1 objective positions cultural and land-based subsistence practices as a central priority overseeing and guiding interpretations of the land claim agreement. Notwithstanding this objective, Dokis (2015) suggests that, in practice, both the SDMCLCA and the associated establishment of regulatory and land use planning institutions under the agreement and enabling MVRMA have been more effective at facilitating resource extraction than encouraging land-based subsistence practices. That is especially the case, given what Dokis (2015) refers to as the "corporatization of lands and land use decision-making" that followed upon settlement of the SDMCLCA.

Section 21 of the SDMCLCA requires an agreement with designated Sahtú organizations (land corporations) to explore, develop, or produce minerals upon or under Sahtú settlement lands; these agreements have come to be known as Access and Benefits Agreements (ABAS). However, Dokis notes "a general apprehension that those who negotiate ABAS are focused on money and business contracts rather than on the needs of the community or appropriate human relationships with the land" (2015). The SDMCLCA also outlines the larger regional co-management framework for environmental decision-making, creating the Sahtú Renewable Resources Board and invoking the MVRMA legislation that subsequently established the regional and cross-regional regulatory boards in the NWT.

The proposal for a natural gas pipeline through the Mackenzie Valley for a second time in 2004 tested the ability of this new EA framework to address the core objectives of the SDMCLCA. The pipeline was proposed by a consortium of multinational corporations led by Imperial Oil Ltd. and including ExxonMobil, ConocoPhillips, and Royal Dutch Shell PLC. The purpose was to tap into immense gas reserves in the Mackenzie Delta and Beaufort Sea. There was also a new player on the team, the Aboriginal Pipeline Group (APG) representing the Inuvialuit, Gwich'in, and Sahtú regions, all of which had settled land claims and had signed a memorandum of understanding for a financial stake in the pipeline (though the Dehcho First Nations, claiming an unsettled area along 44 per cent of the pipeline route, was notably absent from the APG and steadfastly opposed the pipeline). This was in sharp contrast to the pipeline proposal of the 1970s, which was met by near-universal opposition from Indigenous communities and leaders throughout the Mackenzie Valley.

The federally appointed joint review panel established to assess the pipeline project held 115 days of hearings between 2006 and 2007 in twenty-six communities involving 558 presenters, and it delivered its report at the end of 2009. The panel's 679-page report included 176 recommendations and concluded: "The panel is confident that the project as filed, if built and operated with full implementation of the panel's recommendations, would deliver valuable and lasting overall benefits and avoid significant adverse environmental impacts." (2009) The report was handed to the National Energy Board, which held its own hearings in 2010 and issued a report with conditional approval in December of that year. Although the project achieved the final level of approval by the federal cabinet in 2011, it was cancelled by the proponents in 2017 due to high costs and low gas prices. In her analysis of the five-year EA process, Dokis (2015) critiqued the EA framework established by the MVRMA, suggesting that the EA helped to minimize financial risks for companies by securing community consent. The EA recognized and mitigated impacts on subsistence, but it did not centre on Dene ways of life.

A new crucible for Indigenous governance came in 2012 with a shale oil exploration boom that, for a couple of years, led some to refer to the region as "the Dubai of the North." The two communities closest to the Canol shale oil play were Tulít'a and Tłegǫ́hłį,two very different communities within the Tulít'a District along Dǝgho (the Mackenzie River). Tulít'a is a small community of about 500. During the peak of the shale oil play, Dene and Métis comprised about 90 per cent of the community, the unemployment rate was approximately 14 per cent, and the average income was less than $40,000.[1] In contrast, the regional industrial and government hub Tłegǫ́hłį ("where the oil is"), or Norman Wells, had a population of nearly 800 in 2012-14, of which about 37 per cent were Indigenous, with unemployment at a rate of about 7 per cent. Average income was $100,000—well over twice that of Tulít'a and the highest in the NWT.[2] The other three communities of the Sahtú, specifically Délįnę, Fort Good Hope, and Colville Lake, resembled Tulít'a in being primarily Indigenous and reliant on a mixed economic base. They were located at the fringes of the shale oil play. There was strong awareness, however, that they would also be greatly affected by both impacts and benefits of the prospective industrial development in their region. These communities actively participated in planning and governance initiatives.

The exploration activities were supported by confidential Access and Benefits Agreements, along with project by project pre-screening processes overseen by the Sahtú Land and Water Board. Concerned about seemingly irreconcilable interests that could be addressed neither by the ABA mechanism nor the narrowly focused pre-screenings, Sahtú community leadership organizations took advantage of opportunities provided by federal and non- governmental organizations to undertake parallel governance initiatives. Three projects described here were initiated in 2014, just before the sudden and unexpected decline of the shale oil play, including one local and two regional projects:

1) The Canadian Northern Economic Development Agency (CanNor) sponsored the Tulít'a Hįdo (Future) Community Readiness Project, a socio-economic assessment and planning initiative with Tulít'a, the community expected to be most impacted by shale oil development (Headwater Group 2015).

2) CanNor also sponsored Best of Both Worlds, a two-year assessment and plan to support a robust traditional economy as part of the region's "mixed economy," facilitated by the Ɂehdzo Got'įnę Gots'ę́ Nákedı (Sahtú Renewable Resources Board

– SRRB) in partnership with local Renewable Resources Councils
(Harnum et al. 2014; Headwater Group 2015).

3) People felt that they had few opportunities to learn about the
    implications of resource extraction for the land and
    communities from independent (non-industry) voices of
    experience that they could relate to. As a result, the SRRB also
    worked with community leaders to support the Sahtú
    Community Story-telling Tour funded by the Dragonfly Fund
    (Tides Canada), featuring guests from the Fort Nelson First
    Nation sharing stories of shale gas development in their
    traditional territory (Morgan 2014).

Prospects for industrial development turned to bust once again with
collapsing oil prices in 2014, and the three major corporations operating in the region—Husky, ConocoPhillips, and MGM Energy—have
since shut down operations. Nevertheless, the community governance
initiatives propelled the region forward in reconceptualizing benefits
in relation to contemporary Indigenous aspirations. Dene and Métis
women did not have a distinct presence in discussions about shale oil
extraction through these planning processes or through the regulatory
reviews. As individuals, however, women played key roles in community debates about the impacts and benefits of resource extraction.
Debates were highly polarized because community economies have
become more complex in the Sahtú Region. Indigenous corporations
hoped to gain traction and promote employment through the shale
oil play, while other community organizations and families were concerned about social and environmental impacts.

1. Tulít'a Hįdo (Future) Community Readiness Project

This project was a two-year initiative undertaken by the community
of Tulít'a in 2014-15. Six of the nine members of the Tulít'a Readiness
Advisory Committee were women. Key goals of the project were to
assess changes, effects, and opportunities for the community into the
future based on potential development activity, and to develop a community readiness map and implementation plan. Women were engaged
in all phases of the project, including as researchers. The community
study included a survey and workshop series, including specific workshops to provide youth and Elders with spaces to develop ideas

independently. Unfortunately, a similar space was not provided for women. The project gave rise to five key community goals with associated actions and recommendations, all of which were broad enough to include women, including: effective local governance and partnerships; strong community, youth, and well-being; strong culture and relationship to the land; support work in the modern economy; and to engage Tulít'a in development planning (Headwater Group 2015a).

The Tulít'a Hįdo project served as a planning process during the peak of the shale oil boom. One of the principal outcomes of the process was the realization that social issues were preventing Dene and Métis from being in the drivers' seat with respect to the shale oil boom. Though women were involved in the research and planning, explicit gender analysis remained marginal and incomplete. Possible reasons for this omission are that the baseline information used for the project did not include gender-specific data except for employment rates, and that gender has not been a standard theme in Sahtú socio-economic or environmental discourse. However, the emphasis on family health, well-being, and youth, as well as on both traditional and modern economies, indicates a holistic understanding of the intersections between well-being, economy, and governance. It also creates space for interpolating discussions about gender relations.

2. Best of Both Worlds Project

The two-year regional Best of Both Worlds project explored the nature of the mixed economy in the Sahtú Region as the basis for a regional action plan in the context of the shale oil boom. The project followed upon earlier research with the community of Délįne, sponsored by the Social Economy in Northern Canada (SERNNOCA) program (Simmons et al. 2015). The Best of Both Worlds project involved: 1) a literature review; 2) interviews with key traditional economy champions and program managers within the Sahtú communities and at the Government of the NWT's head offices in Yellowknife; and 3) a regional workshop in Délįne. The workshop brought together five women and men delegates from each of the five communities, including youth, adults, and Elders sponsored by local Renewable Resources Councils, to discuss people's visions for the traditional economy or "way of life" (echoing the language of the land claim agreement), while taking into consideration the potential shale oil development. The project took special care to seek out documentation and elicit discussions about aspects of the traditional economy that are considered the domain of women. These often remain invisible because hunting and trapping activities tend to be equated with the traditional economy.

While the resulting discussion paper and action plan documents do not specifically address gender issues, they do account for components of the economy that have been important for women. The report, titled "Best of Both Worlds: Sahtú Gonę̀nę̀ T'áadets'enı̨tǫ – Depending on the Land in the Sahtú Region," notes that "In all communities except Norman Wells, ts'éku kə (women) have higher employment rates than deneyu kə (men). However, the definition of 'employment' used in the report does not include those who are active in traditional pursuits, the majority of whom are deneyu kə (men) (Harnum et al. 2015). The Best of Both Worlds report recognizes that more work would be required for a fulsome gender analysis of the traditional economy, recommending such an analysis for a future phase of the Best of Both Worlds project. However, a gender lens is evident in the report's discussion of arts and crafts, reflecting on comments made by women in the workshop:

> A number of ts'éku kə [women] expressed concern that the importance of náats'enelu hə́ ası̨́ı̨ yáts'íhtsı̨ [arts and crafts] is under-estimated. Náats'enelu hə́ ası̨́ı̨ yáts'íhtsı̨ contributes to a strong Dene/Métis identity and a sense of pride, and creates much-needed deneghágǫ́t'á (opportunity) for the intergenerational transmission of not only skills for producing articles, but also deneghágǫ́t'á to share dene náoweré (traditional knowledge) with young ts'éku kə in particular. For example, a local "sewing circle" involves not only dene gháonetę (teaching) skills for náats'enelu (sewing), but also creates a venue for passing on dene náoweré about producing raw materials such as sinew and hides, traditional ʔewə́ t'áadenakwı̨́ (hide-tanning) methods, child rearing, caring for the sick and ʔǫhda kə (elderly), roles of deneyu kə (men) and ts'éku kə and related customs, kinship, denewá ná rídı̨ı̨ (traditional medicines), and so on (Harnum et al. 2015).

The report's discussion of the undervaluation of arts and crafts has broad implications for community planning by highlighting aspirations of women participants in the workshop for a cultural resurgence that involved a revaluation of women's traditional activities in the regional mixed economy.

3. Sahtú Community Story-telling Tour

The third initiative exemplifying environmental governance processes
in response to the shale oil play was the Sahtú Community Story-telling
Tour in March 2014. Two women leaders from the Fort Nelson First
Nation (FNFN), Chief Sharleen Gale and Lana Lowe, travelled to Tłegǫ́hłı̨,
Tulít'a, and Fort Good Hope. The tour was sponsored by the SRRB in
partnership with local Renewable Resources Councils and the K'áhsho
Got'ı̨nę Community Council (Fort Good Hope) to facilitate two-way
exchange about the FNFN's past and ongoing experience with shale gas
development, to initiate a dialogue about the role that Sahtú communities could play in governing future development. There were strong
turnouts in each of the three communities. The following theme question
emerged repeatedly and was discussed in detail: "What does it mean
and what does it look like, to be self-governing and take charge of land
and water management in relation to oil and gas development?"

The discussion addressed concerns and questions about water;
cumulative impacts; the economy, including hunting and trapping;
and governance. Sahtú community members realized that they had
more power through instruments in their land claim than they thought
compared with the FNFN, and they were also impressed with how
proactive the FNFN had been in asserting their rights and setting up
their own monitoring and land stewardship programs (Morgan 2014).
The community members who turned out to see the two FNFN women
leaders were a notably mixed group, including both women and men,
and women actively participated in asking questions and sharing their
concerns from the audience. This cross-gender discussion provided a
space for considering linkages between cumulative impacts on homeland and way of life, and erosion of community well-being. The events
showcased the critical role of women in asserting strong governance
with respect to external pressures for resource extraction.

*

The three Sahtú initiatives described here highlight the importance of
creating ethical space for Indigenous people to discuss plans for
resource extraction considering their aspirations to maintain their
traditional ways of life. Where previously discussions about development had been characterized by false polarization and conflict, these
initiatives offered a means for women and men to develop conceptual
tools arising from their own histories and the histories and experiences

of other Indigenous peoples to navigate the evolving mixed economy of the Sahtú. Women often took the lead in these discussions by challenging men to question their acceptance of external drivers for development. This was in stark contrast to the formal meetings related to development plans composed of mostly male leaders and with agenda items narrowly reflecting concerns such as environmental screenings, industry engagement activities, or IBA negotiations. The larger community initiatives highlighted here all deliberately encouraged women's participation. These initiatives created ethical spaces for diverse participants to express concerns with socio-economic and cultural impacts of resource extraction, and to put forward plans for shaping development to support healthy communities that are strong in their identities and ways of life.

NUNATSIAVUT – THINKING THROUGH WOMEN'S PARTICIPATION

In contrast with the community-driven initiatives in the Sahtú Region, environmental decision-making pertaining to the Voisey's Bay Mine in Nunatsiavut was confined to formalized EA and IBA processes. Women's groups were formally involved in the environmental assessment process and gender was considered in the negotiation of the IBA between Vale Inco Newfoundland and Labrador (VINL) and the Labrador Inuit Association, which was signed in 2002. Ten years after the mine began production, the Nunatsiavut Government convened with the researchers to initiate discussions about the gendered impacts of the mine with Inuit women. Unlike the Sahtú, Nunatsiavut women were formally involved in discussions about resource extraction in their communities. The Voisey's Bay case also diverged from the Sahtú since initial discussions about resource extraction followed state mandated EA processes and were not Inuit-led.

Nunatsiavut resembles other regions in Northern Canada, in that Inuit aspirations for rights over lands, resources, and self-governance were intimately tied to resource extraction. The resolution of the Labrador Inuit Association's comprehensive claim was originally filed in 1977, but it was only done so after the Labrador Inuit Association (LIA) asserted territorial rights to stop road construction at the site through a court injunction and protests held conjointly with the Innu Nation. These activities ensured that an EA was conducted and that an IBA was signed prior to further construction. It also speeded progress on the negotiation of their land claim. In 1997, the Labrador Inuit Association, the Innu Nation, and the federal and provincial governments established a joint EA process for the Voisey's Bay Mine and its associated mill development. That process culminated in 1999.

Women's groups were active participants in the EA process. Four groups representing the voices of Innu and Inuit women participated in the EA: 1) Postville Women's Group; 2) the Tongamiut Inuit Annait (TIA); 3) the Labrador Inuit Women's Association; and 4) the Ad Hoc Committee on Aboriginal Women and Mining and Labrador Legal Services. Women's contributions to the EA process highlighted broad concerns about the effects of the mine on traditional harvesting, on community well-being, and on the health of individuals and families. Women were also concerned that the benefits of development, namely employment, and especially better-paid jobs, would not be equally shared by women. Though some of these concerns influenced the language of the IBA signed in 2002, notably attention to women's employment at Voisey's Bay, the influence on employment practices and on the land claim agreement signed in 2005 were less apparent. Women, however, were involved in the negotiation of the IBA and the land claim (Cox and Mills 2015). As described by Cox and Mills (2015), women's early involvement in environmental decision-making did not result in improved employment outcomes for women: women's experiences at Voisey's Bay resembled those of women in other Canadian mines in terms of number, experiences of discrimination, and concentration in the lowest paid occupations such as housekeeping and catering.

The Labrador Inuit Land Claims Agreement (LILCA) established the Nunatsiavut Government stipulation that IBAS must be negotiated before any major development can proceed on Inuit lands. Though economic benefits are centred throughout LILCA and the discussion of IBAS, the agreement states that "the benefits must be consistent with and promote Inuit cultural goals" (LILCA 2005). There is no explicit mention of gender in relation to economic development or to Voisey's Bay in the land claim. Notably, the land claim also separates economic development from subsistence and commercial harvesting activities, which are discussed in Chapters 12 and 13 (wildlife and plants and fisheries) (LILCA 2005). LILCA also established a Regional Planning Authority charged with drafting a Regional Land Use Plan separate from the Voisey's Bay development. The land use planning process involved hearings and consultation in Inuit communities run by the planning authority, which is jointly appointed by the Nunatsiavut Government and the Government of Newfoundland and Labrador. However, the planning process took liberties with Inuit governance by imposing state structures that were not sensitive to Inuit world-views. For example, planning meetings were overly formal, and planning processes understood Inuit interests in lands narrowly as economic. According to Proctor and Chaulk (2012), the process imposed structural and political constraints that inhibited Inuit-led processes, and as of 2012, no plan had been ratified.

Discussions with Inuit women in 2015 aimed to look beyond the experiences of individuals to discuss the role of the mine in community development and gender relations more broadly. Over two days of focus groups, women described their perceptions of the changes that had taken place over the past decade in their communities. Themes that emerged repeatedly in these discussions were: economic well-being, sharing, and economic inequality. Most of the women described visible signs of wealth in their communities and observed that many people were better off. One participant said that the changes had resulted in "more money in the community and less people on income support." Several women described how the flow of wealth into their communities also resulted in an increase in consumer goods such as "... more buildings, homes, and boats," and "more variety of foods in the stores ...." Many people felt that there were more opportunities to obtain a postsecondary education, a sentiment one woman summarized: "more women are now earning a formal education and getting certified and trained in occupations that they've not sought out for before the mine at Voisey's Bay began." Women also pointed to grant opportunities that resulted from funds flowing into the community from land claim and IBA funds. One participant described:

> ... there is an increased availability of funding programs in the region. We have our local trust, Ulapitsaijet, that provides dollars to each Nunatsiavut community for social programs ... We have the regional trust, Tasuijatsoak Trust Fund, in which individuals and organizations can apply for monies for social programs or economic programs, all of which did not exist prior to the mine of Voisey's Bay.

Overall, focus group participants felt that mining had increased wealth in their communities and was providing more opportunities for many individuals locally.

Notwithstanding these positive observations, the women also articulated their concerns about growing inequality that resulted from the wealth generated by Voisey's Bay. They observed mounting relative poverty and a growing schism between those who worked at the mine or for the Nunatsiavut Government and those who did not. One woman explained how living in poverty also created a feeling of disconnection: "I think a lot of people are struggling though. I know I'm struggling ... We can't afford to buy a lot of things now ... it's like having internet, if you don't have it then you're left out of the communication, when you've got it you're contacting the whole world."

Food prices were a flash point for discussions about inequality. Women described how their grocery stores were stocking more expensive food and how all food was costing more: "I can see how the lower income families could have it even harder, with the prices increasing because of Voisey's Bay in stores for food and groceries." Some participants also described the psychological impact of this rising inequality, stating that it "lowered self-worth because some people cannot afford high-price products and services." Therefore, though average incomes may have risen as a result of the mine, people's ability to maintain the average standard of living within their community was declining. Women highlighted the psychological effects of unequal food access: "Imagine what that can do to someone on low income, who for the most part is buying the basics of food, flour, sugar, Carnation milk, canned goods, and then standing next to someone else who is purchasing all these wonderful healthy ingredients, healthy vegetables and fruit." According to some, this growing inequality was creating new divisions among individuals and families:

It kind of started to divide the community too, I think, eh. Like I remember before [my spouse] started to work [at Voisey's Bay], the people that worked in Voisey's Bay were real lucky, lucky ol' people, and real privileged ... And now since we grew up and we're adults now and we got our own family, [my spouse] works at Voisey's Bay ... and we get some comments and are talked about and that, and people think we think we're too good and all that ... We've even actually talked about moving away.

Women also cited regional disparities describing how Inuit living in coastal communities, without road access or cellphone service, were not able to benefit from the mine to the same degree as Inuit living in nearby Goose Bay.

Several women also noted a decline in the level of sharing in their communities. One woman observed, "There is less sharing of foods than there used to be," while another remarked, "I find that less people help each other within the community." Specifically, focus group members felt that people in their communities expected monetary compensation for helping each other with tasks that they previously would have done out of goodwill, such as cutting wood, or giving someone a ride in a car. As one woman in the group explained, "If you want help you gotta pay for it; every little thing." Women highlighted how Elders were particularly affected by the decline in sharing. As one woman said, "The younger people seem to think more of just immediate family and don't notice that Elders are less fortunate people, they don't offer free help anymore, and they expect to get paid."

Mirroring the concerns highlighted by women's groups who participated in the EA, women who participated in the focus groups felt that alcohol and drug use as well as crime had also increased over the past ten years. Some of the participants saw rising inequality as a main driver of increased substance abuse.

Finally, some participants identified a growing sense that communities on the coast were isolated from the larger centre of Goose Bay and from the mine, despite their proximity. For example, some participants felt that over the past ten years there has been less information and communication on the coast about the mine. There was consensus among the women that once the company closed their office in Nain (after the mine was in operation), the flow of information to the community dwindled. Other women felt that the link between Nain and Voisey's Bay was not as strong as it should be because many mine workers from the coast had moved to Goose Bay so that they could more conveniently travel to and from the mine. Relocation was attributed to the strict fly-in/fly-out work arrangements that did not allow workers to organize their own travel between Voisey's Bay and their home communities.

Though Indigenous women's groups were involved in pre-project decision-making processes concerning Voisey's Bay, efforts to mitigate the negative social impacts that the women had identified prior to the mine construction, such as the provision of grants to organizations, were few. Though questions about subsistence activities were raised in the focus groups, many of the women did not have much to say about shifts in subsistence activities. Overall, women were concerned with how the mine related to social well-being in their communities in a broad sense, and they articulated ideas about how to foster development unrelated to mining. These ideas included tourism and crafts. For example, one woman suggested that they develop sustainable tourism initiatives and that they also become a port stop for small cruise ships. The focus groups, in effect, offered a way to begin to re-centre discussions about community well-being and development outside of extractive resource development.

DISCUSSION

During the EA process in the lead up to establishment of the mine at Voisey's Bay, women were provided with both space and resources to self-organize and make submissions. Women expressed fears about environmental impacts and the erosion of the traditional economy, and intervened in discussions about employment and training opportunities, leading to a directive to the company to develop a plan for women's employment (Cox and Mills 2015). Women's submissions aimed to shift the EA process from considering gender as a variable to establishing gender as a framework to assess the full range of impacts. Ten years after production began, women described mixed results in their own retrospective experiences of impacts. Notwithstanding the increased overall wealth in the region, women communicated fears about growing income disparities, increasing crime and addiction, and declines in traditional sharing practices. These observations parallel the concerns articulated by women's organizations through the EA. Additionally, there was a sense that the governance role of Innu/Inuit women and the ability of their formal government to uphold the IBA were undermined by the power of the giant mining company.

In the Sahtú case, women's participation as individual community members in a series of three environmental governance initiatives gave rise to findings that are remarkably like those from the Voisey's Bay EA experience. All three Sahtú initiatives discussed here pointed to a risk that resource extraction would be at the expense of the traditional economy and maintenance of Dene and Métis identities and ways of life. Each of the three projects highlighted a distinct facet of the strategies needed to mitigate these risks. The Tulít'a Hдdo project concluded that social issues including violence and addictions were a barrier for self-determination and effective governance, and that they foregrounded the need for strategic initiatives in individual and community wellness. The Best of Both Worlds project considered the nature of a healthy mixed economy in the Sahtú Region, asserting the need for strategic supports to maintain a robust traditional economy, including vital roles for women. The Sahtú Community Storytelling Tour shed new light on the potential for strong governance involving women and using instruments provided in the Sahtú land claim agreement.

The two cases reviewed highlight the analysis of northern Indigenous women in environmental governance and planning for resource extraction. Both the Sahtú and Nunatsiavut cases illustrate the ongoing challenges in creating ethical space for gender informed planning to support proactive Indigenous governance for healthy communities, and in their efforts to maximize benefits and minimize or mitigate negative impacts from extractive development.

CONCLUSION

In his recently published book *Indigenous Empowerment through Co-management*, Graham White undertakes a series of case studies to explore the question of whether institutions arising from modern treaties "have been effective in ensuring substantial Indigenous influence over policies affecting the land and wildlife of traditional territories" (2020). White concludes that overall, the answer is "yes." However, he does not apply a gender lens to his analyses. Here, we suggest that if a gender lens were applied to this question, the answer would almost certainly be quite different. The two contrasting but complementary cases reviewed here are important in that they highlight the role Indigenous women play in supporting gender analysis as a frame for northern resource extraction planning and environmental governance, regardless of whether they are participating explicitly "as women."

Notwithstanding their different histories, the Sahtú and Nunatsiavut have much to learn from each other. Each contributes important lessons for thinking forward about Indigenous environmental governance and gender justice. There is no doubt that the Sahtú can learn much from the self-organized collective leadership and voices of Nunatsiavut women speaking consciously as women about their knowledge and concerns, where gender is recognized as a frame in a formal EA process, and where there is an opportunity to reflect on lessons learned over time. Conversely, the Sahtú community and regional efforts arguably point the way forward by planning ethical spaces outside the formal scope of co-management and an EA process, and by directly addressing key social barriers to Indigenous governance. The result is more effective learning and strategic action, and in the process, also provides openings for gender to be recognized and consciously affirmed as a critical factor in assessing impacts and benefits of resource extraction. The Sahtú and Nunatsiavut regions alike are challenged to consider the role of gender-informed planning as a vehicle for channelling both fears and aspirations related to resource extraction into proactive Indigenous governance for healthy communities that appropriately benefit from economic development initiatives.

NOTES

1 NWT Statistics 2012-2014 (www.statsnwt.ca).
2 *Ibid.*

REFERENCES

Archibald, Linda and Mary Crnkovich. 1999. *If Gender Mattered: A Case Study of Inuit Women, Land Claims and the Voisey's Bay Nickel Project.* Ottawa: Status of Women Canada.
Brandth, Berit. 1995. "Rural Masculinity in Transition: Gender Images in Tractor Advertisements." *Journal of Rural Studies* 11, no. 2: 123-133.
Brandth, Berit and Marit Haugen. 2000. "From Lumberjack to Business Manager: Masculinity in the Norwegian Forestry Press." *Journal of Rural Studies* 16, no. 3: 343-355.
Brockman, Aggie and Marsha Argue. 1995. "Review of NWT Diamonds Project Environmental Impact Statement: Socio-Economic Impacts on Women." Ottawa: Status of Women Canada, 1-41.
Coen, Stephanie E., John L. Oliffe, Joy L. Johnson, and Mary T. Kelly. 2013. "Looking for Mr. PG: Masculinities and Men's Depression in a Northern Resource-based Canadian Community." *Health & Place* 21: 94-101.
Cox, David and Suzanne Mills. 2015. "Gendering Environmental Assessment: Women's Participation and Employment Outcomes at

Voisey's Bay." *Arctic* 68, no. 2 (June): 246-260. Doi: 10.14430/arctic4478.

Czyzewski, Karina, Frank J. Tester, Nadia Aaruaq, and Sylvie Blangy. 2014. *The Impact of Resource Extraction on Inuit Women and Families in Qamani'tuaq, Nunavut Territory*. Vancouver: School of Social Work, University of British Columbia. https://www.pauktuutit.ca/wp-content/uploads/Quantitative-Report-Final.pdf.

Dokis, Carly A. 2015. *Where the Rivers Meet: Pipelines, Participatory Resource Management, and Aboriginal-State Relations in the Northwest Territories*. Vancouver: UBC Press.

Dorow, Sara. 2015. "Gendering Energy Extraction in Fort McMurray." In *Alberta Oil and the Decline of Democracy in Canada*, edited by Meenal Shrivastava and Lorna Stefanick, 275-292. Edmonton: Athabasca University Press.

Filteau, Matthew R. 2014. "Who Are Those Guys? Constructing the Oilfield's New Dominant Masculinity." *Men and Masculinities* 17, no. 4 (October): 396-416.

Gibson, Robert. 2002. "From Wreck Cove to Voisey's Bay: The Evolution of Federal Environmental Assessment in Canada." *Impact Assessment and Project Appraisal* 20, no. 3 (September): 151-159.

Green, Joyce. 2017. "Introduction." In *Making Space for Indigenous Feminism*, 2nd ed., edited by Joyce Greene. Halifax, NS: Fernwood Publishing.

Hallett, Vicki and Gail Baikie. 2011. "Out of the Rhetoric and Into the Reality of Local Women's Lives." Submission to the Environmental Assessment Panel on the Lower Churchill Hydro Development. FemNorthNet. Criaw-icref.ca/sites/criaw/files/Out%20of%20the%20Rhetoric-FNN-March%2028%202011.pdf.

Harnum, Betty, Joseph Hanlon, Tee Lim, Jane Modeste, Deborah Simmons, and Andrew Spring. 2014. *Best of Both Worlds: Sahtú goneį ́ę t'áadets'enįt ǫ (Depending on the land in the Sahtú region)*. Tulı́t'a: Ɂehdzo Got'ın ̨ ę Gots'é Nákedı (Sahtú Renewable Resources Board).

Headwater Group. 2015a. *Tulı́t'a Readiness Plan*. Tulı́t'a: Tulı́t'a Readiness Advisory Committee.

–2015b. *Best of Both Worlds: An Action Plan to Build Readiness for a Mixed Economy*. Tulít'a: Ɂehdzo Got'ı ̨n ̨ę Gots'ę́ Nákedı (Sahtú Renewable Resources Board).

Indigenous Circle of Experts. 2018. *We Rise Together: Achieving Pathway to Canada Target 1 Through the Creation of Indigenous Protected and*

Conserved Areas in the Spirit and Practice of Reconciliation. Ottawa:
Government of Canada. www.conservation2020canada.ca/ice.

Joint Review Panel for the Mackenzie Gas Project. 2009. *Foundation for
a Sustainable Northern Future: Report of the Joint Review Panel for
the Mackenzie Gas Project.* Ottawa: Government of Canada. Apps.
cer-rec.gc.ca.

Kennedy Dalseg, Sheena, Rauna Kuokkanen, Suzanne Mills, and Deborah
Simmons. 2018. "Gendered Environmental Assessments in the
Canadian North: Marginalization of Indigenous Women and
Traditional Economies." *The Northern Review* 47: 135-166.
Thenorthernreview.ca/nr/index.php/nr/article/download/764/775.

Kuokkanen, Rauna. 2011a. "From Indigenous Economies to Market-
Based Self-Governance: A Feminist Political Economy Analysis."
Canadian Journal of Political Science 44, no. 2 (June): 275-297.

–2011b. "Indigenous Economies, Theories of Subsistence, and Women:
Exploring the Social Economy Model for Indigenous Governance."
American Indian Quarterly 35, no. 2 (April): 215-240.

–2019. *Restructuring Relations: Indigenous Self-Determination,
Governance, and Gender.* New York, NY: Oxford University Press.

Lahiri-Dutt, Kuntala. 2012. "Digging Women: Towards a New Agenda for
Feminist Critiques of Mining." *Gender, Place and Culture* 19, no. 2
(April): 193-212.

LILCA Labrador Inuit Land Claims Agreement. 2005. "Land Claims
Agreement Between the Inuit of Labrador and Her Majesty the Queen
in Right of Newfoundland and Labrador and Her Majesty the Queen in
Right of Canada." *Nunatsiavut.* Accessed April 3, 2020. https://www.
nunatsiavut.com/wp-content/uploads/2014/07/Labrador-Inuit-Land-
Claims-Agreement.pdf.

Miller, Gloria. 2004. "Frontier Masculinity in the Oil Industry: The
Experiences of Women Engineers." *Gender, Work and Organization* 11,
no. 1 (January): 47-73.

Mills, Suzanne. 2006. "Segregation of Women and Aboriginal People
Within Canada's Forest Sector by Industry and Occupation." *Canadian
Journal of Native Studies* 26, no. 1 (January): 147-171.

Moreton-Robinson, Aileen. 2000. *Talkin' Up to the White Woman:
Aboriginal Women and Feminism.* St. Lucia, QLD: University of
Queensland Press.

Morgan, Shauna. 2014. *Sahtú Story-telling Tour with Fort Nelson First
Nation Guests.* Tulıt'a: Ɂehdzo Got'ın ę Gots'ę́ Nákedı (Sahtú
Renewable Resources Board and Pembina Institute).

Natcher, David C. 2013. "Gender and Resource Co-Management in Northern Canada." *Arctic* 66, no. 2 (June): 139-245. www.jstor.org/stable/23594686.

Nightingale, Elana, Karina Czyzewski, Frank Tester, and Nadia Aaruaq, 2017. "The Effects of Resource Extraction on Inuit Women and Their Families: Evidence from Canada." *Gender and Development* 25, no. 3 (September): 367-385. Doi:10.1080/13552074.2017.1379778.

O'Faircheallaigh, Ciaran. 2007. "Reflections on the Sharing of Benefits from Australian Impact Benefit Agreements (IBAS)." May 2-7, 2007. Retrieved on 21 of February 2012. http://landkeepers.ca/images/uploads/reports/Reflections_on_sharing_of_benefits_from_Australian_I Bas_Ofaircheallaigh.pdf.

–2011. "Indigenous Women and Mining Agreement Negotiations: Australia and Canada." In *Gendering the Field: Towards Sustainable Livelihoods for Mining Communities,* Volume 6, edited by Kuntala Lahiri-Dutt, 87-110. Canberra: Australian National University.

Peterson St-Laurent, Guillaume and Philippe Le Billon. 2015. "Staking Claims and Shaking Hands: Impact and Benefit Agreements as a Technology of Government in the Mining Sector." *The Extractive Industries and Society* 2, no. 3 (August): 590–602.

Proctor, Andrea and Keith Chaulk. 2013. "Chapter 19: Our Beautiful Land: The Challenge of Nunatsiavut Land-Use Planning." In *Reclaiming Indigenous Planning,* edited by Ryan Walker, Ted Jojola, and David Natcher, 435-453. Montreal & Kingston: McGill-Queen's University Press.

Sharma, Sanjay. 2010. "The Impact of Mining on Women: Lessons from the Coal Mining Bowen Basin of Queensland, Australia." *Impact Assessment and Project Appraisal* 28, no. 3 (September): 201-215.

Simmons, Deborah, Walter Bayha, Ingeborg Fink, Sarah Gordon, Keren Rice, and Doris Taneton. 2015. "Gúlú Agot'ı T'á Kə Gotsúhʔa Gha (Learning about Changes): Rethinking Indigenous Social Economy in Déline, Northwest Territories." In *Northern Communities Working Together: The Social Economy of Canada's North,* edited by Chris Southcott, 254-274. Toronto: University of Toronto Press.

Staples, Kiri, and David C. Natcher. 2015. "Gender, Decision Making, and Natural Resource Co-management in Yukon." *Arctic* 68, no. 3 (September): 356-366.

Weitzner, Viviane. 2006. *Dealing full force: Lutsel K'e Dene First Nation's Experience Negotiating with Mining Companies.* The North-South Institute.

White, Graham. 2020. *Indigenous Empowerment through Co-management: Land Claims Boards, Wildlife Management, and Environmental Regulation.* Vancouver: UBC Press.

8

"It's a precarious situation": Situating Housing First Within the Resource Dependent Economy in Yellowknife, Northwest Territories

Lisa Freeman and Julia Christensen

In 2019, the Canada Mortgage and Housing Corporation (CMHC) released its latest *Northern Housing Report*. The findings were not a surprise for residents of Yellowknife, Northwest Territories (NWT). Rents continued to rise, making Yellowknife the most expensive rental market in Canada for several consecutive years. Complaints of housing costs were commonplace and grumbling could be heard throughout the city in grocery store and post office lineups, local news outlets, and on social media. Meanwhile, anxieties were high around the potential economic fallout that awaits the city and the territory, as the three local diamond mines gear down their production. The high costs of northern living, including housing affordability, mean that such an economic shift could have dire consequences for many. The northern economy, highly dependent as it is on resource extraction, is a precarious one. Yellowknife's housing landscape, then, exists within the context of this precarious economy. But what is the relationship between the two? How exactly does extractive resource-based economy impact housing in Yellowknife?

Mining has a long history in Yellowknife. Though the area was a traditional, seasonal hunting ground for the Yellowknife's Dene, it became a destination for settlers in the 1930s during the gold rush. Gold mining sustained the community's expansion for decades, alongside the emergence of the city as a centre for northern administration.

When the last of the gold mines closed in 2004, resource extraction in the area shifted primarily to diamond mining. Yellowknife was situated as the northern service centre for this activity; though the diamond mines are not as close to the city as the gold mines were, they still provide jobs for locals and migrant workers alike. Diamond mining in NWT has propelled Canada to be the third largest diamond producer globally (Government of Canada 2017). The territorial economy is largely dependent on non-renewable resource development, with its attendant booms and busts

The ways in which resource-dependent economies shape housing geographies in the Canadian North are poorly understood. There have been studies of housing inaccessibility and unaffordability in the Canadian North (Abele, Falvo, and Haché 2012; Canadian Polar Commission 2014; Christensen 2012, 2016; Falvo 2011; Turner Strategies 2017), but the relationship between extractive resource development and northern housing markets has not been widely examined. To fill this gap, we argue that a shift in our viewpoint, towards one that puts housing first, is required. Instead of focusing on how a resource-dependent economy impacts housing, we draw attention to particular dynamics within Yellowknife's current housing landscape and discuss how they may respond to fluctuations in the resource-based economy, and in turn, impact the city's more marginalized residents.

The difficulties many Yellowknifers face in finding affordable housing are of significant social concern. Housing affordability remains elusive in the face of a northern rental housing landscape that is highly constrained. The limited Yellowknife rental housing spectrum, that we have argued elsewhere (Freeman and Christensen, forthcoming), leads to greater vulnerability for low-income tenants in need of accessible and affordable housing due to reliance on a small set of public and private housing options; options that can be easily removed due to policy change, discrimination, or rent hikes. Just one primary property management company provides the majority of rental accommodations in the city and the NWT Housing Corporation provides the public housing in the territory. Thus, a lack of competition in the private housing market and not enough diversity in the public housing supply has led to a semi-monopoly situation that underpins affordable housing supply.

This very limited housing spectrum presents serious obstacles to supportive and transitional housing programs and renders such programs vulnerable to ongoing fluctuations in the local economy, itself precariously pinned to non-renewable resource development. By way of its impacts on the housing security of low-income Yellowknife residents, the resource-dependent economy poses significant impacts to individual and community well-being—impacts that are explored further by Parlee in Chapter 12. In an effort to address the city's housing crisis, in 2017, the City of Yellowknife endorsed a 10-Year Plan to End Homelessness. This plan built upon previous programs in the city (such as Lynn's Place and the Bailey House). It emphasized the Housing First approach, which is a widely used housing program that prioritizes securing housing for homeless people.

The plan proposed an expanded Housing First project to be implemented through engagement with private-sector business, including developers, builders, and landlords, and stressed that "ultimately, government will not solve the affordable housing crisis without partnerships and innovation for a made-in-the-North approach with the private sector" (Turner Strategies 2017).

In this chapter, we argue that the lack of housing choices in Yellowknife complicates the goals of Housing First. In particular, we look at how the implementation of Housing First rests upon the participation of the private rental housing market, which itself is largely reliant upon one major private rental company: Northview Apartment Real Estate Investment Trust. We argue that this reliance on the private rental housing market creates a precarious landscape for the delivery of supportive housing programs. That is particularly the case when these programs are within a resource-dependent economy, such as the one found in Yellowknife. We identify a set of concerns related to the delivery of Housing First in a context of limited public and private housing options, where one is left vulnerable to policy shifts by both the territorial government and Northview. We highlight significant opportunities in bolstering the affordable housing market and housing security through collaborations between the non-profit sector, private rental housing providers, and the territorial government. Bringing these recommendations together, we re-envision the objectives and implementation of Housing First in Yellowknife, addressing and working with the challenges of a resource-dependent economy.

BACKGROUND: LANDSCAPE OF HOUSING INSECURITY IN YELLOWKNIFE

The current state of housing and homelessness in Yellowknife, and across the NWT, is the result of multiple factors including but not limited to government policy, colonization, unemployment, and the resource-based economy.

Resource extraction has negatively impacted social infrastructure for local communities throughout Canada and the United States. The relationship between the socio-economics of oil and gas industries, for example, and social infrastructures relating to community health and housing have been well-documented (Gillingham et al. 2015; Goldenberg et al. 2010; Van Hinte, Gunton, and Day 2007). Housing, however, has primarily been discussed as a by-product of extractive resource-based economies, often framed in terms of how boom and bust cycles and industry-purchased housing for labourers decimate local housing sectors (Asselin and Parkins 2009; Brabant and Gramling 1997; The Canadian Research Institute for the Advancement of Women 2016; Waegemakers Schiff, Schiff, and Turner 2016). Even though the impact of resource extraction on homelessness and housing has not (yet) been explored at length in the NWT, it is embedded within government policies and societal responses to unaffordable housing and homelessness throughout the territory.

Yellowknife is the administrative, economic, and political hub for the Northwest Territories, so many people are attracted there for work, social and health services, social networks, and housing (Abele, Falvo, and Haché 2010; Abele et al. 2010; Levan et al. 2007; Community Partnership Forum 2016). However, when they arrive, they face housing challenges similar to those they experienced elsewhere in the territory (Christensen 2012; Christensen 2017; Canadian Mortgage and Housing Corporation 1997; Dawson 2003; Inuit Tapiriit Kanatami (ITK) 2004; Nunavut Housing Corporation 2004; Stern 2005; Tester 2009). In northern urban or regional centres, the increasing unaffordability in the private rental market, combined with a limited number of public housing units, has led to a particularly bleak picture of low-income housing availability (Falvo 2011). Schmidt et al. (2015) identify the shortage of housing as a critical factor affecting homelessness in Canada's North, specifically citing the very low rental vacancy rates in larger centres like Yellowknife.

Visible homelessness first began to emerge in Yellowknife in the late 1990s (Christensen 2012). Point-In-Time (PIT) Counts and Yellowknife's Homelessness Report Cards state that 3 to 5 per cent of the community's residents stay in emergency shelters or sleep outside. The vast majority of people experiencing homelessness in Yellowknife are single adults (or adults whose children are not currently in their care) and youth (Turner Strategies 2017). Christensen (2017) underlined the chronic nature of many single adults' experiences living in emergency shelters, with many citing stays of more than five years. Yellowknife's 10-Year Plan to End Homelessness similarly emphasized the prevalence of chronic homelessness (and long shelter stays) as a dimension in the overall spectrum of homelessness in the city. The large number of individuals who stay in emergency shelters for years in Yellowknife, as Christensen (2017) describes, suggests that housing inaccessibility and inadequate social supports perpetuate chronic homelessness.

While a lack of affordable, adequate housing across northern communities is behind the rise in homelessness in Yellowknife, a wider-reaching northern housing crisis has been ongoing since the first modern housing programs were established in the Canadian North in the 1950s and 1960s (Tester 2006). A federal northern resettlement policy was implemented shortly after World War II, when the Canadian government embarked upon a period of welfare state reform that included a more interventionist approach to northern Indigenous peoples (Tester and Kulchyski 1994; Bone 2003; Northwest Territories Housing Corporation 2016). Not surprisingly, these policies of centralization and housing provision were significant drivers of social and cultural change. They also increased vulnerability for northern Indigenous peoples by creating reliance on the governments for shelter that was and remains inadequate, both in terms of quality and quantity (Tester 2006; Tester and Kulchyski 1994).

The geographic isolation of northern rural settlements is compounded by the fact that there is a critical shortage of employment opportunities, as most were not formed around a sustainable economy (Collings 2005; Tester 2009; Bone 2003). By contrast, most jobs in Yellowknife reflect the structure of the territorial economy that is reliant on jobs in resource development and government. Abele (2006) argues that the uneven development of the northern economy is made worse by the boom and bust nature of major resource development projects, such as diamond mines and gas pipelines. Thus, northern resettlement policy, combined with non-renewable resource development, has shaped a geography of economic and social disparity between rural and urban settlements in the North.

This rural-urban disparity is reflected in the geography of northern housing. For example, Yellowknife is one of only five NWT communities with functional housing markets (Christensen 2012; Government of the Northwest Territories 2005). Housing stock is much more diverse, with private ownership and private rental housing being the main forms of housing tenure. Meanwhile, public housing comprises the bulk of the housing stock in most northern rural settlements, though it is also the primary source of affordable housing in regional centres like Yellowknife. Thus, not only is there a rural-urban disparity in housing types, but there is also a deep divide within Yellowknife, with affordable housing provided largely by public housing, with little to no regulation of the private rental market to ensure its affordability and accessibility.

Similar rural-urban geographies are also reflected in the geographies of visible homelessness within the territory. As Christensen (2012, 2017) has shown, Yellowknife is a place of economic and social opportunity in the resource economy of the territory. However, the city has few affordable housing options, and many single adults experience homelessness. Life in Yellowknife is characterized by high-cost living, unaffordable housing, inadequate public housing, and insecure employment (Christensen 2012). As Christensen argues, "there is a tremendous need for not only public housing units for single adults, but also for additional supportive housing programs that combine needed social supports, such as counselling or skills development, with housing" (2012, 2017). Greater collaboration and coordination between the various levels of government in the territory—municipal, territorial, Indigenous, and federal—is required.

Recently, the federal government has prioritized funding Housing First programs through the Homelessness Partnering Strategy (HPS), which allows smaller, remote cities, like Yellowknife, to initiate their own Housing First program. The city launched its program in September 2016. While transitional and supportive housing programs have been available in Yellowknife since the Bailey House opened its doors to single men in 2009, the launch of Yellowknife's inaugural Housing First initiative was welcomed by housing providers, government workers, and politicians alike (Beers 2018). It was a joint project, funded through the HPS, administered through the Community Advisory Board (CAB), and (initially) coordinated through the

Yellowknife Women's Society (and other non-profit organizations) intended to provide affordable housing for Yellowknife's most vulnerable individuals.

Housing First was established by Sam Tsemberis and Pathways to Housing in 1992 (Tsemberis, Gulcur, and Nakae 2004). It is based on the premise that it is necessary to secure long-term housing for chronically homeless individuals before addressing other issues such as addictions and mental health that often underlie chronic homelessness (Kirst et al. 2015; Tsemberis et al. 2004). This model is premised upon the idea that providing a person with housing first creates a foundation on which recovery can begin and build (Padgett, Gulcur, and Tsemberis 2006). The core elements of Housing First aim to provide immediate permanent housing for chronically homeless individuals in scattered sites. They also offer harm reduction with respect to mental health and addiction treatment, and ongoing access to wraparound support (Padgett et al. 2006).

By December 2017, Yellowknife's Housing First project reached its goal of housing twenty formally homeless individuals (Canadian Mortgage and Housing Corporation & Government of Canada 2018). That same year, Yellowknife endorsed its 10-Year Plan to End Homelessness, which built on the previous experiences at Lynn's Place, the Bailey House, and the Yellowknife Housing First program, to emphasize a Housing First approach to addressing homelessness in the community. The plan proposed expanding the Housing First project to include further engagement with private-sector business, including developers, builders, and landlords. The plan offered eighty-five scattered site rental units to be accessed through existing private rental housing stock (primarily through Northview) in order to immediately address a number of acute cases of homelessness in the community. Then, an additional eighty new units of Permanent Supportive Housing will be constructed.

While the initial success of Yellowknife's Housing First project is promising, it is not without its limitations. Like most housing in the city, it is dependent on a good working relationship with the private rental property management company, Northview. While this suggests potential productive collaborations between the public, private, and non-governmental sectors, it also highlights the precarious nature of supportive housing programs in a restricted housing market like the one in Yellowknife. There, the private rental housing market is vulnerable to economic fluctuations due to the boom and bust nature of extractive industries.

METHODS

This chapter emerged from our initial collaboration with Alternatives North (AN), a social justice coalition in Yellowknife that addresses poverty, health, climate change, and oil and mining development across the NWT. From our conversations with AN members, we chose two methods for our research: a gap analysis literature review and semi-structured interviews. We began our extensive literature review in 2016, which led us to focus on research on the relationship between housing and resource development in Yellowknife.

With the help of two detail-oriented research assistants, we conducted a thorough analysis of grey and academic sources that focused on northern housing policy, resource extraction, the NWT, and rural and remote homelessness. Nearly 300 articles from multiple academic data bases were scanned, with approximately fifty of them focusing on our key areas of resource extraction/development, housing/homelessness policy, and/or rurality/Northern Canada. In total, nineteen articles that mention all of the key terms searched were downloaded, and seventeen of those that were most relevant to our research were studied. Overall, thirty-one peer-reviewed articles, two magazine articles, two scholarly books, nine government and NGO reports, and two factsheet documents were reviewed. That was forty-seven sources in total.

After conducting our literature review, it became apparent that there were few studies directly focused on the relationship between housing in Yellowknife and the resource-based economy in the NWT.

From November to May 2018, we conducted semi-structured interviews with thirty housing and service providers, advocates, municipal politicians, city staff, members of the NWT legislative assembly, Government of the Northwest Territories (GNWT) staff, and other government officials. The purpose of this work was to talk to housing providers, non-governmental organizations (NGOs), housing advocates, and policymakers within government about housing security and affordability in the context of extractive resource development. However, after the majority of interviews were conducted, it was clear that the relationship between extractive resource development and housing in Yellowknife was not straightforward or even readily visible. Many workers fly-in to the mine from all over Canada and live in on-site housing temporarily. Fluctuations in vacancy rates were a concern, and it was noted that the diamond mining industry made a (rare) donation to supportive housing initiatives. However, it was very clear that everyone interviewed viewed the non-renewable resource-dependent economy as a challenging context within which critical policy, funding, service, and programming dynamics vis-à-vis housing affordability were taking place.

In particular, they were concerned with the ways in which economic uncertainty exacerbates the vulnerability of low-income northerners. That vulnerability is being intensified because of a reliance on both industry and on the private rental sector for the delivery of critical programs for homeless Yellowknifers. At the time of our research, there were mounting questions surrounding the sustainability of Housing First programming which relies on the private rental market, itself highly sensitive to the dynamics of the non-renewable resource economy. These new threads in our work directed us to explore the impacts that public and private housing monopolies have on housing advocates, service providers, and decision-makers, and the constraints that this housing landscape presents for the implementation of Housing First programs.

PUBLIC AND PRIVATE HOUSING MONOPOLIES

Public Rental Housing Monopoly

The lack of a housing spectrum in Yellowknife can be understood as the result of a lack of competition and therefore choices in public and private housing markets. The Northwest Territories Housing Corporation (NWTHC), a territorial governmental department, is the main provider of public or social housing. Meanwhile, Northview Apartment, a real estate investment trust (REIT), is the primary private property management company in Yellowknife offering rental apartments. These two housing providers represent two ends of the Yellowknife rental housing market—social housing and private rentals. Unfortunately, there are few other options for low-income tenants. The limited choices in rental housing present serious obstacles to supportive and transitional housing programs. Such programs are vulnerable to ongoing fluctuations in the local, resource-dependent economy and to changes in the overall local housing market.

The NWTHC was created to oversee the construction, maintenance, and governance of housing throughout communities and regional hubs in Yellowknife. It receives and manages funding from CMHC, and operates several housing programs throughout the territory, including but not limited to Yellowknife. "Technically, the NWTHC has a mandate to provide opportunities for suitable and affordable housing in the territory. [They] have operations and programs and services across the housing continuum, including support for homeless persons, all the way up to ... transitional housing, to public housing, market housing, affordable housing in kind of a leased-owned manner, all the way up to rental incentives for developers to develop more market housing." (Interview with NWTHC representative 1, 2017) As such, the NWTHC is one of the primary providers of affordable housing in Yellowknife.

Despite the extensive role of the NWTHC in providing public housing to territorial residents, several housing advocates we interviewed were critical of the Housing Corporation in terms of housing conditions and governance. They provided detailed accounts of houses that were not adequately built for the northern climate, and described homes in disrepair with problems such as major mould infestations. Many noted that the design of NWTHC houses did not meet the needs for multi-family dwellings or the cultural needs of Indigenous communities, such as space for the preparation of country foods or proper ventilation for the boiling of foods. Further, several housing advocates described the NWTHC as having a reputation for being inflexible, punitive, and unsupportive of tenants. Some noted that "there is no alternative community voice or Indigenous voice to present an alternative model" (Interview with Yellowknife housing advocate 1, 2017). Many of our research participants were critical of NWTHC policy changes around housing subsidies and welfare. They view those as detrimental to marginalized tenants.

Rental arrears with the NWTHC, in particular, were viewed as a barrier to accessing affordable housing in Yellowknife. There are multiple reasons why people go into rental arrears in public housing. According to our research, some tenants refuse to pay rent for political reasons in response to (colonial) policies and the fiduciary responsibility the Crown has towards Indigenous peoples. Others take leave to pursue seasonal traditions such as hunting, fishing, or trapping, and do not pay their rent while away. Moreover, we also heard accounts of women who were stuck paying full rent when trying to escape an abusive partner and falling behind. That experience was also described in Christensen (2017). Regardless of how arrears are accrued, once they are on record, it is an uphill battle to find any other housing in Yellowknife. Since individuals with rental arrears can no longer access public housing, they are forced to look to private market rentals. However, in Yellowknife, few affordable rental housing options are available.

Private Rental Housing Monopoly

The other primary affordable rental housing option is Northview, a REIT with a significant presence across Northern Canada. It is the largest property management company renting apartments in Yellowknife. Northview dominates the private rental housing market in Yellowknife, owning a range of properties from low-rise apartments to townhomes. In fact, August (2020) argues that Northview is one of a small number of REITs that focus their rental interests in non-renewable resource-based economies, chasing the speculative growth of resource extraction. Indeed, it is precisely because of resource extraction that Northview has been interested in the North. As August (forthcoming) illustrates, Northview has been clear in its aim to rent to creditworthy professional tenants in Northern Canadian markets that have rapid economic growth and where there are significant structural impediments to new competition, such as short building seasons, barriers to the transportation of construction materials, and labour shortages.

In a forthcoming article by August (2019), she describes how in the early 1990s, real estate investment trusts (REITS) cropped up across Canada, allowing real estate investors to buy up scattered rental properties and fold them into large asset management firms. Since the late 1990s, these financialized landlords enjoyed incredible growth, particularly in areas of Canada where there is little government regulation of the private rental housing market. The Northwest Territories is one of many areas of Canada where the local economy is dependent on non-renewable resource development. Thus, there is a significant geography to the prevalence of REITS across Canada, suggesting common links in housing affordability and accessibility challenges across regions with non-renewal resource-based economies. The emergence of a private rental monopoly in Yellowknife is reflective of a nation-wide trend, with REITS representing nine of the top ten biggest landlords in Canada (August, forthcoming). This centralization of private rental ownership in the hands of a small number of REITS is a new phenomenon in Canada, and has had a significant impact on the diversification of rental housing in many communities, small and large.

When one considers that Northview holds 85 per cent of all private rental structures in Yellowknife (August, forthcoming), it is clear that the REIT has significant influence over rental rates and conditions. That effect has been clearly illustrated in Yellowknife in recent years. For example, in 2014, Northview (then Northern Properties) announced that it would no longer provide housing to income assistance recipients in Yellowknife (Falvo 2014). Following an outcry, the company reversed this position; however it retained de facto power to rent selectively (August, forthcoming). For example, drawing on interview participants' firsthand experiences, Christensen (2017) described how Northview is known for blacklisting tenants based on reputation or racialized discrimination, leaving them with few private rental housing options.

Several research participants noted a complex yet challenging relationship with Northview, saying that the company has blatantly discriminated against low-income and homeless individuals in the past by refusing to rent to them. One advocate noted, "… the private market landlord that is a monopoly has declared they won't rent to anyone on income support in spite of the fact that it's against human rights. No one's challenged this so they still don't rent to [people on disability]" (Interview with Yellowknife housing advocate 1, 2017). Another housing advocate called it a policy, "one major landlord … who rents most of the [rental] stock, and stopped taking income support tenants" (Interview with Yellowknife housing advocate 2, 2017). They went on to explain that because their organization is now operating a Housing First program, they have more access to this rental housing, but was very clear in stating that Northview did have an actual policy where they would "not accept income support for their rentals, because it was, unpredictable about when they would receive payment … Yeah, so, they would have units sitting empty rather than take tenants [that were on income support]" (Interview with Yellowknife housing provider 1, 2017).

The dominant presence of a single private rental company means that it is easy for tenants to burn their bridges if they have a negative experience with the REIT. Christensen (2017) has described the ways in which the Yellowknife housing market is highly precarious for low-income residents. For example, she describes several instances where tenants were evicted during a personal crisis in their lives. Later, when their personal situations were resolved, they had difficulty renting again from the company (Christensen 2017). Similarly, public housing tenants described several instances where they were not given a second chance to access public housing due to a negative first experience. Our research corroborates these observations, further illuminating the difficult housing situation for low-income individuals in Yellowknife.

Recent economic change in Northern Alberta due to the collapse in global oil markets has seen vacancy rates rise, putting Northview in a position where apartment buildings that were once at maximum capacity now struggle to find tenants (August, forthcoming). In Yellowknife, there is a similar dynamic as local diamond mine production has begun to taper off after hitting a peak in 2017 (Blake 2018). All three operating diamond mines in the territory—Ekati, Diavik, and Gahcho Kué—will be closed before 2030 (Blake 2018). The vacancy rates in Yellowknife are already reflecting these shifts, as people begin to leave the territory in search of employment prospects elsewhere (Dow 2018). The Canada Mortgage and Housing Corporation's 2019 *Northern Housing Report* cited a vacancy rate in Yellowknife of 5.2 per cent, up from 1.5 per cent in 2011. Importantly, these economic shifts form the context within which we must situate, at least in part, Northview's engagement in supportive housing and in the implementation of Housing First programming in Yellowknife. In other words, Northview may well be willing to participate in Housing First when vacancy rates are higher, and then reject further engagement once demand from professional tenants resumes.

The creation of supportive housing and the implementation of programs like Housing First become increasingly important in a city with virtual monopolies in public and private rental housing. The challenges of relying on one private rental company seem insurmountable, but this is the reality of housing affordability in Yellowknife. Though there were many challenging situations with Northview, there were also positive stories related to the Housing First project, as we describe in the next section, where housing providers collaborated and felt supported by Northview.

Housing First: Building Relationships with Northview

The limited housing market in Yellowknife is a primary challenge faced by the city's Housing First program. Several research participants spoke about the precariousness involved in relying on one private market company for the provision of housing units for Housing First programs. For example, a tenant evicted from Northview would have a very difficult time finding another housing option because there are so few options. Building relationships with private rental management staff was seen to be integral to the success of the Housing First program. According to one research participant, "It has been a nice relationship to have with [the] major landlord. It's kind of a three-way tenant program—tenant, support, landlord triangle—it kind of all works to keep them housed" (Interview with City of Yellowknife staff member 1, 2017).

This type of working relationship highlights the important role that support providers play in securing housing for Housing First clients. It demonstrates the extra work required by support providers. That also includes wraparound services for Housing First clients, to ensure that they are securely housed. It also demonstrates the challenges of implementing a Housing First project in a remote, smaller northern city. Housing First initially started in New York City and has been successful in major cities across North America, such as Toronto and Vancouver (Hwang et al. 2012; Patterson, Moniruzzaman, and Somers 2014; Shan and Sandler 2016). Thus, many foundational components of Tsemberis' Housing First model rely heavily on resources found in larger cities, such as a diversified housing stock and a large support service network (Tsemberis 2011). Even so, the Housing First model has been successful in rural and suburban areas and in smaller regional cities, by adapting the urban model to local needs (Cloke, Widdowfield, and Milbourne 2000; Stefancic and Tsemberis 2007; Waegemakers Schiff, Schiff, and Turner 2016; Waegemakers Schiff and Turner 2014).

A study of twenty-two rural and remote Canadian regions/cities outlined some of the challenges of implementing Housing First programs (Waegemakers Schiff, Schiff, and Turner 2016; Waegemakers Schiff and Turner 2014). The authors premised their analysis on the important distinctions between rural and urban homelessness: homelessness in rural municipalities looks different than its urban counterpart, and illustrates "distinct dynamics from urban regions, particularly related to the availability of social infrastructure, the impacts of macro-economic shifts, housing markets, and migrations" (Waegemakers Schiff and Turner 2014). Rural and remote communities often experience underdeveloped social service infrastructure, unqualified staff, limited emergency shelter space, and limited availability of rental housing stock. In Newfoundland and Labrador, Housing First programs found challenges in outreach workers' isolation from peers, and in rural New Brunswick, Housing First participants had a choice of moving to a different city with more housing options or staying where there are few. Unfortunately, not all adaptations are long-term solutions, but some do help in the immediate short term.

In the context of Yellowknife's limited housing options and the dominance of one major rental landlord, building healthy and collaborative relationships with Northview properties is essential for a successful Housing First program. The case of a support provider who worked with at-risk youth was a key example of how good relationships with property management staff kept tenants housed. During our interview with the support provider, they kept mentioning how "brokering" with property management really worked in maintaining secure housing for hard-to-house individuals. They told us a story of a tenant who "trashed a unit," resulting in $20,000 in damages, but due to the excellent working relationship they had with property management, the tenant did not get evicted but got re-housed. This tenant had to be re-housed three times. The housing worker allocated this success to the good relationship they had with Northview: "They're a profit business. And maybe because it's [name of worker]. He's new and he's so personable that is there's an issue I'll go over to his office and problem-solve. I don't think its pure economics. Because he really likes what he's doing" (Interview with support provider 1, 2018). Thus, brokering and building relationships is critical to the sustainability of the Housing First program in Yellowknife.

Relationships between housing providers and property management, then, are essential to maintain housing for Housing First tenants. The potential precarity of these relationships with the only private property management company in town, however, raises serious questions about the longevity of the program:

> It's really person-based, and that's the unfortunate thing. And, I mean, right now, we're having a—we have a great relationship with the manager, and when we were initially starting up, the

manager ... [who] was advocating for Housing First in
Yellowknife. But, if someone comes in who's not supportive of
our program, then we can very easily lose the units. And it's not
sustainable because of the funding model. It has to rely—we have
to rely on funding coming in to support our tenants and support
our staff.

(Interview with housing provider 2, 2017).

Unfortunately, the status of this relationship could change without
notice. "That's kind of out of our hands. They do what they need to
do. We're not in the business of competing with them ... But we can't
tell anyone what they can set their prices at" (Interview with NWTHC
representative 1, 2018). At the moment, Northview is willing to provide apartments for use within the Housing First program, but as
another interview participant indicated, it is not necessarily sustainable: "having all your eggs in one basket [is a] difficult thing"
(Interview with City of Yellowknife representative 1, 2017).

Challenges to Yellowknife's Housing First Program

Though successful and thriving, Yellowknife's Housing First program faces three significant challenges: funding, support for wraparound services, and lack of rental options. These challenges are
reflective of obstacles faced by other rural and remote communities
implementing Housing First programs across Canada (Waegemakers
Schiff and Turner 2014). Yellowknife, like other smaller cities, has
a limited network of social service providers, a significant lack of
affordable rental options, and precarious funding arrangements.

The allocation of funding was considered a major concern by many
of our research participants. In particular, those working for non-
profit organizations or the municipal government had concerns about
the centralization of federal funding within the territorial government, and how that impacted policy and programming decisions.
Several non-profit support providers expressed the desire to see the
municipal government in greater control of funding, suggesting it
would allow for more contextually-appropriate programming.

As mentioned above, it is difficult for non-profit housing providers to push back against funding agencies, for they are in a precarious situation as being dependent on government funding while

also wanting to advocate on their client's behalf. "I'm both an advocate and a service provider," one informant told us. "And that's really hard because we don't have an advocacy office here ... I'm going to lose my funding sometime because I'm a mouthpiece and saying too much against the hand that feeds you" (Interview with support provider 1, 2018).

However, the territorial and federal governments are not the only funding providers of relevance here. Since the beginning of diamond mining in the NWT, the presence of this industry in various aspects of Yellowknife community life has been palpable. To varying degrees, and at various times, the parent companies of the three diamond mines—Ekati, Diavik, and Gaucho Kué—have been involved in sponsoring initiatives aimed at providing supports to low-income Yellowknife residents, including those experiencing homelessness. In particular, the mines have significantly funded such initiatives as the Bailey House transitional home for men, Lynn's Place transitional home for women, the day shelter, the Side Door Youth Centre, and other organizations directed at caring for people living without housing in the city. Most recently, in 2019, De Beers (Gaucho Kué's parent company) partnered with the NWTHC and the Yellowknife Women's Society to help fund needed renovations on the women's shelter.

The extractive resource industry has therefore played a considerable role in the delivery of key social envelope programs in Yellowknife over its lifespan. At the same time, homelessness has not decreased. In fact, it has increased over those years, and continues to do so. This reality has left many of our interview participants wondering what lies ahead for funding such initiatives in the future, since all the diamond mines are set to close within the next decade. Industry-sponsored programming has been an important source of funding with more flexible requirements than those imposed by government, and has in many ways, been a funding course that the municipal government and local NGOs have been able to access directly.

SUSTAINABILITY, SOCIAL INFRASTRUCTURE, AND EXTRACTIVE RESOURCE ECONOMY

Evaluating the challenges within Yellowknife's inaugural Housing First program provides some tangible lessons for the future of housing policy in Yellowknife and the NWT. Both private and non-profit providers play a significant role in creating and maintaining affordable housing in Yellowknife. However, they do not have consistent funding for the operation of Housing First, nor are they represented in the decision-making of territorial or federal governments. "We're very vulnerable in that way. So, you know, a couple of things that were noted and we're mindful of, and we're just seeing if we can advocate our way forward" (Interview with City of Yellowknife staff member 1, 2017). This highlights an important thread in our examination of this precarious housing landscape, that is, vulnerability.

In fact, "vulnerable" is a key descriptor for the ways in which housing and support services for low-income Yellowknifers are situated within the context of extractive resource development. They are vulnerable to booms and busts and the impacts they have on affordability and accessibility; vulnerable to the closure of entire industries, upon whom key social programming has come to depend; and finally, vulnerable in that those organizations who are tasked with addressing and alleviating homelessness on the ground are not in positions of decision-making or policy setting. As a result, an additional layer of precariousness is created through a reliance on building and brokering personal relationships with the private and public sectors in order to access funding, to provide key services and, ultimately, to implement programs like Housing First.

Our research presents some key concerns about the sustainability of social infrastructure in a context like Yellowknife's, where programming like Housing First is dependent on private rental market participation and public/private funding arrangements. How can the city best cope with these vulnerabilities? There are many ways in which the private sector works with NGOs and government to ensure much-needed social programming, and yet these relationships are highly precarious. Working to better engage NGOs and housing advocates in setting funding and policy agendas would ensure a greater institutional memory, reinforcing relationships that exist between advocates and the public sector. Meanwhile, pooling industry contributions to social infrastructures would provide more dependable funding and would enable support providers and advocates to apply for grants on a consistent basis, rather than the current project by project fashion.

CONCLUSION

In this chapter, we have explored the implementation of Housing First in Yellowknife as a means to illustrate how the extractive resource economy sets the context for a constrained and precarious housing landscape. In particular, we highlighted how a reliance on the private rental market, largely pinned upon major private rental company, Northview REIT, makes housing programs and low-income housing highly sensitive to the ebbs and flows of the local extractive resource economy. Programs like Housing First are heavily dependent upon limited public and private rental housing markets, and are vulnerable to policy shifts within both. Moreover, we have aimed to advance the call in this volume for a more nuanced and comprehensive understanding of the socio-economic impacts of non-renewable resource development on northern individuals and communities.

The history of Northview indicates that there can be quick policy changes on the part of the REIT. After all, it was only in 2014 that Northview announced it would not be renting to tenants on income support ("Northern Property tightens rules for renters on income support," 2014). The 2019 CMHC *Northern Housing Report* indicated that Yellowknife's housing market is cooling, as the local diamond mines approach the end of their lifespan and the territory's economic future looks increasingly bleak. Providing units for Yellowknife's Housing First initiative may be seen as a productive use of Northview's resources now, but what might happen during a time of relative economic boom? These types of concerns support Petrov et al.'s call in Chapter 13 for socio-economic indicators based on the cultural and contextual priorities of northern communities in assessing the impacts of non-renewable resource development.

The eventual closure of diamond mining will bring to a close an era of industry financing of key social support services for people experiencing homelessness in the city. Unfortunately, the future for affordable and accessible housing in Yellowknife is not looking good, unless systemic change occurs and governments and extractive resource industry stakeholders are held to account. Moreover, without a policy emphasis on the well-being of all northerners, as Parlee suggests in Chapter 12 of this volume, the social costs of development, particularly for low-income residents, remain high.

Though the situation seems rather dire, and it would be difficult to ensure that benefits of resource development continue to support housing initiatives and affordable housing in Yellowknife, many local housing advocates and policymakers we interviewed had hopeful suggestions for mitigating the negative impacts of resource development. Their ideas included urging diamond mining companies to renovate and move on-site labourer's housing into sustainable long-term housing in Yellowknife, and to invest in a more clear and consistent way in housing infrastructure in the city and territory. As our engagements with policymakers and support providers has demonstrated, there is significant potential for partnerships and collaborations between governments, NGOS, and industry to ensure all northerners enjoy the benefits of extractive resource development. The best and most sustainable way forward to ensure that Yellowknife has a future filled with affordable and accessible housing is to continue to build relationships between all levels of government, to continue establishing new pathways for housing programs, and to hold the key players in the housing market and extractive resource-based economy to account.

REFERENCES

Abele, Frances, Nick Falvo, and Arlene Haché. 2010. "Homeless in the Homeland: A Growing Problem for Indigenous People in Canada's North." *Parity* 23, no. 3 (November): 21-23.

Asselin, Jodie and John Parkins. 2009. "Comparative Case Study as Social Impact Assessment: Possibilities and Limitations for Anticipating Social Change in the Far North." *Social Indicators Research* 94, no. 3 (December): 483–497.

August, Martine. 2020. "The Financialization of Canadian Multi-family Rental Housing: From Trailer to Tower." *Journal of Urban Affairs* 42, no. 7 (February): 975-997.

Beers, Randi, 2018. "Yellowknife's Housing First Eyes Expansion and Graduating 1st Clients Out of Program." *CBC News North*, June 25, 2018. https://www.cbc.ca/news/canada/northyellowknife-housing-first-update-1.4717121.

Blake, Emily. 2018. "Northwest Territories Economic Future 'Grim,' Says Report." *CBC North*, May 1, 2018. https://www.cbc.ca/news/canada/north/nwt-economic-forecast-2018-1.4643792.

Bone, Robert. 2003. *The Geography of the Canadian North: Issues and Challenges*. Oxford: Oxford University Press.

Brabant, Sarah and Robert Gramling. 1997. "Resource Extraction and
    Fluctuations in Poverty: A Case Study." *Society & Natural Resources* 10,
    no. 1 (January): 97–106. https://doi.org/10.1080/08941929709381011.
Canadian Mortgage and Housing Corporation. 1997. *Housing Need
    Among the Inuit in Canada, 1991* No. 35. Ottawa.
–2018. *Northern Housing Report.*
Canadian Polar Commission. 2014. *Housing in the Canadian North:
    Recent Advances and Remaining Knowledge Gaps and Research
    Opportunities.* http://www.polarcom.gc.ca/sites/default/files/housing_
    summary_1.pdf
Canadian Research Institute for the Advancement of Women. 2016. *Housing
    Market Fluctuations in Resource-Based Towns.* Retrieved from http://
    fnn.criaw- icref.ca/images/userfiles/files/HousingMarketFluctuations.pdf
Christensen, Julia. 2012. ""They Want a Different Life": Rural Northern
    Settlement Dynamics and Pathways to Homelessness in Yellowknife and
    Inuvik, Northwest Territories." *The Canadian Geographer / Le
    Géographe Canadien* 56, no. 4 (December): 419–438. https://doi.
    org/10.1111/j.1541-0064.2012.00439.x.
–2016. "Indigenous Housing and Health in the Canadian North:
    Revisiting Cultural Safety." *Health & Place* 40: 83–90. https://doi.
    org/10.1016/j.healthplace.2016.05.003.
–2017. *No Home in a Homeland: Indigenous Peoples and Homelessness
    in the Canadian North.* Vancouver: UBC Press.
Cloke, Paul, Rebekah C. Widdowfield, and Paul Milbourne. 2000. "The
    Hidden and Emerging Spaces of Rural Homelessness." *Environment
    and Planning A: Economy and Space* 32, no. 1: 77–90. https://doi.
    org/10.1068/a3242.
Collings, Peter. 2005. "Housing Policy, Aging, and Life Course
    Construction in a Canadian Inuit Community." *Arctic Anthropology* 42,
    no. 2 (January): 50–65.
Community Partnership Forum. 2016. *Homeless in Yellowknife:
    Community Partnership Forum April 26-27, 2016.* Northwest
    Territories Housing Corporation. http://nwthc.gov.nt.ca/sites/default/
    files/2016_homelessness_in_yellowknife_-_community_partnership_
    forum_-_summary_0.pdf.
Dawson, Peter C. 2003. "Examining the impact of Euro-Canadian archi-
    tecture on Inuit families living in Arctic Canada." Paper presented at
    International Space Syntax Symposium. London.

Dow, Keven. 2018. "Yellowknife Vacancy Rate Rises." *My Yellowknife Now*, November 29, 2018. https://www.myyellowknifenow.com/33854/ yellowknife-vacancy-rate-rises/.

Falvo, Nick. 2011. *Homelessness in Yellowknife: An Emerging Social Challenge*. The Homeless Hub.

– 2014, May. "10 Things to Know About Yellowknife's Northern Property Conundrum." *Northern Public Affairs*. Accessed October 31, 2019. http://www.northernpublicaffairs.ca/index/2014/05/.

Gillingham, Michael, Greg Halseth, Chris Johnson, and Margot Parkes. 2015. *The Integration Imperative: Cumulative Environmental, Community and Health Effects of Multiple Natural Resource Developments*. New York, NY: Springer International Publishing.

Goldenberg, S. M., J. A. Shoveller, M. Koehoorn, and A.S. Ostry. 2010. "And They Call This Progress? Consequences for Young People of Living and Working in Resource-extraction Communities." *Critical Public Health* 20, no. 2 (June): 157–168. https://doi. org/10.1080/09581590902846102.

Government of the Northwest Territories. 2005. *Homelessness in the NWT: Recommendations to Improve the GNWT Response*. Yellowknife, NWT.

Hwang, Stephen W., Vicky Stergiopoulos, Patricia O'Campo, and Agnes Gozdzik. 2012. "Ending Homelessness Among People with Mental Illness: The At Home/Chez Soi Randomized Trial of a Housing First Intervention in Toronto." *BMC Public Health* 12, no. 1 (September): 787.

Inuit Tapiriit Kanatami (ITK). 2004. *Backgrounder on Inuit and Housing*. Presented at the *Housing Sectoral Meeting, Ottawa, November 24-25*. Ottawa: ITK.

Kirst, Maritt, Suzanne Zerger, Vachan Misir, Stephen W. Hwang, and Vicky Stergiopoulos. 2015. "The Impact of a Housing First Randomized Controlled Trial on Substance Use Problems Among Homeless Individuals with Mental Illness." *Drug and Alcohol Dependence* 145: 24–29. https://doi.org/10.1016/j. drugalcdep.2014.10.019.

Levan, Mary Beth, Judie Bopp, Gillian McNaughton, and Mira Hache. 2007. *Being Homeless is Getting to be Normal: A Study of Women's Homelessness in the Northwest Territories*. YWCA Yellowknife and the Yellowknife Women's Society. http://ywcacanada.ca/data/publica- tions/00000011.pdf.

"Northern Property Tightens Rules for Renters on Income Support." 2014. *CBC News North*, May 9, 2014. https://www.cbc.ca/news/canada/north/

northern-property-tightens-rules-for-renters-on-income- support-1.2638266.

Nunavut Housing Corporation. 2004. *Nunavut Ten Year Inuit Housing Action Plan: A Proposal to the Government of Canada.*

NWT Housing Corporation. 2016. *Homelessness in Yellowknife: Community Partnership Forum April 26-27.* Yellowknife: Government of the Northwest Territories.

Padgett, Deborah, Leyla Gulcur, and Sam Tsemberis. 2006. "Housing First Services for People Who are Homeless with Co-occurring Serious Mental Illness and Substance Abuse." *Research on Social Work Practice* 16, no. 1 (January): 74–83.

Patterson, Michelle L., Akm Moniruzzama, and Julian M. Somers. 2014. "Community Participation and Belonging Among Formerly Homeless Adults with Mental Illness After 12 Months of Housing First in Vancouver, British Columbia: A Randomized Controlled Trial." *Community Mental Health Journal* 50, no. 5 (July): 604–611.

Shan, LeeAnn and Matt Sandler. 2016. "Addressing the Homelessness Crisis in New York City: Increasing Accessibility for Persons With Severe and Persistent Mental Illness." *Columbia Social Work Review* 14, no. 1 (June): 50–58. https://doi.org/10.7916/D82J6C8K.

Stefancic, Ana and Sam Tsemberis. 2007. "Housing First for Long-term Shelter Dwellers with Psychiatric Disabilities in a Suburban County: A Four-year Study of Housing Access and Retention." *Journal of Primary Prevention* 28, no. 3 (July): 265–279.

Stern, Pamela. 2005. "Wage Labor, Housing Policy, and the Nucleation of Inuit Households." *Arctic Anthropology* 42, no. 2 (January): 66–81.

Tester, Frank J. 2006. *Iglutaq (In My Room): The Implications of Homelessness for Inuit. A Case Study of Housing and Homelessness in Kinngait, Nunavut Territory.* Kinngait, Nunavut: The Harvest Society.

–2009. "Iglutaasaavut (Our New Homes): Neither "New" nor "Ours": Housing Challenges of the Nunavut Territorial Government." *Journal of Canadian Studies/Revue d'études canadiennes* 43, no. 2: 137–158.

Tester, Frank J. and Peter K. Kulchyski. 1994. *Tammarniit (Mistakes): Inuit relocation in the Eastern Arctic, 1939-63.* Vancouver, BC: UBC Press.

Tsemberis, Sam. 2011. "Housing First: The Pathways Model to End Homelessness for People with Mental Illness and Addiction." *European Journal of Homelessness* 5, no. 2.

Tsemberis, Sam, Leyla Gulcur, and Maria Nakae. 2004. "Housing First, Consumer Choice, and Harm Reduction for Homeless Individuals with a Dual Diagnosis." *American Journal of Public Health* 94, no. 4 (April): 651–656.

Turner Strategies. 2017. *Everyone is Home: Yellowknife's 10 year Plan to End Homelessness.* https://www.yellowknife.ca/en/living-here/resources/Homelessness/EVERYONE-IS-HOME—YELLOWKNIFE-10-YEAR-PLAN-TO-END-HOMELESSNESS-FINAL-REPORT-JULY-2017.pdf.

Van Hinte, Tim, Thomas I. Gunton, and J. C. Day. 2007. "Evaluation of the Assessment Process for Major Projects: A Case Study of Oil and Gas Pipelines in Canada." *Impact Assessment and Project Appraisal* 25, no. 2 (June): 123–137. https://doi.org/10.3152/146155107X204491.

Waegemakers Schiff, Jeannette, Rebecca Schiff, and Alina Turner. 2016. "Rural Homelessness in Western Canada: Lessons Learned from Diverse Communities." *Social Inclusion* 4, no. 4 (October): 73. https://doi.org/10.17645/si.v4i4.633.

Waegemakers Schiff, Jeannette and Alina Turner. 2014. *Housing First in Rural Canada: Rural Homelessness & Housing First Feasibility Across 22 Canadian Communities.* Government of Canada.

9

Environmental Legacies: Mine Remediation Policy and Practice in Northern Canada

Anne Dance, Miranda Monosky, Arn Keeling, and John Sandlos

In 2013, a journalist, attempting to clarify who was responsible for reclaiming Nunavut's recently abandoned Jericho Diamond Mine, threw up her hands in frustration, characterizing the case as a jurisdictional "hot potato" (Herman 2013). The mine's two former owners went bankrupt in 2010 and 2012, leaving twenty-three workers unpaid and millions of dollars of remediation work incomplete; the federal government assumed responsibility for its cleanup in 2014 (Gregoire 2014; Nunatsiaq News 2017; Rohner 2016).

Jericho is not unique. There are thousands of abandoned and orphaned mines and exploration sites scattered across Canada, many of them in the north (National Orphaned/Abandoned Mines Initiative [NOAMI] 2010). Their legacies include heavy metal leaching, wildlife habitat loss, polluted rivers and undrinkable water, the disruption of animal migratory routes and other traditional food sources, and long-term socio-economic strains in northern communities (Wynn 2007; Office of the Auditor General of Canada [OAGC] 2002; Gibson and Klinck 2005; Rodon and Lévesque 2015; Sandlos and Keeling 2016b). Remediation (also known as reclamation in other parts of Canada, and rehabilitation in both Quebec and in Newfoundland and Labrador) has the potential to help lessen these negative impacts and ensure that development does not leave northerners worse off (Séguin and Larivière 2011; Department of Natural Resources 2010; Government of the Northwest Territories 2013). Given that the federal government must spend billions of dollars to reclaim existing contaminated sites and that two abandoned northern mines (Giant and Faro) top this list, the magnitude of these problems is appreciated even in distant Ottawa (Parliamentary Budget Office 2014).

Jericho Diamond Mine's recent troubles are particularly disturbing given that politicians, bureaucrats, and industry representatives often differentiate between two periods of Canadian mining: an earlier era of regulatory laxness and limited reclamation, and its modern, enlightened counterpart from the 1980s onwards (Nahir and David 2007; Monosky 2021). The former spawned a disastrous inheritance: unreclaimed legacy mines, including orphaned or abandoned mines whose owners' bankruptcy or dissolution ensures that ownership, and the incumbent responsibilities, reverts to the Crown (Castrilli 2007; Worrall et al. 2009). The latter embraces technological innovations and cradle to grave planning, cementing remediation's central role.

However, more recent mine abandonments, like Yukon Zinc Corporation's Wolverine and Keno Hill mines, blur the line between these periods without a corresponding recognition of the problem by the government and the industry (CBC News 2015). Much to northerners' frustration, abandoned mines and exploration sites continue to cause problems. For example, Vancouver-based Snowfield Development Corp. abandoned a diamond exploration site near Drybones Bay (near Yellowknife, NWT) in 2014. Not only did Snowfield's directors choose to go ahead with advanced exploration without adhering to the environmental review process, but the security they posted was a fraction of the total anticipated remediation costs (CBC News 2014). Attempts to resolve such challenges are complicated by a constantly evolving jurisdictional landscape that is, in turn, shaped by modern land claims agreements and devolution.

The environmental legacies of these extraction activities have a significant impact on the long-term sustainability of northern communities. Not only do these impacts have to be adequately repaired, but mitigation activities must be informed and shaped by communities to support their future well-being. As part of the Resources and Sustainable Development in the Arctic (ReSDA) project, we looked at changes in policy relating to mine remediation in Canada. In this chapter, we bridge literature on contaminated site management policies (Adams et al. 2010; Castrilli 2002; De Sousa 2001, 2002, 2003, 2006, 2008) and historical scholarship that has called attention to the complicated ways Indigenous peoples have navigated employment, knowledge production, and cultural change in the face of large-scale resource projects (Sistili et al. 2006; Tester and Irniq 2008; Piper 2009). Through an analysis of policy documents, legislation, research reports, and newspaper articles, we discuss the factors shaping remediation regimes in the territorial and provincial North. We explore how policymaking has responded to the challenges of remediating mines and mineral exploration sites in the North, and we ask what this means for the region by examining past and current closure plans.

From this analysis, it is clear that no overarching vision informs remediation planning in the North. Jurisdictional overlap explains some of the disaggregated policy regime, as responsibility for policy-making in many cases sits with some combination of federal, provincial, territorial, and regional governments, along with local land and water boards. In Chapter 8 of this volume, Freeman and Christensen show that northern housing policies necessitate collaboration between a confusing number of government departments and other stakeholders; northern mine remediation efforts are similarly constrained and complicated. Efforts to mitigate the impacts of new and legacy mines are complicated by the highly site- and case-specific nature of remediation; the lack of a clear, ambitious technical and regulatory definition or vision of remediation; and the jurisdictional overlap and governance issues associated with cleanup. Particularly in the territorial North, both already and yet-to-close sites pose significant challenges for Indigenous organizations and co-management bodies. Addressing these wider policy challenges in the North is crucial to meet the expansive, expensive demands of mine remediation and the long-term sustainability goals of northern communities.

The problems are significant, but not hopeless. Remediation efforts, particularly those that draw on Indigenous knowledge and encourage local involvement, can mitigate and manage some of the worst impacts of northern resource development. Remediation policy reform, such as strengthened regulations and more rigorous government enforcement, will help facilitate this. However, it is worth remembering that remediation can also exacerbate inequality and environmental problems, both in terms of how (and by whom) remediation is undertaken and the outcomes of the cleanup itself. (Keeling and Sandlos 2017). Effective remediation demands more than a particular technological fix or planning strategies; it involves a candid discussion of the goals and limitations of remediation projects, both past and present.

CANADIAN REMEDIATION IN CONTEXT

Contemporary remediation practice generally involves planning, engineering, and management strategies undertaken to help monitor, mitigate, and in some cases remove disturbances and pollution in areas affected by mining and mining-related activities. In the Canadian North, this can be tremendously difficult. Revegetating, removing buildings and equipment, covering physical hazards, backfilling mined out pits, stabilizing waste piles, and containing hazardous materials for decades or even centuries make for daunting technical challenges. Remote locations, logistical barriers, limited government funding, and skilled labour shortages, combined with limited oversight and difficulties enforcing regulations, further complicate these efforts (Hart et al. 2012).

Realistically, remediation cannot restore a site to its previous, pre-mining state. Some remediation activities may actually remobilize contaminants, deepening community concerns about ongoing environmental and health impacts. As Sandlos and Keeling write, potential redevelopment projects, as well as the ongoing toxic legacies of mineral extraction, can act to revive the North's historic "zombie" mines so that true restoration of a site to its pre-mine state remains elusive (Keeling and Sandlos 2013; 2017). Furthermore, the *meaning* of remediation has changed over time. Rather than a task to be under-taken after operations cease, modern progressive remediation strat-egies ensure that mitigation begins before an operation's official opening. At its best, remediation includes ambitious, holistic land-use strategies, habitat planning, and the comprehensive treatment of contaminants beyond those in the mine's immediate vicinity. Technical solutions are only the start: effective remediation entails sound environmental stewardship, planning, and policy spanning decades and even centuries.

In addition, each remediation project is unique, differentiated by disparate environments; industrial processes and by-products; com-munity expectations and input; and time scales stretching over months, years, and decades—all within complicated histories of disruptive resource development. These factors are layered upon the unique compositions of local communities (depending on their proximity) and governance arrangements. Thus, reclaiming Yukon's Faro Mine, with its 320 million tonnes of acid-generating waste rock and 70 mil-lion tonnes of acidic tailings demands considerably more time, money, and expertise than reclaiming a small mine exploration site with a half-dozen abandoned fuel barrels and a small area of contaminated soil. The Faro Remediation Project is particularly challenging because of perpetual water management on site; there is a significant potential for this large amount of acid-generating waste to enter the water (Faro Mine Remediation Project 2019; Kuyek 2019). Like the ore bodies miners exploit, remediation is a strongly local, site-specific process. As Keske (Chapter 11) demonstrates, local communities are often left to contend with the long-term management of environmental problems created by extractive industries, making their involvement in cleanup and remediation planning imperative.

When surveying the multitude of contemporary remediation requirements, it is worth remembering that, until the late 1970s and early 1980s, mining companies operated within a Canadian legal framework that was "open, straightforward, democratic, and encouraging," which is to say, deeply favourable to them and largely inattentive to environmental concerns (McPherson 2005). But by the late 1980s, sustained Indigenous and settler activism, often in opposition to specific projects, led to more proactive (albeit still fragmentary) environmental policymaking, as well as a greater recognition of Indigenous land rights and the duty to consult with Indigenous communities (Laforce et al. 2012). As McAllister and Alexander (1997) show, public unease about the social and environmental impacts of mining across Canada spurred companies to seek a social licence for their extractive activities, most notably through the 1992 multi-stakeholder Whitehorse Mining Initiative (WMI). In their comprehensive study of the WMI, they demonstrate that the initiative led to several changes for both legacy and start-up mines, including more active engagement in the oversight process from Natural Resources Canada and the federal government in general. Collaborative research and development initiatives such as the Mine Environmental Neutral Drainage (MEND) program, from 1989 onwards, the Canada Centre for Mineral and Energy Technology (CANMET), and the Green Mining Initiative complemented these efforts (OAGC 2002; Natural Resources Canada 2018).

WMI's implementation and reach was uneven and fragmented by the same jurisdictional silos it was created to address. Newfoundland and Labrador was not a signatory to the agreement, which meant that Labrador's Indigenous groups had little input, while Yukon's government only adopted WMI provisions that were favourable to industry. Junior mining companies (including exploration companies and small producers) were largely left out of the process (McAllister and Alexander 1997). Efforts to improve regulations for newer mines also had their limitations: new legislation in the 1980s and 1990s requiring mine remediation plans were stymied by a cap on the number of required securities, leading to liabilities outstripping submitted funds (Nahir and David 2007).

THE CURRENT REGULATORY REGIMES

Multiple levels of government now inform northern remediation policies in Canada: legislative bodies (provincial and territorial legislatures and the Canadian Parliament); the executive (provincial, territorial, and federal cabinets, departments, and agencies); and the judiciary (courts, including the Supreme Court of Canada and tribunals). Remediation projects on Indigenous lands are also shaped by regional land and water boards, land claims agreements, and Impact and Benefit Agreements (Natural Resources Canada 2015a). On paper, national guidelines for pollution releases and health standards crafted by the collaborative federal and provincial Canadian Council of Ministers of the Environment [CCME], as well as federal laws setting out pollution release standards, actuate remediation efforts (Skogstad and Kopas 1992; Fowler 2007; CCME 1996, 1997). However, the provinces, and increasingly the territories, wield the most regulatory power over natural resource management, including mining and remediation policy. Rather than a single remediation "regime," there are multiple arrangements for mine site remediation. These efforts are shaped by a mixed bag of legislation, regulations, permit and licensing systems, environmental review processes, guidelines, and site-specific initiatives (Castrilli 2007). Which regulations and programs are relevant for particular mines and how. quickly remediation will occur depends on the characteristics at each site. Remediation programs and policies for abandoned or orphaned sites, for instance, are distinct from requirements for new or future projects, and specific geographical features, such as lakes and rivers, determine which licensing requirements and environmental standards apply. Additionally, sites presenting major health hazards or environmental damage are sometimes subject to an environmental review process and reclaimed more quickly.

Whatever their origins, modern remediation policies share two common elements:

1   the requirement to file a mine closure plan as a condition of development approval and

2   tax-deductible financial sureties (also known as bonds or securities) to guarantee money will be available for remediation activities should the mining proponent go bankrupt.

Companies' contributions to mine remediation trusts are tax deductible in every Canadian jurisdiction except for Nova Scotia and the Yukon (Natural Resources Canada 2015b). Mine remediation includes everything from managing large tailings ponds, to dismantling mine workings and processing facilities, to removing leftover explosives and fuel stockpiles. For newer mines, operators must keep abreast of these requirements and send updated remediation plans to the appropriate authority. Abandoned mines are assessed using the CCME's classification system for contaminated sites (CCME 2014).

Embedded within numerous Canadian environmental laws and planning guidelines, the "polluter pays" principle requires that parties responsible for pollution must fund and carry out remediation. In practice, however, these requirements have not always been met (FCSAP 2019b). When bankrupt mining companies abandoned the remains of exploration activities and mining operations on northern Crown land, the remediation of these sites fell to federal departments and taxpayers. Newer land use planning and mine licensing instruments rarely apply to these sites.

Federal leadership on northern mine reclamation was largely nonexistent throughout the twentieth century; national initiatives to address orphaned and abandoned sites only began in the late 1980s. From 1989 to 1995, the CCME and the federal government's National Contaminated Sites Remediation Program [NCSRP] paid for the basic cleanup of forty-five high-risk orphaned sites, created a classification scheme based on environmental and human health impacts, and funded industry technology trials. Subsequent efforts and broader enforceable legislation for contaminated sites were derailed by federal-provincial friction, weak leadership, and funding cuts. (OAGC 1995, 1996; Cowan and Mackasey 2006; FCSAP 2019a) The North was also largely an afterthought: since NCSRP funding was allocated on a per capita basis, this meant that support for projects in the Yukon and NWT combined was less than the share for Prince Edward Island (NCSRP 1994).

Other, more specialized, initiatives were subsequently launched for northern sites. In 1991, the federal Department for Indigenous and Northern Affairs, then titled Indian and Northern Affairs Canada [INAC], began a waste management program for assessing and remediating contaminated sites, with some limited funding for abandoned mines (CCSG Associates 2001). The same year, the federal government's Green Plan created contaminants programs under the Arctic

Environmental Strategy. These programs led to the removal of barrels from several exploration sites and remediation at two Yukon mines. By 1995, the Contaminated Sites Management Working Group (CSMWG) had developed an expansive interdepartmental strategy to address contaminated sites across the country (FCSAP 2019a). Despite this, there was still no substantial funding for abandoned mines and other contaminated sites.

By the mid-1990s, the scale of remediation challenges could no longer be ignored: collapsing mineral prices had hastened a wave of private sector bankruptcies and mine abandonment, including the massive Faro and Giant operations (INAC 2010). Changes to federal accounting systems put contaminated sites on the books as contingent liabilities, a move that the Auditor General of Canada had repeatedly called for during the previous decade. Finally, Ottawa was overtly acknowledging its responsibility for thousands of contaminated sites across the country, many of which were created by mining activities. The government's liabilities translated into hefty additions to federal debt numbers, as did growing future care and maintenance costs. These could only be mitigated through remediation (OAGC 1995, 1996; Fowler 2007). Yet many of these sites were in the North, where abandoned mine strategies frequently amounted to little more than "crisis management" (Van Dijken 2001). In 2002, the federal environment commissioner questioned why federal authorities knew little about the health risks posed by hundreds of abandoned mines and contaminated sites. The commissioner stressed the government had failed to adequately study, monitor, and remediate the sites, or secure stable funding to fulfill these responsibilities (OAGC 2002).

The federal government responded with the Federal Contaminated Sites Inventory [FCSI], the Federal Contaminated Sites Action Plan [FCSAP], and the multi-stakeholder National Orphaned/Abandoned Mines Initiative [NOAMI]. An advisory committee made up of mining groups, First Nations, NGOs, and governments, NOAMI began studying mine abandonment and associated Canadian regulatory reform in 2002 and has since produced guidance documents and reports on abandoned mines. The federal government alone funds FCSI and FCSAP. While FCSI compiles data on thousands of contaminated sites, FCSAP supplies money and expertise to Crown corporations and responsible departments to remediate them (Treasury Board Secretariat 2019). Established in 2005 with dedicated funding for an initial fifteen years, FCSAP is a rare example of a substantive financial commitment for

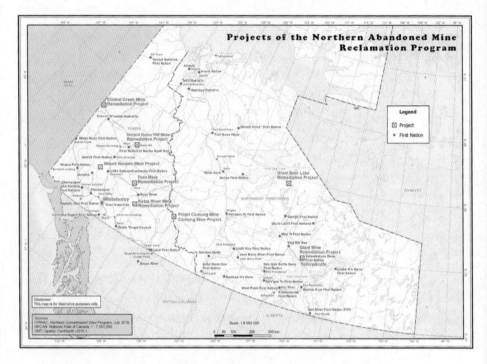

Figure 9.1 Projects of the Northern Abandoned Mines Reclamation Program

contaminated sites across Canada. Under FCSAP, the federal government accepts full responsibility for sites where pollution levels exceed national guidelines and pose health hazards to humans and the environment (OAGC 2012). After the relevant government department (in the North, generally Crown-Indigenous Relations and Northern Affairs Canada) determines the degree of site contamination, FCSAP provides training and funding for remediation. FCSAP has made abandoned mines and other contaminated sites visible to federal departments and agencies beyond those regulating natural resource extraction (Environment Canada Evaluation Project Team 2014). Funding allocated to FCSAP has simultaneously bolstered older programs like the Northern Contaminated Sites Program [CSP], which has funded assessments of hundreds of sites in the territories. FCSAP money has helped the CSP shift from basic care and maintenance to more ambitious remediation planning, consultation, and monitoring (INAC 2010; Aboriginal Affairs and Northern Development Canada [AANDC] 2014).

In the 2019 federal budget, FCSAP was renewed for another fifteen years with guaranteed funding through 2024; the federal environment department now administers it with support from the Treasury Board (Environment and Climate Change Canada [ECCC] 2019; Crown-Indigenous Relations and Northern Affairs Canada [CIRNAC] 2019). The budget also created the Northern Abandoned Mine Reclamation Program (NAMRP) to remediate or manage the highest risk sites in the Northwest Territories (Cantung Mine, Giant Mine; the Great Bear Lake Remediation Project) and the Yukon (Clinton Creek Mine; Faro Mine; Ketza River Mine; Mount Nansen Mine; United Keno Hill Mines) over the next fifteen years (see figure 9.1).

REMEDIATION POLICIES IN THE NORTH

A handful of federal laws and regulations apply to all new and orphaned mines across Canada (see table 9.1). The federal government's Minerals and Metals Policy acknowledges differing jurisdictional competencies and emphasizes sustainability, Indigenous involvement in planning, and safe mining practices for companies operating within Canada and abroad (Natural Resources Canada 2013). Each northern jurisdiction has a handful of standard regulatory elements that can affect remediation planning for new mines. Mining legislation generally focuses on assigning mineral rights and has little to say about environmental protection other than the requirement for remediation plans and financial securities (Hart et al. 2012). Separate environmental acts, regulations, and guidelines establish pollution limits, discharge permit processes, initiate cleanup authorities, and authorize specific boards to carry out environmental assessments of projects and exploratory work (Castrilli 2007). Water and land use acts also set out license and permit requirements for remediation-specific mining concerns. Regionally developed guidelines, such as the Nunavik Inuit Mining Policy (discussed by Schott et al. in Chapter 3 of this volume) also increasingly inform extractive polices. Workplace safety and land planning legislation likewise affect exploration, development, and site cleanup operations.

Territorial and provincial programs intersect with the federal ones noted above. Yukon's devolution transfer agreement has an entire section devoted to Type II mines, which are remediated by the federal government (through NAMRP) and the Yukon, with the engagement of Yukon First Nations (Yukon Territorial Government 2021).

Abandoned mines strategies like those in Quebec and Manitoba, while not exclusively northern focused, are very active in the provincial norths (Ministère de l'Énergie et des Ressources naturelles 2021; Government of Manitoba 2021). Some territorial and provincial governments have also collaborated through special inter-jurisdictional or public-private collaborative agreements and programs to manage these sites. In 2007, for example, Quebec mining companies partnered with the provincial government, Inuit, First Nations, and regional governments to develop a program for remediating abandoned mines in Nunavik, called the Fonds Restor-Action-Nunavik (FRAN); this fund inspired a similar initiative in the James Bay area, created by the Cree Nation of Eeyou Istchee (Dea and Gaumand 2016).

Underlying the many policies, regulations, and programs for mine remediation is the reality that Canada's historic dependence on natural resource development and close ties between industry and government have produced a weak record of environmental regulation and enforcement (Boyd 2003). The most robust components of remediation policymaking have been developed in response to environmental groups' complaints and community advocacy. In 2009, for example, MiningWatch Canada, Ecojustice Canada, and Great Lakes United successfully sued for toxic mining tailings releases to be included within the National Pollutant Release Inventory (MiningWatch Canada 2014). As Wenig and O'Reilly persuasively argued in their 2005 review of NWT remediation regulations, there is no true Canadian remediation "regime," as this word implies a coherence wholly absent from current requirements.

Instead, remediation requirements are undermined by feeble enforcement, monitoring, and review capacities, and available data is fragmented according to the source of pollution or jurisdiction authority. Thus, the relative success of remediation efforts depends less on policy or legislation than on the case-by-case practices of mine stakeholders and regulators. That said, NOAMI's multi-stakeholder approach has led to substantive legislative reform in most jurisdictions across Canada; it has also created a toolkit based on best practices to help parties clean up abandoned mines (Tremblay and Hogan 2016).

REMEDIATION REFORM IN A NORTHERN CONTEXT

Scholars and communities have called for an enhanced remediation regime that considers all environmental impacts through improved cumulative impact assessment standards (Wenig and O'Reilly 2005;

Duinker and Greig 2006; Ehrlich 2010; Beckett et al. 2020). Legislative reforms could also lead to stronger enforcement and funding (Castrilli 2010). Another consideration for reform involves all levels of government affirming (or reaffirming) the goals of reclamation. National environmental quality guidelines informing remediation are incomplete and far from "holistic benchmarks" for human health and the environment, so how can remediation itself be trusted (Wenig and O'Reilly 2005)? The North would be better served by embedding the precautionary principle (operating in ways that prevent harm, be it health-related, psychological, or environmental) into remediation regulations.

Effective northern remediation reform also involves tackling broader difficulties facing northern communities and governments. Frances Abele (2009) stresses that northern peoples in Canada confront three major challenges: finding ways to make their evolving governance arrangements work, developing their economy in sustainable ways, and dramatically improving local well-being. The limitations and failures of current remediation policy and practice exemplify all these challenges.

Devolution and modern treaty making have created new political and institutional co-management arrangements in the North, many of which directly affect mining policy and provide leverage for increased recognition of Indigenous land use rights—and Indigenous control over their lands (Monosky and Keeling 2021). Northern-staffed agencies and co-management boards like the Nunavut Impact Review Board or territorial land and water boards can monitor or assess the federal authorities responsible for remediation projects, including the ongoing cleanup of Jericho Diamond Mine and other sites. Modern land claims processes could potentially provide openings for Indigenous expectations to be interpreted not simply as mine remediation *guidelines,* but as enforceable, non-negotiable standards. But these agreements are still an incomplete patchwork. The absence of a modern treaty or land claim led the Ross River Dena Council to successfully challenge free entry (mining proponents' unrestricted access to their lands for exploration activities) in their traditional territory in the courts in 2012 (Hoogeveen 2015).

Devolution and land claims are not a comprehensive panacea for improved remediation standards: new authorities charged with issuing licenses and assessing mine closure planning suffer from major capacity deficits in keeping with limitations across the North. In autumn 2014, for instance, a quarter of positions within the Government of

Nunavut were unfilled (Nunatsiaq News 2014). In other words, there are not enough trained people to plan, carry out, and oversee remediation enforcement, monitoring, and reviews. Nor is there enough money to support these efforts.

Additionally, the responsibilities of new territorial regulatory authorities sometimes overlap, leading to further difficulties (Wenig and O'Reilly 2005). Territorial-federal relationships can be fraught, as illustrated by Ottawa's 2014 attempt to merge all regional land and water boards in the NWT, which was successfully resisted by northerners (Cohen 2018). And although devolution has created new arrangements for natural resource management, these are not necessarily effective. Northern agencies and governments struggle with inadequate staff and funding, and continued devolution might only reinforce this pattern. Nor is support from Ottawa always forthcoming (Hart et al. 2012; MacNeil 2014). NOAMI, too, has struggled to overcome fragmentary jurisdictional divisions (NOAMI 2009). Despite new commitments such as the NAMRP, such long-term initiatives are always vulnerable to government cuts and changing policy priorities in Ottawa (Tremblay and Hogan 2005; Kuyek 2011). Northern regions within provincial borders face additional jurisdictional barriers. Multiple regional authorities in Nunavik are responsible for overseeing the activities of mining companies, but these organizations act more as advisers for the provincial and federal government, and have little power to create and enforce their own remediation regulations (Monosky and Keeling 2021b).The challenges of territorial devolution also complicate the other key element of remediation practices for new mines: financial securities (funds set aside by mining companies for remediation activities). Across the North and much of southern Canada, securities are weakly enforced, incorrectly calculated, and inconsistently collected; they provide inadequate protection in cases of mine abandonment, and measures in place to track these funds are few and largely inaccessible to the public (Wenig and O'Reilly 2005; Hart et al. 2012). First Nations and organizations such as MiningWatch Canada continue to advocate for more rigorous securities regimes, and some jurisdictions such as Quebec have responded by requiring companies to provide 100 per cent of cleanup liability costs up front (Pollon 2019; Lavoie 2019). However, in the absence of stringent regulations, it still falls to the government to assume liability for remediation costs that can rise significantly over time (Parliamentary Budget Office 2014). While the federal government has committed funds for abandoned mines through initiatives such as FCSAP and now the NAMRP, where does this leave more recent mines? Devolution could lead to increased financial burdens for territorial governments in the form of unremediated mines—and increased risk for northern communities given the territories' limited capacity to finance the remediation of newly abandoned mines.

The North's second challenge, achieving sustainable economic development, also resonates with the limitations and opportunities inherent in remediation policymaking. It is difficult to overstate the importance placed on mineral development in northern regions (Thompson 2019). For those expressing reservations about the social, economic, and environment legacies of increased mineral exploration, remediation challenges loom large, especially since mining is far from a sustainable activity (Fonseca et al. 2014; McAllister and Alexander 1997). Effective remediation involves monitoring and treating tailings that may leach toxicants for decades or even centuries after mine closure. Despite this, management and governance regimes in Canada generally lack laws and regulations for perpetual care. Closure plans typically only contain commitments for five to ten years of post-closure monitoring, while a 100-year lifespan is the standard expectation for waste containment facilities. Stewardship and emergency response plans are inadequate, and environmental assessment processes inconsistently address broader concerns about community safety and environmental quality (Hart and Coumans 2013). According to critics, modern remediation approaches like adaptive management have become "euphemism[s] for stumbling along, and keeping costs to a minimum" (Kuyek 2011). The narrowness of the current regimes is particularly evident at sites with substantive long-term problems. The Giant Mine Remediation Project epitomizes the most extreme challenges. The gold mine operations left behind 237,000 tons of toxic arsenic trioxide dust buried underground at the site. The technological solution advanced through the environmental review process involves freezing the arsenic underground, a strategy that has exacerbated the Yellowknives Dene's anxiety about this toxic site on their traditional lands (Castrilli 2007; Sandlos et al. 2014). As remediation practices continue to evolve, policymakers must consider how such legacies will affect future generations (Kuyek 2011; Sandlos and Keeling 2016b).

One way to ensure northern communities actively benefit from these troubling intergenerational impacts is to guarantee them jobs, training, and lead roles in remediation. The community of Faro, for example, has reoriented itself from mineral extraction to mine closure, with the Faro Mine closure plan stressing that mitigation activities and monitoring would take centuries, thereby ensuring local employment for decades (Faro Mine Complex 2010). That said, although Faro is often cited as an example of remediation projects providing employment opportunities for First Nations, these opportunities were highly localized and bypassed the Ross River Dena until early 2019 (Rudyk 2019). Similarly, Northerners only made up 20 per cent of the workforce for recent Giant Mine remediation activities (Brockman 2019). Achieving sustainable economic resource development in the North means more than hiring northerners for remediation; it also involves incorporating traditional knowledge into policy and practice. At least nominally, initiatives like NOAMI and Parks Canada remediation efforts (funded through FCSAP) incorporate local knowledge provisions (NOAMI 2002). Different forms of local knowledge have expedited and improved remediation. This was the case during a Parks Canada cleanup at Ivvavik, where former Inuvialuit employees of the DEW Line site helped scientists distinguish between natural features and potentially contaminated land (Edwards 2010).

That said, the mandated integration of local knowledge into mine planning has not systematically addressed local concerns, particularly those related to loss of "spirit of place" and connections to land incurred by contamination and pollution (Kuyek 2011). Too often, Traditional Knowledge (TK) is pigeonholed as cultural or ecological information within policymaking rather than incorporated into extractive activities that create the need for remediation in the first place (Keeling et al. 2019). For instance, during the environmental assessment of the Giant Mine Remediation Project, Dene and Métis knowledge tended to be confined to issues related to historic local biotic resources and land-based practices, marginalizing local input on technical and procedural matters related to the remediation plan. Nor were Indigenous claims related to environmental injustice for past harms related to mining pollution (and their ongoing legacies beyond the mine site) addressed (Sandlos and Keeling 2016a).

A review of eight northern closure plans from currently operational mines across the territorial and provincial North demonstrates how increasingly regionalized governance structures are not necessarily resulting in better, more democratic and locally appropriate remediation planning (Agnico Eagle Mines Limited 2014, 2015; Alexco Resource Corp 2018; Baffinland Iron Mines Corporation 2018;

Canadian Royalties 2016, 2019a, 2019b; Diavik Diamond Mines 2017; Dominion Diamond Mines 2018; Vale Newfoundland & Labrador Limited 2016). The flexibility allowed through devolution, combined with effective community consultation, *could* help mining companies develop more locally appropriate closure objectives in consultation with impacted communities. However, consultation requirements for closure planning are inconsistent across jurisdictions and the methods and outcomes of engagement are left vague in closure plans. Of the eight closure plans reviewed, all but one explicitly referred to some form of public consultation, but most emphasized the importance of *future* or *ongoing* consultation with relevant Indigenous groups, while neglecting to detail the consultations that have already occurred and what community members' expectations are.

A notable exception to this is the closure plan for the Diavik Diamond Mine, which includes summary reports from their Traditional Knowledge Panel and other community engagement activities. Alternatively, the closure plan for Keno Hill (owned by Alexco Resource Corp. and operated in Yukon) simply indicates that closure objectives were "developed with and agreed to by Alexco/ERDC, Canada, ... FNNND [First Nation of Na-cho Nyäk Dun], and Yukon," and that the relationship between these actors "provides for significant consultation and collaboration on closure objectives and final options for the development of the District Closure Plan" (Alexco Resource Corp 2018). But no clear consultation methods or results are included in the Keno Hill closure plan. This latter example is much more representative of the standard approach to writing closure plans, where Indigenous communities and the need for consultation may be briefly acknowledged, but community expectations, concerns, knowledge, and perspectives are not presented in any meaningful way (Monosky and Keeling 2021a). The lack of transparency means it is impossible to know how community knowledge is being collected, interpreted, and applied to closure planning, and evaluations of these plans can be challenging as a result.

In the absence of community input and expectations, much of the structure, contents, and guiding principles of these documents are oriented to meeting the minimal or baseline requirements written in territorial and provincial policy documents and guidelines governing the closure of new mines. In most cases, the objectives and goals of the documents are copied directly from mine closure guidelines; both the Ekati and Diavik closure plans refer to the Mackenzie Valley Land and Water Board's closure principles as their own guiding principles, for example. This is also true for much of the post-closure land use planning written in closure plans, where the documents cite specific guidelines to justify how landscapes will look at relinquishment. Meeting baseline requirements, though, does not necessarily result in social acceptance and the restoration of cultural land uses (Rosa et al. 2020a; 2020b), especially when those requirements are from governments that may not have the human and financial capacity to develop and enforce effective policy. These minimal requirements typically focus on relatively short-term technical remediation fixes and do not require companies to engage with conversation about colonialism, historical and ongoing environmental injustice, (in)equity, and community consent (Beckett 2020; Gregory 2021).

Sound and enduring remediation means re-establishing functioning ecosystems that are valued, safe, and accessible for locals. It requires consultation and consent as core features of the environmental review process and integrating TK into site cleanup and mining. At Drybones Bay, NWT, the site of sacred Yellowknives Dene hunting and fishing grounds, burial grounds, and historic villages, TK considerations led to the rejection of a mineral exploration application in 2004 (Vela 2011; Parlee 73). In the North and across southern Canada, there is a need for a conversation about the goals and aims of remediation and how TK should shape them. Does "remediation" refer to a return to pre-extraction landscapes, or is its aim to mitigate the worst pollution problems and ensure specific human activities can go forward? How should remediation planning incorporate Indigenous hunting and fishing needs? Northern residents deserve a greater debate about what has been achieved—and what is achievable.

The final pressing northern challenge involves improving community well-being. This too, mirrors ongoing remediation dilemmas, especially those relating to knowledge and information sharing. A growing number of tools provide guidance for communities grappling with the long-term challenges of remediation in the North. These include plain language overviews of the various agreements and remediation activities (Gibson and O'Faircheallaigh 2010; INAC 2007, 2009). Still, straightforward information on the transportation and management of toxic materials and other aspects of the remediation process is not easy to find in Canada. Endemic institutional "structural secrecy" within government departments responsible for managing and regulating contamination further exacerbates public anxieties (Kuyek 2011). The mandate of the Giant Mine Oversight Board reflects this history: it was created in part to ensure that contemporary remediation efforts are transparent and consider the safety of local communities through a health advisory committee which engages communities and all government stakeholders (Giant Mine Oversight Board 2019).

Community well-being likewise resonates with the environmental justice dilemmas underlying mine remediation histories. Mineral development in Canada is based on free mining, a system that privileges proponents' goals over community rights and equality (Laforce 2012). No matter how sophisticated they may seem, remediation strategies must confront the profound historical injustices that marginalized northern people and created enduring social, economic, and environmental problems on their lands (Keeling and Sandlos 2009; Gosine and Teelucksingh 2008; Horowitz et al. 2018; Beckett and Keeling 2018). Those who laud the cost-effectiveness and engineering finesse of certain technological solutions often miss the point. Reclaiming Giant Mine, for example, involves more than addressing asbestos, soils contaminated with hydrocarbons, and abandoned buildings, or even managing hundreds of thousands of tons of arsenic trioxide. Locals ask that other aspects of Giant's legacy be considered, especially the Yellowknives Dene children poisoned by the mine's operations (Sandlos and Keeling 2011).

While approaches to modern mine closure in the North continue to emphasize technical solutions to physical problems (tailings containment, removal of infrastructure, etc.), recent research points to more complex, long-term post-closure socio-economic and cultural impacts experienced by communities, as well as more nuanced understandings of remediation and perpetual care (Bainton and Holcombe 2018; Beckett 2020; Bowes-Lyon et al. 2010; Rodon and Lévesque 2015). In addition to the lasting harm caused by environmental degradation and pollution, the closure of a mine can leave communities without the income, infrastructure, and services once provided by the company. In cases of mine abandonment, these changes occur abruptly and without warning. Despite this knowledge, closure planning does not adequately address these challenges and relies on narrow definitions of closure and remediation. It is common for closure plans to either only mention the possible *benefits* of mine closure for communities, or to acknowledge potential negative outcomes while providing no details or clear commitments for mitigating those outcomes.

For example, the closure plan for Baffinland's Mary River Mine states that "no significant adverse residual effects are predicted to occur to VSECS [valued socio-economic components] identified within the socio-economic environment, and the Project is expected to have significant positive effects for most of the VSECS" (Baffinland Iron Mines Corporation 2018). Addressing negative socio-economic impacts is also not included in closure cost estimates: while closure plans always contain a breakdown of estimated costs for closure, none of those reviewed for this chapter dedicated any money to addressing social, economic, or cultural challenges of mine closure. Cost estimates are limited to the physical removal and/or containment of waste, equipment, and infrastructure and post-closure monitoring of physical and environmental integrity. The cost estimates present in closure plans are used to determine the financial security amount required from the proponent, and so the limited scope of these cost estimates means communities are more likely to be left with few resources after closure. In order to ensure that remediation strategies address wider understandings of post-closure health and well-being, and that they successfully mitigate negative impacts to both land and people, André Moura Xavier argues that "more than good intentions are required" (2013).

Reconsidering remediation in terms of well-being reflects the reality that site cleanup is only one part of a complicated northern mining history. While heritage organizations nostalgically portray former mine sites as the harbinger of northern modernity, Indigenous communities reflect upon a past that was rife with discord, shaped as it was by the long reach of colonialism (Sandlos and Keeling 2013; Cooke 2013). Scholarship like Boutet's study (2014) of iron mining near Schefferville illustrates the complexity of Indigenous experiences: local Innu and Naskapi people adapted their land-use practices and took part in the wage labour economy, often on their own terms. Nevertheless, there is still a tendency to characterize northern mining as a completely transformative activity affecting empty and unproductive landscapes (Samson and Cassell 2013; Desmarteau 2014). As many Indigenous communities have argued, remediation can provide an opening to challenge assumptions about the value of land, knowledge production, and development's legacies. In Rankin Inlet, Qallunaat residents reassessed their community history after the closure of the North Rankin Nickel Mine, while Arctic Bay residents sought an apology from a mine operator about changing ice conditions during the remediation of the Nanisivik Mine (Cater and Keeling 2013; Midgley 2015; Matimekush Lac-John Innu Nation Council 2001).

Mine remediation policy must undergo a philosophical shift to reflect this potential. More than two decades ago, Brian Bowman and Doug Baker (1998) underscored how remediation meant different things to different people, and such ambiguity persists to this day. Many contemporary remediation policies and laws exist because concerned community members challenged assertions and assumptions about mine closure. For example, when the Waswanipi Cree First Nation fought for the cleanup of abandoned mines in their territory in the late 1990s, they were frustrated that "remediation work" did not involve monitoring wet tailings or closing open shafts, while budget limitations prevented more extensive work (CCSG Associates 2001). Mines continue to affect locals even after their closure, but there remains a lack of technical knowledge about the longer-term, more complex impacts of mine closure on communities that can be applied to policy and closure planning (Bainton and Holcombe 2018). Current scientific and legal definitions of closure and remediation only superficially account for the lived experiences of northerners (LeClerc and Keeling 2015). Parlee's contribution to this volume (Chapter 12) provides a valuable approach for incorporating the loss of place brought about by large-scale natural resource extraction into well-being indicators. Such considerations are essential to effectively assess and remediate the impacts of northern mines.

CONCLUSION

This chapter sought to unpack current Canadian mine remediation policies and practices for both legacy and new mines. Every northern territory has its own remediation policies, programs, and authorities; determining responsibility for a site's cleanup involves asking a series of questions about the site's location and history. As it stands, anemic linkages between different authorities and the shaky incorporation of cumulative environmental impacts obstruct effective mine remediation planning. Divergent remediation programs and guidelines in each province and territory means that sites outside of one jurisdiction are effectively invisible to regulators in other places. This is especially problematic because reclaiming mine sites in places like the Labrador Trough necessitates a cross-border approach. Remediation projects in the North face various additional challenges, including integrating local knowledge, capacity building, and strengthening government oversight and enforcement. Efforts to improve these processes through environmental reviews and other means take time, energy, and money, but are essential given that some remediation works require management in perpetuity.

Pushed by legal challenges and Indigenous efforts to control their own lands, the policy landscape has evolved to respond (however imperfectly) to some of the most pressing difficulties of remediation. Changes include reforms to existing mining legislation, more prescriptive remediation guidelines, and programs for addressing liability issues and mitigating the health and environmental concerns at abandoned mines. However, these many policies only address some of the inherent challenges of northern mine remediation for both new and old mines. In general, they fail to adequately appreciate and respond to the complex socio-environmental problems those living near the mines have confronted for decades. To be truly effective, remediation requires considering environmental harm and community health while recognizing the uneven power dynamics underlying mine siting, operations, and remediation protocols.

If the goal of remediation policymaking in northern Canada has been to remediate polluted sites and prevent the creation of new ones, then it has patently failed. This is in keeping with international trends. Despite the industry's focus on sustainable development principles, remediation around the world has simply not kept pace with extraction (Worrall et al. 2009). Canada's failure can be attributed to an unwillingness to embed and enforce strict remediation standards and requirements, not only within environmental policies, but also within mining acts and regulations first drafted decades ago. With few exceptions, federal, territorial, and provincial governments have proven reluctant to strengthen remediation policy or ensure industry compliance with existing rules and regulations (Hart et al. 2012). Mainstream calls for reclamation in the 1990s and early 2000s, such as the National Roundtable's urban brownfield strategy and the federal Environmental Commissioner's reports on federal contaminated sites—were often centred on other public policy issues, such as urban redevelopment or reducing federal liabilities (NRTEE 2003; OAGC 2002).

Recent mine abandonments, like those of Jericho and Wolverine, illustrate one of the current regime's greatest limitations: the distinction it draws between legacy mines and new ones. Not only is redevelopment "reanimating" legacy mines like Keno Hill, but new mines are

Table 9.1
Remediation programs and policies in the Canadian North

Laws, regulations, policies, and governance for new mines or mineral exploration

Yukon	Northwest Territories	Nunavut	Nunavik	Nunatsiavut
Environment Act	Environmental Protection Act	Canadian Environmental Protection Act, 1999	Commission de la qualité de l'environnement Kativik	Department of Environment and Climate Change
Placer Mining Act	Environmental Rights Act	Impact Assessment Act	Guide de préparation du plan de réaménagement et de restauration des sites miniers au Québec	Environmental Protection Act
Quartz Mining Act	Guidelines for Closure and Reclamation Cost Estimates for Mines	Mine Site Reclamation Policy for Nunavut, 2002	Loi sur la qualité de l'environnement, 2018	Labrador Inuit Land Claims Agreement
Reclamation and Closure Planning for Quartz Mining Projects	Guidelines for the Closure and Reclamation of Advanced Mineral Exploration and Mine Sites	Nunavut Impact Review Board	Loi sur les mines	Mineral Act
Territorial Lands (Yukon) Act	Mackenzie Valley Environmental Impact Review Board	Nunavut Mining Regulations	Nunavik Inuit Mining Policy	Mineral Exploration and Quarrying Standards Act (Nunatsiavut)
Waters Act	Mackenzie Valley Resource Management Act	Nunavut Planning and Project Assessment Act	Nunavik Regional Board of Health and Social Services	Mining Act and Mining Regulations
Yukon Environmental and Socio-Economic Assessment Act	Mine Site Reclamation Policy for the Northwest Territories	Nunavut Planning Commission -Nunavut Land Use Plan (Draft)		Nunatsiavut Land Use Plan (Draft)
Yukon Environmental and Socio-Economic Assessment Board	Northwest Territories Lands Act (Mining Regulations)	Nunavut Tunngavik Inc. (NTI)		
Yukon Mine Site Reclamation and Closure Policy	Waters Act	Nunavut Water Board		
Yukon Surface Rights Board Act		Nunavut Waters and Nunavut Surface Rights Tribunal Act		
		Territorial Land Use Regulations		
		Territorial Lands Act		

Inuvialuit Settlement Region

Environmental Impact Screening
Environmental Impact Review Board

Authorities, policies, and programs for abandoned and legacy mines

Department of Energy, Mines and Resources Northern Abandoned Mine Reclamation Program Yukon Northern Affairs Program Devolution Transfer Agreement, Chapter 6	Cooperation Agreement Respecting the Giant Mine Remediation Project Délı̨nę-Canada Remediation Management Committee and Operations Committee Giant Mine Remediation Project Environmental Agreement Northern Abandoned Mine Reclamation Program Northwest Territories Devolution Agreement, Chapter 6	Nunavut Waters and Nunavut Surface Rights Tribunal Act	Loi sur la qualité de l'environnement Loi sur les mines Fonds Restor-Action Nunavik Politique et plan d'action de protection des sols et de réhabilitation des terrains contaminés Programme gouvernemental des sites contaminés de l'État	Department of Industry, Energy and Technology (Mineral Development Division)

Other authorities, programs, and legislation relevant to new or abandoned mines

• Land claims and self-government agreements; impact and benefit agreements
• *Federal Policies*: Canadian Environmental Protection Act, 1999; Environmental Damages Fund; Federal Contaminated Sites Action Plan; Federal Contaminated Sites Inventory; Fisheries Act; Impact Assessment Act; Metal and Diamond Mining Effluent Regulations; Minerals and Metals Policy; Northern Contaminated Sites Program; Species at Risk Act; Toxic Substances Management Policy

sometimes just as vulnerable to collapsing mineral prices and unenforced environmental regulations as their predecessors. Remediation policies must address this reality and move beyond applying the most cost-effective technological solution and operating within the appropriate regulatory framework. Rather, it necessitates northern involvement through employment, regulation, transparency, and the integration of TK and Indigenous governance into every aspect of decision making, including more locally appropriate closure objectives. Gladstone and Kennedy Dalseg's discussion of increasing meaningful Indigenous local governance of resource development in Chapter 11 of this volume resonates for reclamation policymaking: effective remediation requires greater transparency and ongoing conversations with communities about their expectations. In Chapter 7 of this volume, Mills and Simmons provide a valuable example of how this could occur through the creation of "ethical space" for Indigenous communities to develop "conceptual tools" shaped by individual and community histories; such tools would be useful when revisiting the goals and processes informing site cleanup. Remediation planning also invites renewed conversations about long-term sustainability and land use goals, as well as comprehensive strategies for achieving these goals and prioritizing positive outcomes for local communities. In Chapter 3 of this volume, Schott et al. show that mining policies like the Nunavik Inuit Mining Policy can be regionally driven; remediation policies should likewise be informed by local concerns, needs, and intergenerational equity.

Instead of characterizing remediation projects as technological undertakings wholly divorced from social and economic well-being, existing political mechanisms should incorporate objectives of reparation and environmental justice. In this way, some of the most damaging impacts of historical northern resource development can be at least partially mitigated. Such an approach will reduce the cumulative costs of resource development in the North. Nevertheless, legacy mine programs have yet to expand to cover more recent abandonments, while newly arranged devolution agreements transfer authority and responsibility for newer mines to territorial or regional governments, regardless of these governments' limited capacity to fund these responsibilities. The current policy regime threatens to leave us with the same problems as before: massive public liabilities, enormous mitigation challenges, and communities trying to reconcile promises made and promises broken.

NOTES

An earlier version of this chapter was published in *The Northern Review*. Resources and Sustainable Development in the Arctic (RESDA) and a Social Sciences and Humanities Research Council (SSHRC) of Canada post-doctoral fellowship generously funded this research. Several government, industry, and NGO representatives took the time to meet with the authors and answer their questions, and we are grateful for their assistance. Thanks also to Jean-Sébastien Boutet, Deanna McLeod, Scott Midgley, Catherine Mills, Kevin O'Reilly, Chris Southcott, John Thistle, Theresa Wallace, and two anonymous reviewers at *The Northern Review* for their helpful comments on earlier versions of this chapter. We would like to express a special thanks to Caitlynn Beckett for her time and thoughtful comments, without which this chapter would not be what it is.

REFERENCES

AANDC. 2014. *Northern Contaminated Sites Program Project Report: 2012-2013*. https://www.rcaanc-cirnac.gc.ca/eng/1406904610234/1537372620664.

Abele, F. 2009. "Northern Development: Past, present and future." In F. Abele, T. Courchene, F. St-Hilaire, & L. Seidle (Eds.), *Northern Exposure: Peoples, powers, and prospects in Canada's North* (pp. 19-65). Institute for Research on Public Policy.

Adams, David, Christopher De Sousa, and Steven Tiesdell. 2010. "Brownfield Development: A Comparison of North American and British Approaches." *Urban Studies* 47, no. 1 (January): 75-104.

Agnico Eagle Mines Limited. 2014. *Meadowbank Closure Plan*.

–2015. *Meliadine Preliminary Closure and Reclamation Plan*.

Alexco Resource Corp. 2018. *Reclamation and Closure Plan: Keno District Mine Operations*.

Baffinland Iron Mines Corporation. 2018. *Mary River Interim Closure and Reclamation Plan*.

Bainton, Nicholas and Sarah Holcombe. 2018. A Critical Review of the Social Aspects of Mine Closure. *Resources Policy* 59: 468-478.

Beckett, Caitlynn. 2020. Beyond Remediation: Containing, Confronting, and Caring for the Giant Mine Monster. *Environment and Planning E: Nature and Space* (doi: 10.1177/2514848620954361).

Beckett, Caitlynn, Elizabeth Dowdell, Miranda Monosky, and Arn Keeling. 2020. "Integrating socio-economic objectives for mine closure

and remediation into impact assessment in Canada." Report for the
SSHRC Knowledge Synthesis Grant: Informing Best Practices in
Environmental and Impact Assessment. https://www.sshrc-crsh.gc.ca/
society-societe/community-communite/ifca-iac/evidence_briefs-donnees_
probantes/environmental_and_impact_assessments-evaluations_envi-
ronnementales_et_impacts/beckett_dowdall-eng.aspx.

Beckett, Caitlyn, and Arn Keeling. 2018. "Rethinking Remediation: Mine
Reclamation, Environmental Justice, and Relations of Care." *Local
Environment* 24, no. 3 (March): 216-230.

Boutet, Jean-Sébastien. 2014. "Opening Ungava to Industry: A Decentring
Approach to Indigenous History in Subarctic Québec, 1937-54."
Cultural Geographies 21, no. 1 (January): 79-97.

Bowes-Lyon, Léa-Marie, Jeremy P. Richards, and Tara M. McGee. 2010.
"Socio-Economic Impacts of the Nanisivik and Polaris Mines, Nunavut,
Canada." In *Mining, Society, and a Sustainable World,* edited by Jeremy
P. Richards, 371-396. Heidelberg: Springer Berlin Heidelberg.

Bowman, Brian and Doug Baker. 1998. "Mine Reclamation Planning in
the Canadian North, Northern Minerals Program." *Northern Minerals
Program.* Working Paper 1. Canadian Arctic Resources Committee.

Boyd, David R. 2003. *Unnatural Law: Rethinking Canadian
Environmental Law and Policy.* Vancouver, BC: UBC Press.

Brockman, Alex. 2019. "Northerners Aren't Ready to Cash in on $1B
Giant Mine Cleanup, Oversight Board Says." *CBC News North,* April
24, 2019.

Canadian Royalties Inc. 2016. *Projet Nunavik Nickel: Plan de restauration pour le Site de la Fosse Puimajuq du Projet Minier Nunavik Nickel.*

– 2019a. *Projet Nunavik Nickel: Plan de restauration pour le Site de la
Fosse Expo du Projet Minier Nunavik Nickel.*

–2019b. *Projet Nunavik Nickel: Plan de restauration pour le Site Minier
Allammaq du Projet Minier Nunavik Nickel.*

Castrilli, Joseph F. 2002. *Barriers to Collaboration: Orphaned/Abandoned
Mines in Canada.* National Orphaned/Abandoned Mines Initiative.

–2007. *Report on the Legislative, Regulatory, and Policy Framework
Respecting Collaboration, Liability, and Funding Measures in Relation
to Orphaned/Abandoned, Contaminated, and Operating Mines in
Canada.* National Orphaned/Abandoned Mines Initiative.

–2010. "Wanted: A Legal Regime to Clean Up Orphaned/Abandoned
Mines in Canada." *McGill International Journal of Sustainable
Development Law and Policy* 6, no. 2: 130-135.

Cater, Tara and Arn Keeling. 2013. ""That's Where Our Future Came From": Mining, Landscape, and Memory in Rankin Inlet, Nunavut." Études/Inuit/Studies 37, no. 2: 59-82.

CBC News. 2014. "Snowfield Fined $40K for Abandoned NWT Mining Camp." *CBC News North*, August 21, 2014. https://www.cbc.ca/news/canada/north/snowfield-fined-40k-for-abandoned-n-w-t-mining- camp-1.2742008.

–2015. "Wolverine Mine Flooding, Yukon Zinc Not Complying with Agreements." *CBC News North*, March 24, 2015. https://www.cbc.ca/news/canada/north/wolverine-mine-flooding-yukon-zinc-not- complying-with-agreements-1.3007520.

CCME. 1996. *A Framework for Ecological Risk Assessment: General guidance*. CCME.

–1997. *A Framework for Ecological Risk Assessment: Technical appendices*. CCME.

–2008. *National Classification System for Contaminated Sites: Guidance document*. CCME.

CCSG Associates. 2001. *Financial Options for the Remediation of Mine Sites: A preliminary study*. MiningWatch Canada.

CIRNAC. 2019. *Backgrounder: The Northern Abandoned Mine Reclamation Program*. https://www.canada.ca/en/crown-indigenous-relations-northern-affairs/news/2019/08/the- northern-abandoned-mine-reclamation-program.html.

–2019. *Northern Abandoned Mine Reclamation Program*. https://www.rcaanc- cirnac.gc.ca/eng/1565968579558/1565968604553.

Cohen, Sidney. 2018. "Feds Move to Scrap N.W.T.'s Planned Land and Water Superboard, Keep 4 Regional Boards." *CBC News North*, November 9, 2018.

Cooke, Lisa. 2013. "North Takes Place in Dawson City, Yukon, Canada." In *Northscapes: History, Technology, and the Making of Northern Environments*, edited by Dolly Jørgensen and Sverker Sörlin, 223-246. Vancouver, BC: UBC Press.

Cowan, W.R. and W.O. Mackasey. 2006. *Rehabilitating Abandoned Mines in Canada: A Toolkit of Funding Options*. NOAMI.

Dea, Nancy and André Gaumand. 2016. *Le leadership du Québec en restauration minière: Le Fonds Restor-Action Nunavik*. https://mern.gouv.qc.ca/wp-content/uploads/S04-02_Gaumond.pdf.

Department of Natural Resources. 2010. *Guidebook to Exploration, Development and Mining in Newfoundland and Labrador*. https://www.findnewfoundlandlabrador.com/files/2017/03/Guidebook.pdf.

Desmarteau Raymond. 2014. "Fonds Restor-Action Nunavik: Une humble fierté pour André Gaumond des Mines Virginia." *Radio Canada International*, July 3, 2014.

De Sousa, Christopher A. 2001. "Contaminated Sites: The Canadian Situation in an International Context." *Journal of Environmental Management* 62, no. 2: 131-154.

–2002. "Brownfield Redevelopment in Toronto: An Examination of Past Trends and Future Prospects." *Land Use Policy* 19, no. 4: 297-309.

–2003. "Turning Brownfields into Green Space in the City of Toronto." *Landscape and Urban Planning* 62, no. 4: 181-198.

–2006. "Urban Brownfields Redevelopment in Canada: The Role of Local Government." *Canadian Geographer* 50, no. 3: 392-407.

–2008. *Brownfields Redevelopment and the Quest for Sustainability*. Elsevier.

Diavik Diamond Mines Inc. 2017. *Diavik Closure and Reclamation Plan – Version 4.0*.

Dominion Diamond Mines. 2018. *Ekati Mine Interim Closure and Reclamation Plan*.

Duinker, Peter N. and Lorne A. Greig. 2006. "The Importance of Cumulative Effects Assessment in Canada: Ailments and Ideas for Redeployment." *Environmental Management* 37, no. 2: 153-161.

ECCC. 2019. *Action Plan for Contaminated Sites*. https://www.canada.ca/en/environment-climate-change/services/federal-contaminated-sites/action-plan.html.

Edwards, Frank B. 2010. *Parks Canada's EAP Contaminated Sites Cleanup*. http://www.hedgehogproductions.com/pdfs/Parks_Canada_Clean_Up.PDF.

Ehrlich, Alan. 2010. "Cumulative Cultural Effects and Reasonably Foreseeable Future Developments in the Upper Thelon Basin, Canada." *Impact Assessment and Project Appraisal* 28, no. 4: 279-286.

Environment Canada Evaluation Project Team. 2014. *Evaluation of the Federal Contaminated Sites Action Plan: Final Report*. https://ec.gc.ca/ae-ve/82F2991C-8730-41D1-A321- F6314509C1D5/Evaluation%20of%20the%20Federal%20Contaminated%20Sites%20Action%20Plan.pdf.

Faro Mine Complex. 2010. *A Plan for Closure*. Yukon Government.

Faro Mine Remediation Project. 2019. *Remediating Faro Mine in the Yukon.* https://www.rcaanc-cirnac.gc.ca/eng/1480019546952/15375549 89037.

FCSAP. 2019a. *History of Federal Contaminated Sites.* Contaminated Sites. https://www.canada.ca/en/environment-climate-change/services/federal-contaminated- sites/history.html.

—2019b. *Managing the Sites.* Federal Contaminated Sites Portal. https://www.canada.ca/en/environment-climate-change/services/federal-con-taminated- sites/managing.html.

Fonseca, Alberto, Mary L. McAllister, and Patricia Fitzpatrick. 2014. "Sustainable Reporting Among Mining Corporations: A Constructive Critique of the GRI Approach." *Journal of Cleaner Production* 84 (Special Issue): 70-83.

Fowler, R. 2007. "Site Contamination Law and Policy in Europe, North America and Australia: Trends and Challenges." Paper presented at the *8th Meeting of the International Committee on Contaminated Land, Stockholm, September.*

Giant Mine Oversight Board. 2019. *Current Status.* GMOB. https://www.gmob.ca/remediation/.

Gibson, Ginger, and Jason Klinck. 2005. "Canada's Resilient North: The Impact of Mining on Aboriginal Communities." *Pimatisiwin* 3, no. 1: 116–39.

Gibson, Ginger and Ciaran O'Faircheallaigh. 2010. *IBA Community Toolkit: Negotiation and Implementation of Impact and Benefit Agreements.* Summer 2015 ed. Toronto: Walter and Duncan Gordon Foundation.

Gosine, Andil, and Cheryl Teelucksingh. 2008. *Environmental Justice and Racism in Canada: An Introduction.* Toronto: Emond Montgomery.

Government of Manitoba. 2021. *Orphaned and Abandoned Mines.* https://www.gov.mb.ca/sd/environment_and_biodiversity/mines/index.html.

Government of the Northwest Territories. 2013. *Northwest Territories Mineral Development Strategy.* NWT & Nunavut Chamber of Mines and Government of the Northwest Territories Industry, Tourism and Investment.

Gregoire, Lisa. 2014. Abandoned Nunavut mine now belongs to Crown. *Nunatsiaq News,* December 19, 2014. https://nunatsiaq.com/stories/article/65674abandoned_nunavut_diamond_mine_now_belon gs_to_the_crown/.

Gregory, Gillian H. 2021. Rendering Mine Closure Governable and
    Constraints to Inclusive Development in the Andean Region. *Resources
    Policy* 72, no. 3 (August): 102053.
Hart, R., and C. Coumans. 2013. *Evolving Standards and Expectations
    for Responsible Mining: A Civil Society Perspective.* MiningWatch
    Canada.
Hart, R., MiningWatch Canada, and Dawn Hoogeveen. 2012.
    Introduction to the Legal Framework for Mining in Canada.
    MiningWatch Canada.
Herman, L. 2013. "Playing the Blame Game at Jericho." *Northern News
    Services Online*, December 21, 2013. https://archive.nnsl.com/2013-12/
    dec23_13jer.html.
Hoogeveen, Dawn. 2015. "Sub-surface Property, Free-entry Mineral
    Staking and Settler Colonialism in Canada." *Antipode* 47, no. 1
    (January): 121-138.
Horowitz. Leah S., Arn Keeling, Francis Lévesque, Thierry Rodon,
    Stephan Schott, and Sophie Thériault. 2018. "Indigenous Peoples'
    Relationships to Large-Scale Mining in Post/Colonial Contexts: Toward
    Multidisciplinary Comparative Perspectives." *The Extractive Industries
    and Society* 5, no. 3 (July): 404-414.
INAC. 2007. *A Citizen's Guide to Cumulative Effects.* https://www.aadnc-
    aandc.gc.ca/DAM/DAM-INTER-NWT/STAGING/texte-text/ntr_pubs_
    CEG_1330635861338_eng.pdf.
–2009. *A Citizen's Guide to INAC's Environmental Stewardship Roles in
    the NWT.* Government of Canada.
–2010. *INAC Northern Contaminated Sites Program: Progress Report
    2005-2010.* Minister of Public Works and Government Services
    Canada.
Keeling, Arn and John Sandlos. 2009. "Environmental Justice Goes
    Underground? Historical Notes from Canada's Northern Mining
    Frontier." *Environmental Justice* 2, no. 3: 117-125.
–2017. "Ghost Towns and Zombie Mines: The Historical Dimensions of
    Mine Abandonment, Reclamation, and Redevelopment in the Canadian
    North." In *Ice Blink: Navigating Northern Environmental History*,
    edited by Stephen Bocking and Brad Martin, 377-420. Calgary, AB:
    University of Calgary Press.
Keeling, Arn, John Sandlos, Jean-Sébastien Boutet, Hereward Longley, and
    Anne Dance. 2019. "Knowledge, Sustainability, and the Environmental
    Legacies of Resource Development in Northern Canada." In *Resources

and Sustainable Development in the Arctic, edited by Chris Southcott, Frances Abele, Dave Natcher, and Brenda Parlee, 187-203. Routledge.

Kuyek, Joan N. 2011. *The Theory and Practice of Perpetual Care of Contaminated Sites.* Alternatives North.

—2019. *Unearthing Justice: How to Protect Your Community from the Mining Industry.* Between the Lines.

Laforce, Myriam. 2012. "Régulation du projet minier de Voisey's Bay au Labrador: Vers un rééquilibrage des pouvoirs dans certains contextes politiques et institutionnels." In *Pouvoir et régulation dans le secteur minier. Leçons à partir de l'expérience canadienne*, edited by Myriam Laforce, Bonnie Campbell, and Bruno Sarrasin, 157-189. Quebec, QC: Presses de l'Université du Québec.

Laforce, Myriam, Ugo Lapointe, and Véronique Lebuis. 2012. "Régulation du secteur minier au Québec et au Canada." In *Pouvoir et régulation dans le secteur minier. Leçons à partir de l'expérience canadienne*, edited by Myriam Laforce, Bonnie Campbell, and Bruno Sarrasin, 9-50. Quebec, QC: Presses de l'Université du Québec.

Lavoie, Judith. 2019." BC First Nations Should Require Full Clean-up Costs Up-front for Mines: New Study." *The Narwhal*, November 8, 2019. https://thenarwhal. ca/b-c-first-nations-should-require-full-clean-up- costs-up-front-for-mines-new-study/.

LeClerc, Emma and Arn Keeling. 2015. "From Cutlines to Traplines: Post-Industrial Land Use at the Pine Point Mine." *Extractive Industries and Society-an International Journal* 2, no. 1: 7-18.

MacNeil, Robert. 2014. "Canadian Environmental Policy Under Conservative Majority Rule." *Environmental Politics* 23, no. 1 (January): 174-178.

Matimekush Lac-John Innu Nation Council. 2001. "Site Restoration and Mining Facilities of the Schefferville IOC Company." Paper presented at *Workshop on Orphaned/Abandoned Mines in Canada, Winnipeg, June*.

McAllister, Mary L. and Cynthia J. Alexander. 1997. *A Stake in the Future: Redefining the Canadian Mineral Industry.* Vancouver, BC: UBC Press.

McPherson, Robert. 2005. *New Owners in their Own Land: Minerals and Inuit Land Claims.* Calgary, AB: University of Calgary Press.

Midgley, Scott. 2015. "Contesting Closure: The Science, Politics, and Community Responses to Closing the Nanisivik Mine." In *Mining and Communities in Northern Canada: History, Politics, and Memory*, edited by Arn Keeling and John Sandlos, 293-314. University of Calgary Press.

MiningWatch Canada. 2014. *2013 Annual Report.* MiningWatch Canada.

Ministère de l'Énergie et des Ressources naturelles. 2021. *Restauration des sites miniers sous la responsabilité réelle de l'État*. https://mern.gouv.qc.ca/mines/restauration-miniere/restauration-des-sites-miniers-abandonnes/.

Monosky, Miranda. 2021. *Social and Community Engagement Mine Closure: An Exploration of Mine Closure Governance and Industry Practices in Northern Canada*. Master's thesis. Memorial University of Newfoundland.

Monosky, Miranda and Arn Keeling. 2021a. Planning for Social and Community-Engaged Closure: A Comparison of Mine Closure Plans from Canada's Territorial and Provincial North. *Journal of Environmental Management* 277, no. 1 (January): 111324.

−2021b. Social considerations in Mine Closure: Exploring Policy and Practice in Nunavik, Québec. *The Northern Review* 52: 29-60.

Nahir, Michael and Claudia David. 2007. *Abandoned Mines in Northern Canada: Program Challenges and Case Studies*. INAC.

Natural Resources Canada. 2013. *The Minerals and Metals Policy of the Government of Canada*. Government of Canada.

−2015a. *Interactive Map of Aboriginal Mining Agreements*. Government of Canada. http://www2.nrcan.gc.ca/mms/map-carte/MiningProjects_cartovista-eng.html.

−2015b. *Tables on the Structure and Rates of Main Taxes*. Government of Canada. http://www.nrcan.gc.ca/mining-materials/taxation/mining-taxation-regime/8890.

−2018. *Managing Water in the Mining Cycle*. https://www.nrcan.gc.ca/mining-materials/mining/green-mining-innovation/8178.

NCSRP. 1994. *Annual Report 1993-94*. CCME Secretariat.

NAOMI. 2002. *Best Practices in Community Involvement: Planning for and Rehabilitating Abandoned and Orphaned Mines in Canada*. National Resources Canada.

−2009. *2002-2008 Performance Report*. Natural Resources Canada.

−2010. *Programs to Enhance and Sustain Safety and the Quality of the Environment in and around Orphaned and Abandoned Mine Sites*. Natural Resources Canada.

NRTEE. 2003. *Cleaning Up the Past, Building the Future: A National Brownfield Redevelopment Strategy for Canada*. National Round Table on the Environment and the Economy.

Nunatsiaq News. 2014. "Nunavut Government Operates at Three-Quarters of its Human Capacity." *Nunatsiaq News*, November 13, 2014. https://nunatsiaq.com/stories/

article/65674nunavut_government_still_operating_at_three_quarter_
human_capacity/.

–2017. "Nunavut Review Board Keeps Close Watch on Derelict Mine."
Nunatsiaq News, November 30, 2017. https://nunatsiaq.com/stories/
article/65674nunavut_review_board_keeps_close_watch_on_aban-
doned_mine/.

OAGC. 1995. *Report of the Auditor General of Canada, May 1995.*
Minister of Public Works and Government Services Canada.

–1996. *Report of the Auditor General of Canada, November 1996.*
Minister of Public Works and Government Services Canada.

–2002. *Report of the Commissioner of the Environment and Sustainable
Development, Fall 2002.* Minister of Public Works and Services.

–2012. *Report of the Commissioner of the Environment and Sustainable
Development, Spring 2012.* Minister of Public Works and Services.

Parlee, Brenda. 2012. "Finding Voice in A Changing Ecological and
Political Landscape: Traditional Knowledge and Resource Management
in Settled and Unsettled Claim Areas of the Northwest Territories,
Canada." *Aboriginal Policy Studies* 2, no. 1: 56-81.

Parliamentary Budget Office. 2014. *Federal Contaminated Sites Costs.*
Library of Parliament.

Piper, Liza. 2009. *The Industrial Transformation of Subarctic Canada.*
Vancouver, BC: UBC Press.

Pollon, Christopher. 2019. "BC's 'Archaic' Mining Laws Urgently Need
Update: 30 Groups." *The Narwhal*, May 15, 2019. https://thenarwhal.
ca/b-c-s-archaic-mining-laws-urgently-need-update-30-groups/.

Rodon, Thierry and Francis Lévesque. 2015. "Understanding the Social
and Economic Impacts of Mining Development in Inuit Communities:
Experiences with Past and Present Mines in Inuit Nunangat." *The
Northern Review* 41: 13-39.

Rohner, Thomas. 2016. "In Wake of Nunavut's Jericho Mine Fiasco,
Terminated Workers Still Unpaid." *Nunatsiaq News*, September 22,
2016. https://nunatsiaq.com/stories/
article/65674in_wake_of_nunavuts_jericho_mine_fiasco_terminated_
workers_still_unpai/?lang=fr.

Rosa, Josianne Claudia Sales, Davide Geneletti, Angus Morrison-Saunders,
Luis Enrique Sánchez, and Michael Hughes. 2020a. "To what extent
can mine rehabilitation restore recreational use of forest land? Learning
from 50 years of practice in southwest Australia." *Land Use Policy* 90,
no. 1 (January): 104290.

Rosa, Josianne Claudia Sales, Angus Morrison-Saunders, Michael Hughes, and Luis Enrique Sánchez. 2020b. "Planning Mine Restoration Through Ecosystem Services to Enhance Community Engagement and Deliver Social Benefits." *Restoration Ecology* 28, no. 4: 937-946.

Rudyk, Mike. 2019. "Yukon Construction Company to Partner with Ross River Dena Council on Faro Mine Remediation." *CBC News North*, January 17, 2019. https://www.cbc.ca/news/canada/north/pelly-construction-first-nation-1.4981183.

Samson, Colin and Elizabeth Cassell. 2013. "The Long Reach of Frontier Justice: Canadian Land Claims 'Negotiation' Strategies as Human Rights Violations." *The International Journal of Human Rights* 17, no. 1: 35-55.

Sandlos, John and Arn Keeling. 2011. *Giant Mine: Historic Summary.* Abandoned Mines Project.

–2013. "Zombie Mines and the (Over)burden of History." *Solutions* 4, no. 3: 80-83.

–2016a. "Aboriginal Communities, Traditional Knowledge, and the Environmental Legacies of Extractive Development in Canada." *Extractive Industries and Society-an International Journal* 3, no. 2 (April): 278-287.

–2016b. "Toxic Legacies, Slow Violence, and Environmental Injustice at Giant Mine, Northwest Territories." *The Northern Review* 42: 7-21.

Sandlos, John, Arn Keeling, and Kevin O'Reilly. 2014. *Communicating Danger: A Community Primer on Communicating the Arsenic Hazards at Yellowknife's Giant Mine to Future Generations.* Abandoned Mines in Northern Canada.

Séguin, Jean-Marc, and Mylène Larivière. 2011. *Nunavik Guidebook: Mineral exploration, mining development and Nunavik region.* Makivik Corporation.

Sistili, Brandy, Mike Metatawabin, Guy Iannucci, and Leonard J. S. Tsuji. 2006. "An Aboriginal Perspective on the Remediation of Mid-Canada Radar Line Sites in the Subarctic: A Partnership Evaluation." *Arctic* 59, no. 2: 142-154.

Skogstad, Grace and Paul Kopas. 1992. "Environmental Policy in a Federal System: Ottawa and the Provinces." In *Canadian Environmental Policy: Ecosystems, Politics and Process*, edited by Robert Boardman, 43-59. Oxford University Press.

Tester, Frank, and Peter Irniq. 2008. "Inuit Qaujimajatuqangit: Social History, Politics, and the Practice of Resistance." *Arctic* 61, no. 5 (January): 48-61.

Thomson, Jimmy. 2019. "Canadian Taxpayers on Hook for $61 Million for Road to Open Up Mining in Arctic." *The Narwhal*, August 15, 2019. https://thenarwhal.ca/canadian-taxpayers-61-million-road-open- mining-arctic/.

Treasury Board of Canada Secretariat. 2019. *Federal Contaminated Sites Inventory*. December 15, 2019. TBS. http://www.tbs-sct.gc.ca/fcsi-rscf/home-accueil-eng.aspx.

Tremblay, G. A. and C.M. Hogan. 2005. *Initiatives at Natural Resources Canada to Deal with Orphan and Abandoned Mines*. Natural Resources Canada. https://pdfs.semanticscholar.org/7747/bc826424a3c8f1f2b53dc6a6c84acabba8c2.pdf.

–2016. "Managing Orphaned and Abandoned Mines: A Canadian Perspective." Paper presented at *Dealing with Derelict Mines Summit, Singleton, Australia, December*.

Vale Newfoundland & Labrador Limited. 2016. *Rehabilitation and Closure Plan: Voisey's Bay*.

Van Dijken, Bob. 2001. "Mines in the Yukon: Abandoned, Orphaned and in Limbo." Paper presented at *Workshop on Orphaned/Abandoned Mines in Canada, Winnipeg, June*.

Vela, Thandie. 2011. "Review Board Approves Drybones Bay Exploration." *Northern News Services*, November 21, 2011. https://archive.nnsl.com/2011-11/nov21_11db.html.

Wenig, Michael M. and Kevin O'Reilly. 2005. *The Mining Reclamation Regime in the Northwest Territories: A Comparison with Selected Canadian and U.S. Jurisdictions*. CIRL and CARC.

Worrall, Rhys, David Neil, David Brereton, and David Mulligan. 2009. "Towards a Sustainability Criteria and Indicators Framework for Legacy Mine Land." *Journal of Cleaner Production* 17, no. 16: 1426-1434.

Wynn, Graeme. 2017. *Canada and the Arctic North America: An Environmental History*. ABC-CIO.

Xavier, Andre M. 2013. *Socio-Economic Mine Closure (SEMC) Framework: A Comprehensive Approach for Addressing the Socio-Economic Challenges of Mine Closure*. Vancouver, BC: UBC Press.

Yukon Territorial Government. 2021. *Type II Abandoned Mines Sites*. https://yukon.ca/en/type-ii-mines.

10

Facilitating Sustainable Waste Management in a Northern Community and Resource Development Context at Happy Valley-Goose Bay, Labrador

Catherine Keske

This chapter presents a case study of the Happy Valley-Goose Bay, Labrador community to illuminate historical and present-day waste management challenges in northern communities with extractive industry development. The chapter proposes a circular economy model to convert waste biomass into a soil amendment that may improve soil quality and thus food security in the study area. This study supplements other waste management research presented in this volume, including Dance et al.'s work (Chapter 9). Other chapters, written by Rodon et al. and Schott et al. (Chapters 2 and 3), summarize socio-economic studies of impacts and benefits from extractive industry development, as well as highlight issues such as economic rents and employment. This chapter illuminates aspects of resource development that could improve community sustainability through waste management. The chapter discusses the economic feasibility of a proposed project that would convert waste biomass from forests that were clear cut in the construction of the controversial Muskrat Falls hydroelectric dam into a soil amendment to grow local root vegetables. This would provide an opportunity to use waste material to improve the region's food security, and effectively advance a circular economy that converts biological resources and waste streams into value-added products.

The chapter starts out by providing a summary of regional climatic conditions, waste flows, and waste management practices that were compiled during the study's qualitative research phase (Mills and Keske 2017; Keske et al. 2018a; Keske et al. 2018b). The Labrador case study is consistent with a northern community that has endured colonialism, resource extraction, and ecosystem disruption (Hird 2016). Qualitative data were collected using a grounded theory research approach (Charmaz and Belgrave 2007; Glaser and Strauss 1967) supplemented by an archival document review. A grounded theory method was used because community partners from the Town of Happy Valley-Goose Bay (see endnote below) initiated the project out of a desire to promulgate sustainable waste management practices, and they were undecided about how far to expand the study scope due to the complexity of waste management in northern communities. During semi-structured, participatory interviews and discussions with twenty-one community members, participants expressed an interest in conducting a quantitative assessment on how to reduce negative environmental impacts and generate local community benefits. Thus, a mixed-methods research design was used, consisting of a qualitative research phase followed by a quantitative one (Teddlie and Tashakkori 2009). Qualitative data were collected from conversations with partners, a community grassroots organization, and a participatory action meeting. There, at least one investigator asked a series of open-ended questions. Data were transcribed in real time and clustered by themes for improved waste management (Keske et al. 2011; Creswell 2003). Themes included: household recycling, household waste reduction, waste management to improve food security, water quality, illegal dumping, landfill age, mobile labour force waste, and the establishment of landfill tipping fees that were both socially and economically viable.

One group of community partners urged a techno-economic analysis of converting waste forest biomass from the Muskrat Falls hydroelectric dam site into biochar for local agricultural production. Biochar is a substance similar to charcoal with promising properties for soil enriching and stabilizing. It is, however, still considered an experimental soil amendment (Abedin 2018; Abedin 2015; Liu et al. 2013; Field et al. 2013; Nematian et al. 2021). A two–year trial of local beet crops (Abedin 2015; Abedin 2018) showed that biochar, mixed with fish meal, substantially increased yields. The community partners and research team decided to evaluate the idea that the waste biomass, primarily black spruce, could be converted into biochar and used in local root vegetable production in an economically feasible manner.

The project's quantitative research phase consisted of an enterprise budget, a sensitivity analysis, and a stochastic analysis. A summary of findings is presented in the section below dealing with quantitative research results.

Values from Abedin's work and other biochar studies were incorporated into an Excel-based enterprise budgeting tool that was eventually made available to the public through a peer-reviewed, open-source publication (Keske et al. 2018a). The proposed baseline condition utilizes a mobile, slow pyrolysis biochar production unit co-located at the Muskrat Falls hydroelectric dam site. However, the tool's adjustable range of values accommodates project variation and yield and price uncertainty, as well as a slightly amended alternative biochar production process (fast pyrolysis). As a result, the enterprise budgeting tool's parameters can be modified to accommodate more specificity at the local site. It can be adapted to the explicit conditions in other northern communities interested in pursuing a similar venture.

The quantitative economic analysis was then further expanded to consider different risk scenarios that might be useful for anyone considering whether to invest in the proposed venture. Using values from the Excel budgeting tool, a stochastic analysis was conducted to assess the worst, typical, and best-case scenarios for biochar and root vegetable co-production in the Happy Valley-Goose Bay study region (Keske et al. 2019). Break-even and sensitivity analyses of biochar inputs and yield outputs were conducted to accommodate high yield and price uncertainty in the biochar and agricultural production processes, and potential impacts on the economic viability of the proposed project. Beet production was a profitable venture in nearly every scenario evaluated.

A discussion of the integrated case study is presented in the results section. One of the most important findings is that biochar production alone would not be a profitable operation due to the region's remote location and the limited market for biochar. Rather, the increased agricultural yields from the biochar application make the proposed biochar-agricultural co-production venture economically viable. There is an apparent symbiotic relationship between the agriculture and biochar production within the study region. The area's harsh environment makes agriculture extremely difficult. Soil amendment is essential, though it is not always effective. Hence, the case study shows a serendipitous situation for biochar and agricultural co-production. Given the region's dire and widespread food insecurity, this chapter makes the case that a pilot biochar production site is worth the investment.

In summary, it is difficult to roll back deleterious environmental and cultural impacts of large extraction projects in the North, including hydroelectric megaprojects like Churchill Falls and Muskrat Falls. However, this case study illuminates that there is potential to create opportunities for community benefit, like converting waste forest biomass into a soil amendment that can be used to improve food security. If the project proposed in this chapter is implemented successfully, it may be replicated in other extractive communities and improve long-term sustainability in Canada's North.

QUALITATIVE RESEARCH RESULTS: WASTE MANAGEMENT PRACTICES IN STUDY REGION

Study Region

As shown in figure 10.1, Labrador is a large (269,134 km²) coastal region in northeastern mainland Canada, within the nation's easternmost province, Newfoundland and Labrador. The largest population centre of 8,109 people (approximately one-third of Labrador's population) is the Upper Lake Melville area around Happy Valley-Goose Bay. This is in Central Labrador, at the head of the Hamilton Inlet estuary, which drains the Churchill River and several other major watersheds. Most of the region is Indigenous, with 52 per cent of Happy Valley-Goose Bay's population reporting an Indigenous identity (Statistics Canada 2011). Indigenous majorities are in the three outlying communities of North West River (about 550 people) and Sheshatshiu Innu First Nation (a reserve of about 1,300 people), both 35 kilometres to the north, and Mud Lake, a hamlet of about fifty people a few kilometres east and across the Churchill River.

The Labrador climate is predominantly subarctic in inhabited areas, with boreal forests broken by alpine and coastal barrens, giving way to tundra in the north. The area has warm summers, no permafrost, and an average frost-free period of 104 days (St. Croix 2002). It also has abundant sunlight during spring and summer months, all of which facilitate a short but feasible agricultural production season.

Agricultural production is economically and technically challenging in part because the sandy soils have low organic content and little ability to immobilize nutrients (Abedin 2015) or pollutants (Abedin

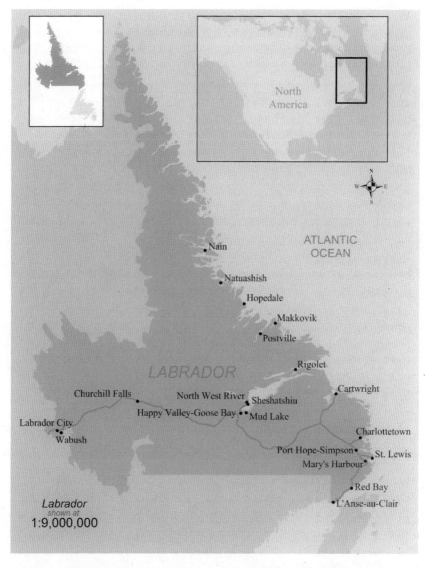

Figure 10.1 Newfoundland and Labrador is Canada's most eastern province, as shown in the upper right inset box. The two land masses that comprise the province are shown to scale in the upper left hand box. Labrador is north of the island of Newfoundland and attached to eastern mainland Canada. The Happy Valley-Goose Bay study region and nearby communities are in Central Labrador.

2017a; Abedin 2017b). In the past, there was small scale agricultural production in the region by Indigenous communities as well as European settlers. The latter included Moravian missionaries who introduced root vegetable crops to the region (Dickenson 2001). However, like many regions across the Canadian North, communities around Happy Valley-Goose Bay are extremely food insecure. They endure high food prices, shortages, and a lack of availability of fresh produce (Schiff and Bernard 2018; Ford et al. 2010).

Along with the change in economy and lifestyles, patterns of food consumption and waste production have correspondingly changed. Several at-risk communities have a higher consumption of pre-packaged foods than other households (Schiff and Bernard 2018). Low-income households frequently don't have access to fresh food or time/resources for food preparation (Keske 2018). High volumes of household food packaging waste have been observed in low-income neighbourhoods by the study's community partners (Keske et al. 2018a). As a province, Newfoundland and Labrador has one of the highest waste production disposal levels per capita in the country, and a high proportion is organic waste. According to the Multi-Materials Stewardship Board (MMSB), it is estimated that more than 400,000 tonnes of municipal solid waste (MSW) materials are generated each year in the province; organic waste comprises as much as 30 per cent of all wastes.

Impacts of Colonialism, Extraction, and Military Defence on Waste Management

Happy Valley-Goose Bay is like many remote northern communities: it is forced to disproportionately shoulder the social costs (including the subsequent waste streams) of projects that propel the economic interests of the nation. Specifically, Hird (2016) notes that natural resource extraction and military developments (such as the Distant Early Warning (DEW) radar installations) designed for the greater good of Canada and its allies have created disproportionately negative impacts on the northern communities where organic and inorganic wastes have been left behind. This is particularly the case beginning in the twentieth century. Figure 10.2 presents a timeline that captures the relationship and coexistence between colonialism, military legacy waste, and resource extraction that is like many other communities across the Canadian Arctic, Alaska (US), Greenland, and Iceland.

As shown in figure 10.2., there was a sharp increase in the population of Happy Valley-Goose Bay, in parallel with the Canadian territories, following the end of World War II. Then, sixty-three temporary military settlements, including Happy Valley-Goose Bay and other Labrador communities, were erected to form the DEW Radar Line to defend against potential Soviet attacks across northern Canada. The construction and operation of a military base at Goose Bay led to rapid population growth and a shift away from Indigenous and pre-industrial economy. The result was a series of disjointed waste management strategies. Population growth also coincided with iron mining development in western Labrador, gold mining in the Northwest Territories, and infrastructure development projects like the Canol Road and Alaskan pipelines. The rapid population growth meant a sharp increase in the number of households, all producing waste. As well, massive infrastructure transformations undertaken in a relatively short period of time created huge amounts of waste.

Population increases and multiple waste sources exerted considerable pressure on Labrador landfills and the surrounding environment. By 1976, both the Goose Bay and Happy Valley landfill sites were identified as unsuitable locations due to potential growing risk of pollution to water and soil. Until 1990, household wastes were discarded in nearby landfill locations, compacted, and bulldozed. These practices and other regulatory shortfalls have left a legacy of contamination from households, industry, and military sources. Lingering contaminants include petrochemicals, polycyclic aromatic hydrocarbons (PAHS), polychlorinated biphenyls (PCBS), pesticides, and other pollutants that were released into local waters and soils. A series of expensive federal government remediation operations have been completed, though many will continue for years due to the large scale and scope of remediation required.

Waste disposal and management practices in Happy Valley-Goose Bay and the surrounding communities have been largely in reaction to defence and extraction projects, as well as dynamic population changes. Many stakeholders, such Nalcor Energy, conduct confidential or sensitive activities and exert sovereignty over their waste management (Mills and Keske 2018; Keske et al 2018a). Hence, it is difficult for communities to coordinate and regulate management over many waste streams from industrial and military sources, for example. It is nearly impossible to attain economies of scale necessary to manage household wastes from temporary or military dwellings

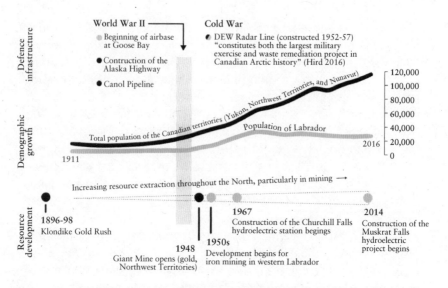

Figure 10.2 Timeline of demographic growth, defense infrastructure, and resource development

in a cost effective and environmentally sustainable manner. There are concerns about several landfills and dumps overflowing near the Churchill River, Goose River, and Lake Melville. Due to logistical barriers, waste in the Mud Lake community continues to be incinerated and buried locally, despite periodic re-examination by provincial regulators (Felsberg 2009). The Canadian Forces Base continues to arrange for disposal of its waste in several sites that are outside of community control.

Shortly after the research project commenced, Nalcor Energy, the Newfoundland Crown Corporation administering the Muskrat Falls hydroelectric project, began openly burning waste on site rather than using the Happy Valley-Goose Bay landfill. The timing also coincided with an increase in the town's waste disposal tipping fees (Barker 2017). The waste management practices may have been motivated, at least in part, by a desire to maintain privacy after years of fervent protests by environmentalists to the dam's siting and construction (Keske et al. 2018), and eventual electricity generation in late 2020. Withdrawal of support of the regional municipal landfill is yet another example of the study region's fractured waste management legacy. Keske, Mills, Tanguay, and Dicker (2018) provide additional historical narrative

about past and present Happy Valley-Goose Bay landfills, while noting
the long-term environmental and economic impacts resulting from the
study region's dispersed waste management practices.

Impacts of Climate on Northern Waste Management Practices

The qualitative research findings illuminated the impacts of harsh
climate on waste management across the Canadian North. Remoteness,
low population densities, limited local capacity, and lack of local
control in public decision-making all complicate waste management,
as pointed out earlier in this volume by Dance et al. For example, it is
well-documented that waste accumulated from extractive industries
and landfilling practices imperils drinking water quality. In studies of
the Canadian Arctic and the territory of Nunavut in particular, Daley
et al. (2015) and Daley et al. (2014) note that most of North America
relies on wastewater buried conveyance systems, but that these systems
are impractical for the Canadian North, due to extremely low
temperatures.

Long winters with a lot of snow accumulation necessitate storage
of bulky household and commercial waste until the spring snow melt.
In May, household wastes are brought to the street curb, for either
repurposing or transporting to landfills in an annual "spring cleaning"
event. Community members also harvest from landfills year round,
and landfill harvesting activity was documented in the study region
during the summer months. A series of photographs are shown in
figure 10.3. These were taken at the Nain, Labrador (Nunatsiavut)
dump, from the same vantage point on 31 May 2017, at the beginning
of summer when there is substantially more waste, and again on 23
August 2017, when volumes of waste were harvested after three
months of summer. In essence, landfills serve as a holding place for
waste to be reclaimed by community members at a later point in time,
though open burning in the rural communities is not uncommon.

Waste Wood from Muskrat Falls Hydroelectric Project

The wood that lies in piles and pits from forest cleared for the construction of the Muskrat Falls hydroelectric dam presents another
example of the waste management complexities created by natural
resource development. The Muskrat Falls megaproject is the most
recent hydroelectric dam installment on Labrador's Churchill River
(Canadian Broadcasting Corporation 2019). This hydroelectric

Figure 10.3 The truck serves as a frame of reference in the first set of photographs. On 31 May 2017, at the beginning of summer, there is substantially more waste than on 23 August 2017. The second photograph also taken from the same vantage point, and at the beginning and end of summer, document the large amounts of residential waste before and after burning.

megaproject was designed to supply electricity to more heavily populated northeastern North America, including Montreal, Quebec, and the northeastern United States. Beginning in 2012, more than 2 million cubic metres—or essentially 1.25 per cent of Canada's 2016 total timber harvest and more than 160 per cent of the province's annual timber harvest (Natural Resources Canada 2016)—was cleared and left to rot. Whole timber logs have rested at the main site, as well as in remote piles cleared for transmission lines.

All that timber sits there because transportation costs are prohibitive due to the region's remote location and harsh climate. It is simply not cost-effective to convert the timber into higher value applications like dimensional lumber and engineered wood products (Moody 1984). For years, the wood has been available for public collection at no cost, and the provincial government even transported the wood to more convenient pick-up areas for household use (Province of Newfoundland and Labrador Forestry and Agrifoods Agency 2015; Pardy 2013). The log piles that remain serve as a poignant example of a commodity that has essentially been left for waste, as there are no sawmills in Labrador with capacity to process the wood and no feasible way to get it to market (Forisk Consulting 2019).

Following the study's qualitative research phase in 2018, community partners expressed a desire to review the economic feasibility of converting the wood into biochar and using it as soil amendment for agricultural production to boost food production. The prevalence of food insecurity is estimated to affect between 62-83 per cent of these households (Schiff and Bernard 2018). An economically viable biochar and agricultural co-production system would show that there is potential to create opportunity within the community by addressing two social goals, namely waste management and improved food security.

QUANTITATIVE RESEARCH RESULTS: ENTERPRISE BUDGET AND TECHNO-ECONOMIC ANALYSIS

Enterprise Budget for Biochar Production

The first step of the techno-economic analysis involved constructing an Excel-based enterprise budget of the various costs for a proposed biochar production site. The budgeting tool and written guideline were published in the appendix of an open source peer review publication (Keske et al. 2018a), which provides additional detailed information on costs and assumptions, with all dollar values expressed in Canadian dollars.

In the development of the enterprise budget, considerable effort was undertaken to secure values relevant to the study region, although this proved difficult at times due to a paucity of publications and biochar commercialization ventures. It was important that the budget reflected a mobile, rather than stationary, pyrolysis unit to burn the wood at high temperatures, and subsequently produce the char. A mobile pyrolysis unit could be set up at various locations of the log piles (pits) that were cleared for the transmission lines, which would reduce start-up costs. Since the largest pit is located next to the Muskrat Falls megaproject site, production there was used as a model, with the business headquarters located in Happy Valley-Goose Bay, near the proposed agricultural production site.

Slow pyrolysis was chosen because it produces more biochar than the fast method (Kung, McCarl, and Cao 2013) and has lower pre-treatment costs (Ahmed et al. 2016). That is an advantage for a small-scale operation, like the one proposed. The biochar enterprise budget included costs from five stages of the production process: 1) transportation to and from production sites; 2) pre-processing; 3) pyrolysis; 4) biochar bagging; and 5) biochar storage. The process begins with transporting an empty trailer from Happy Valley-Goose Bay to Muskrat Falls, and it ends with taking the bagged biochar back to Happy Valley–Goose Bay for storage. The biochar production budget can be fine-tuned or expanded at a future time by anyone interested in exploring biochar production in a rural Canadian community.

Calculation of Fixed and Variable Costs and Break-even Analysis

The enterprise budget presented the costs of establishing a biochar production facility. However, it was determined at the outset that the facility would lose a considerable amount of money because there presently isn't a consistent market for selling the biochar. This is partly because documentation of its soil-enhancing properties is only beginning to emerge. A consistent biochar market for the product is needed to recover fixed (start-up) and variable (operational) costs of a biochar plant. Therefore, the analysis expanded the proposed biochar facility to include agricultural production using biochar as a soil amendment for growing beets (*Beta vulgaris L.*) and potatoes (*Solanum tuberosum L.*) in the Happy Valley-Goose Bay study region, at the site that is now known as the Pye Centre for Northern Boreal Food Systems.

Beets and potatoes were chosen because they can be grown in the region, they are commonly sold at grocery stores, and there is a strong potential for increased local demand for freshly grown vegetables. The region is considered extremely food insecure (Schiff and Bernard 2018). Results from Abedin's (2015, 2018) agronomic beet field trials were used. Potato cultivation was chosen because it is a priority crop of the Newfoundland and Labrador provincial government due to its potential to improve the region's food security. Yield data for local potato production were unavailable, so estimates were used from a related study that looked at a meta-analysis of vegetable yields from biochar application (Liu et al. 2013).

Average net returns over variable and fixed costs were calculated for both beets and potatoes. Breakeven yields per hectare and prices per unit yield, the point at which the business would recover expenses, were estimated to identify when joint biochar-agricultural operation would become profitable.

The value of biochar is calculated as the revenue increase from higher beet and potato yields. In this study, beet production covers total costs per hectare with $2,044 additional profit, and it covers variable costs with $3,758 to spare. Under the baseline scenario, potato production cannot cover total costs, losing on average $1,290 ha^{-1}. However, it can cover average variable costs, at $423 ha^{-1}, in the baseline scenario. More detailed information about the assumptions used in the calculation of these values is presented in Keske, Godfrey, Hoag, and Abedin (2020).

Sensitivity Analysis

Since the data are based on limited field research trials, there is still considerable uncertainty, and there will likely be a range of production yields. Therefore, a sensitivity analysis was conducted to evaluate the impact of three variables on profit, holding all other factors equal. These three variables were: 1) pyrolysis/biochar production rates (Mg day^{-1}); 2) biochar application quantity (Mg ha^{-1}); and 3) biochar application frequency (in years). Yields were also varied through a stochastic process explained in another section.

The sensitivity analysis also assessed the impacts of carbon application and biochar production rates. That analysis demonstrates that applying biochar at either a ten- or twenty-year lifespan is enough to make biochar application on potatoes profitable; otherwise, potato production doesn't cover total costs. Expanding the life of the system to ten years reduces costs to $1,949 ha^{-1}. With a twenty-year biochar application lifespan, costs fell to $1,251 ha^{-1}. Costs don't fall proportionately to the ratio of the increased lifespan due to the cost of borrowing at 6 per cent interest. Costs double, however, if the application rate doubles to 20 ha^{-1}. If the pyrolysis rates could increase to 2.5 Mg day^{-1} instead of 2 Mg day^{-1}, costs would fall to $2,725 ha^{-1}. However, costs would soar to $4,541 ha^{-1} if pyrolysis produces biochar at a rate of only 1.5 Mg day^{-1}.

A best-case scenario would reduce overall production costs to $1,001 ha^{-1} with 10 Mg C ha^{-1} biochar application every twenty years, and 2.5 Mg day^{-1} biochar production rates from a slow pyrolysis system. A worst-case scenario with 20 ha^{-1} application every five years and only biochar production of 1.5 Mg day-1 would cost over $9,000 ha^{-1}. Beet production is profitable under all scenarios except for the 20 Mg C ha^{-1} biochar application rate and worst-case scenario. Achieving the breakeven price and yield for potatoes is unlikely, though there is a low probability that the venture may be highly profitable.

Stochastic Techno-economic Analysis

The best, typical, and worst-case yield scenarios were evaluated through a stochastic analysis using @Risk, an Excel add-in (Palisade 2019), to predict the range of expected returns from the project. The stochastic techno-economic analysis, based upon two years of agronomic field trials in the study region, show that the increased beet and potato yields and subsequent agricultural sales could pay for the construction and operation of a biochar facility, and make use of the discarded timber.

In summary, there is potential to convert waste biomass into a value-added product. Providing concrete quantitative results that connect waste biomass to sustainable biochar and increased food production provides opportunities for communities to engage in informed conversation that may advance sustainable waste management and food production in the North.

DISCUSSION OF RESULTS

The goal of the study was to identify the waste management flows and sustainable waste management strategies that simultaneously reduce negative environmental impacts, yet also provide opportunity to generate benefits for the local community. As others before us have shown, this case study demonstrates that wastes and resources are inextricably linked. Some costs associated with waste management, and specifically waste wood resulting from a large natural resource project in the study region, may be recovered and transformed into opportunities to advance food security. Identification of waste characteristics is a first step in actualizing any potential benefits, and it lays the groundwork for a cost-benefit analysis of specific waste streams. Collaboration with community partners was key to successfully implementing the study and to obtaining grounded values relevant to the study region.

As previously discussed, associated demographic pressures and social changes have transformed local settlement patterns and consumption habits, replacing traditional economies and waste systems with those of Western capitalism. This economic structure is unlikely to be reversed; however, there is opportunity to create value-added product, like biochar, that can facilitate agricultural development opportunities.

The economic analysis presented in this paper illustrates a high likelihood of profitability for biochar and beet co-production in the study area under most conditions. Beet production may generate maximum yield values as high as $9,575 ha^{-1} under a best-case scenario. This is primarily attributable to the high yield potential exhibited in two years of regional beet growing trials where the control condition showed close to zero production capacity. These values may reflect an optimistic scenario, though modelling a more modest midline value of 5.59 Mg ha^{-1} demonstrates strong profitability at $3,240 ha^{-1} net return over total costs. Given certain circumstances, biochar and potato co-production may also be profitable, based upon estimated increases in potato yields extrapolated from Liu et al. (2013). The potential for profitable potato production augmented with biochar supports the supposition that additional potato-biochar field trials are needed to illustrate higher yields, or at least lead to a site-specific range of yield values.

An important finding is that there are increased agricultural production yields from the biochar application that make the proposed biochar-agricultural co-production venture economically viable. This stems from the apparent symbiotic relationship between the agriculture and biochar production within the study region. The area's harsh environmental conditions make agricultural production very difficult, and soil amendment is essential, even though it is not always effective. The case study shows a serendipitous situation for biochar and agricultural co-production. Given the region's dire and widespread food insecurity, this chapter makes the case that a pilot biochar production site is worth the investment.

However, caution should be exercised because biochar field studies show site-specific sensitivity, and there is some general uncertainty throughout the literature about the transferability of results. Local environmental, transportation, and market conditions make every project unique. Combined with a nascent market for biochar demand, any biochar production business venture would be considered highly risky.

In many situations, production in remote areas presents additional costs and risks due to added transportation costs and low market demand. However, in this particular situation, the region's remoteness and harsh climate motivates the pursuit of creative options to improve food security. There is considerable unfilled demand for beet and potato production within the province and across Canada's North.

The proposed operation may facilitate much-needed employment and community income, in addition to increasing local food supplies.

Though the stochastic economic models show that potato production is profitable less than 15 per cent of the time, the province's desire to increase local production, and specifically potatoes, may justify provincial or federal financial support for expanding field trials or breaking ground on a pilot biochar production facility. Specifically, the province of Newfoundland and Labrador has identified provincial potato production as a priority to enhance food security. Potato field trials should reflect crop varieties suited to the region, and document potential differences between potato varieties. Given the potential profitability of beet growing trials, this could diversify costs of potato production, as well, in order to move the province closer to achieving food security.

Ostensibly, from an environmental justice perspective, it could be argued that other Canadian provinces and states within northeastern US should compensate the province of Newfoundland and Labrador for the environmental impacts that the region has sustained as a result of the construction of the Muskrat Falls hydroelectric dam. However, practically speaking, this would be difficult to administer, if not impossible to orchestrate. Although the Churchill River hydroelectric projects have been viewed as contentious and deliver relatively meager financial benefits to the Newfoundland and Labrador government and residents, the courts (including Canada's Supreme Court) have declined to allow contracts to be renegotiated (Harris 2018). There is a case to be made that the Newfoundland and Labrador government should consider compensating Central Labradorians for the resource and associated environmental damage; however, given that the province itself doesn't benefit enough from Muskrat Falls, it may lack the capacity and will to compensate the region's residents.

Using this environmental justice argument, the government of Newfoundland and Labrador could provide the start-up capital required to open the biochar production facility or cover potential downside losses of a critical crop like potatoes, which could feed many households in the region. Extending the environmental and social justice thread a bit further, within Labrador, as well as across the Canadian North, the Canadian federal government does increasingly recognize a mandate for supporting northern food security (and/or food security for Indigenous communities). A subsidy would be consistent with other efforts in that area. Moreover, it would be proper to include Indigenous communities in a joint biochar-agricultural production venture that creates a circular economy in the region. Support from the Canadian government to facilitate this transition into an economically viable enterprise would facilitate economic opportunity and social justice.

In sum, with a nascent market for biochar, two years of field trial data, and preliminary efficacy of biochar field trials, it is difficult to provide a definitive answer to the question of whether to embark on a risky biochar-agricultural co-production venture. Agricultural production has increased within the community. The site where agricultural production field trials were conducted has been established as the Pye Center for Northern Boreal Food Systems. Additional publications have validated the efficacy of applying biochar to the sandy soils at the study site (Altdorff et al. 2019). However, mobile biochar production has yet to be undertaken in the study region. With optimism, the findings from this analysis should encourage local and national governments to fund additional studies to examine the efficacy of biochar on local agricultural production in the quest to pursue improved food security.

Following the publication of qualitative and quantitative findings in the open source academic journals, other northern communities (such as Igloolik, Nunavut) have contacted the authors to report similar waste management concerns, and to request assistance with waste management in their community. As well, other researchers and government agencies expressed interest in advancing biochar and agricultural production in their laboratories and communities. These follow-up discussions suggest that these study findings resonate in other northern communities. It is hoped that other scientists undertaking this type of work will consider using the interdisciplinary mixed methods research approach undertaken here to identify priority waste streams for more detailed techno-economic analysis. The impacts upon human well-being should also be formally assessed, perhaps with an instrument suggested by other authors in this volume, such as Parlee.

In brief, preliminary feedback from published study findings indicates that the dilemma of waste management in the North resonates in other communities, and that continued diligence is necessary to transform these wastes into opportunities that may provide benefits to local communities.

ACKNOWLEDGEMENTS

This study would not have been possible without the help of community partners who prompted the project: Frank Brown, Julianne Griffin, Samantha Noseworthy-Oliver, and Anatolijs Venovcevs at the Town of Happy Valley-Goose Bay; Mike Hickey of Hickey Construction; and Tammy Lambourne and Marina Biasutti-Brown of Healthy Waters Labrador. Credit is also due to Labrador Institute current and former affiliates Dr. Ron Sparkes, project co-PI Dr. Joinal Abedin, and Dr. Nathaniel Pollock, for insights and direction throughout the research process. Project co-PI Morgon Mills was instrumental in project coordination and for qualitative data gathering, interpretation, writing, and publication of qualitative research findings.

REFERENCES

Abedin, Joinal. 2015. "Potential for Using Biochar to Improve Soil Fertility and Increase Crop Productivity in the Sandy Soils of Happy Valley-Goose Bay, NL." Harris Centre, Memorial University of Newfoundland, St. John's, NL, Canada.

–2017a. "Applying Biochar to Reduce Leachate Toxicity & Greenhouse Gas Production in Municipal Solid Waste (MSW)." St. John's, NL: Harris Centre.

–2017b. "Reduction of AMD Generation and Availability of Metals from Sulfidic Mine Tailings Through Biochar Application: A Laboratory Scale Study." Geological Association of Canada/Mineralogical Association of Canada (GAC-MAC) Annual Conference, Kingston, ON.

–2018. "Enhancing Soils of Labrador Through Application of Biochar, Fishmeal, and Chemical Fertilizer." *Agronomy Journal* 110 no.6: 2576-2586. DOI: 10.2134/agronj2018.02.0074.

Ahmed, Mohammad B., John L. Zhou, Huu Hao Ngo, and Wenshan Guo. 2016. "Insight Into Biochar Properties and Its Cost Analysis." *Biomass and Bioenergy* 84: 76-86. DOI://dx.doi.org/10.1016/j.biombioe.2015.11.002.

Altdorff, Daniel, Lakshman Galagedara, Joinal Abedin, and Adrian Unc. 2019. "Effect of Biochar Application Rates on the Hydraulic Properties of an Agricultural-use Boreal Podzol." *Soil Systems* 3, no. 3 (2019): 53.

Barker, Jacob. 2017. "Burning Waste at Muskrat Falls Blasted by Happy Valley-Goose Bay Council." *Canadian Broadcasting Corporation*, April

20, 2017. https://www.cbc.ca/news/canada/newfoundland-labrador/nalcor-controlled-burns-muskrat-falls-1.4077195.

Charmaz, Kathy and Linda L. Belgrave. 2007. "Grounded Theory." *The Blackwell Encyclopedia of Sociology.*

Creswell, John W. 2003. *Research Design: Qualitative, Quantitative, and Mixed Methods Approaches.* 2nd ed. Sage Publications. DOI: 10.12691/jbms-4-4-1.

Daley, Kiley, Heather Castleden, Rob Jamieson, Chris Furgal, and Lorna Ell. 2014. "Municipal Water Quantities and Health in Nunavut Households: An Exploratory Case Study in Coral Harbour, Nunavut, Canada." *International Journal of Circumpolar Health* 73, no. 1. 23843.

−2015. "Water Systems, Sanitation, and Public Health Risks in Remote Communities: Inuit Resident Perspectives from the Canadian Arctic." *Social Science and Medicine* 135: 124-132. DOI: http://dx.doi.org.qe2a-proxy.mun.ca/10.1016/j.socscimed.2015.04.017.

Dickenson, V. 2009. *A Veritable Scoff: Sources on Foodways and Nutrition in Newfoundland and Labrador*, edited by Maura Hanrahan and Marg Ewtushik. St. John's, NL: Flanker Press, 2001. https://www.erudit.org/en/journals/cuizine/2009-v2-n1-cuizine3403/039522ar/abstract/.

Felsberg, S. 2009. "To Burn or Not to Burn: Mud Lake and the Provincial Government Debate the Future of the Village's Incinerator." *Labrador Life* 3, no. 4: 24–27.

Field, John, Catherine M. H. Keske, Greta Lohman Birch, Morgan W. DeFoort, and M. Francesca Cotrufo. 2013. "Distributed Bioenergy and Biochar Co-Production: A Regionally-Specific Case Study of Environmental Benefits and Economic Impacts." *Global Change Biology (GCB) Bioenergy* Special Edition on Bioenergy 5, no. 1: 177-191. http://onlinelibrary.wiley.com/doi/10.1111/gcbb.12032/full.

Ford, James D., Tristan Pearce, Frank Duerden, Chris Furgal, and Barry Smit. 2010. "Climate Change Policy Responses for Canada's Inuit Population: The Importance of and Opportunities for Adaptation." *Global Environmental Change* 20, no. 1: 177-191.

Forisk Consulting. 2019. "20170815 Canada Sawmill Map." https://forisk.com/blog/2017/08/15/sizing-canadas-forest-products-industry/20170815-canada-sawmill-map/.

Glaser, Barney and Anselm Strauss. 1967. "Grounded Theory: The Discovery of Grounded Theory." *Sociology: The Journal of the British Sociological Association* 12, no. 1: 27-49.

Harris, Kathleen. 2018. "Supreme Court Rejects Churchill Falls Corp.'s Bid to Reopen Energy Deal with Hydro-Québec." *Canadian Broadcasting Corporation*, November 2, 2018. https://www.cbc.ca/news/politics/churchill-falls-hydro-quebec-supreme-court-1.4888321.

Hird, Myra J. 2016. "Waste Legacies: Land, Waste, and Canada's DEW Line." *The Northern Review* 42:173–195. http://journals.sfu.ca/nr/index.php/nr/article/view/567.

Keske, Catherine. 2018. *Food Futures: Growing a Sustainable Food System for Newfoundland and Labrador*. St. John's, NL: ISER Books.

Keske, Catherine, Todd Godfrey, Dana Hoag, and Joinal Abedin. 2020. "Economic Feasibility of Biochar and Agriculture Co-production from Canadian Black Spruce Forest." *Food and Energy Security* 9, no. 1. https://doi.org/10.1002/fes3.188.

Keske, Catherine, Dana Hoag, Donald M. McLeod, Christopher T. Bastian, and Michael G. Lacy. 2011. "Using Mixed Methods Research in Environmental Economics: The Case of Conservation Easements." *International Journal of Mixed Methods in Applied Business and Policy Research* 1, no. 1: 16–28.

Keske, Catherine, Morgon Mills, Todd Godfrey, Laura Tanguay, and Jason Dicker. 2018. "Waste Management in Remote Rural Communities Across the Canadian North: Challenges and Opportunities." *Detritus: Multidisciplinary Journal for Waste Resources & Residues* 2: 63-77. https://digital.detritusjournal.com/articles/waste-management-in-remote-rural-communities- across-the-canadian-north-challenges-and-opportunities/106. DOI: 10.31025/2611- 4135/2018.13641.

Keske, Catherine, Morgon Mills, Laura Tanguay, and Jason Dicker. 2018. "Waste Management in Labrador and Northern Communities: Opportunities and Challenges." *The Northern Review* 47: 79-112. DOI: https://doi.org/10.22584/nr47.2018.100.

Kung, Chih-Chun, Bruce A. McCarl, and Xiaoyong Cao. 2013. "Economics of Pyrolysis-Based Energy Production and Biochar Utilization: A Case Study in Taiwan." *Energy Policy* 60: 317-323. DOI: http://dx.doi.org/10.1016/j.enpol.2013.05.029.

Liu, Xiaoyu, Afeng Zhang, Chunying Ji, Stephen Joseph, Rongjun Bian, Lianqing Li, Genxing Pan, and Jorge Paz-Ferreiro. 2013. "Biochar's Effect on Crop Productivity and the Dependence on Experimental Conditions—A Meta-Analysis of Literature Data." *Plant and Soil* 373, no. 1-2: 583-594. DOI 10.1007/s11104-013-1806-x.

Mills, Morgon, and Catherine Keske. 2018. "Industrial Mega-Projects and Waste Management in Northern Communities." *Northern Public Affairs*

Magazine 5, no. 3. http://www.northernpublicaffairs.ca/index/volume-6-issue-3-the-fight-for-our-lives-preventing- suicide/industrial-mega-projects-and-waste-management-in-northern-communities/.

Moody, Douglas. 1984. "An Economic and Technical Analysis of Potential Forest Industry Developments in Labrador." Centre for Newfoundland Studies, Memorial University of Newfoundland. ISBN 0- 315-61771-3. https://research.library.mun.ca/8574/1/Moody_DouglasB.pdf.

Multi-Materials Stewardship Board. 2017. "About MMSB." http:// mmsb. nl.ca/about-mmsb/.

Natural Resources Canada. 2016. Statistical Data. https://cfs.nrcan.gc.ca/statsprofile/overview/nl.

Nematian, Maryam, Catherine Keske, and John N. Ng'ombe. 2021. "A Techno-economic Analysis of Biochar Production and the Bioeconomy for Orchard Biomass." *Waste Management* 135: 467-477. DOI: https://doi.org/10.1016/j.wasman.2021.09.014.

Pardy, Brandon. 2013. "Wood Ya Believe..." *The Newfoundland and Labrador Independent*, May 23, 2013. https://theindependent.ca/2013/05/23/wood-ya-believe/.

Province of Newfoundland and Labrador Forestry and Agrifoods Agency Executive Council. 2015. "Achieving Benefits from the Timber Resource." April 9. https://www.releases.gov.nl.ca/releases/2015/fishaq/0409n02.aspx.

Schiff, Rebecca, and Karine Bernard. 2018. "Food Systems and Indigenous People in Labrador: Issues and New Directions." In *Food Futures: Growing a Sustainable Food System for Newfoundland and Labrador*, edited by Catherine Keske. St. John's, NL: ISER Books.

Statistics Canada. 2011. "National Household Survey Focus on Geography Series: Happy Valley-Goose Bay, Town." http://www12.statcan.gc.ca/nhs-enm/2011/as-sa/fogs-spg/Pages/FOG.cfm?lang=E&level=4&GeoCode=1010025.

–2017. "Census Profile, 2016 Census." https://www12.statcan.gc.ca/census-recensement/2016/dp-pd/prof/details/page.cfm?Lang=E&Geo1=CSD&Code1=1010025&Geo2=CD&Code2=1010&SearchText=Happy%20Valley%20Goose%20Bay&SearchType=Contains&SearchPR=01&B1=All&TABID=1&type=0.

St. Croix, Rick. 2002. *Soils of the Happy Valley East Area, Labrador: Soil Survey Report*. Soil and Land Management Division, Department of Forest Resources and Agrifoods, St. John's, NL.

Teddlie, Charles and Abbas Tashakkori. 2009. *Foundations of Mixed Methods Research: Integrating Quantitative and Qualitative*

Approaches in the Social and Behavioral Sciences. Thousand Oaks, CA: Sage Publications.

"What's the Deal with Muskrat Falls? Answers to a Few Frequently Asked Questions." 2019. *Canadian Broadcasting Corporation*, May 4, 2019. https://www.cbc.ca/news/canada/newfoundland-labrador/ muskrat-falls-whats-the-deal- 1.5083458.

11

The Social Economy and Resource Governance in Nunavut

Joshua Gladstone and Sheena Kennedy Dalseg

Environmental and social economy organizations are often overlooked as actors in the northern political economy (Abele and Southcott 2016), and yet their contributions to resource governance help to guide decision-making towards more sustainable and democratic outcomes (Fitzpatrick, Sinclair, and Mitchell 2008; Diduck and Sinclair 2002). For this reason, it is important to better understand the role these organizations are playing, as well as the roles they could play, in northern and Indigenous jurisdictions grappling with industrial resource extraction. We consider what we present here to be the groundwork of what could, and perhaps should, be a more systematic examination of the full spectrum of actors and interests engaged in northern resource governance in order to better understand the ways in which to mitigate risks and maximize benefits to Inuit and northern citizens and communities from resource development.

In this chapter, we look to Nunavut as a jurisdiction in which environmental and social economy organizations are playing a role in resource governance. Specifically, we examine the range of those organizations that exist in the territory and compare it with those organizations participating in resource governance processes between 2008-18. Our initial findings suggest that only a small number of local environmental and social economy organizations active in the territory are participating in resource governance. Many of those participating organizations are not exclusively Nunavut-based. We discuss this observation in connection with recent insights into Inuit-Qallunaat (an Inuktitut term for non-Inuit) relations in the territory and the moves made by Inuit to formalize Inuit systems of democratic governance, and to advance Inuit interests within Nunavut's non-profit and voluntary sector.

INTRODUCTION

In Canada, the general trend in resource governance since the 1970s has been a shift away from top-down decision-making systems toward more open, participatory processes. These include processes such as environmental assessments that seek to anticipate and prevent environmental harms before they occur. At the same time, they attempt to provide a greater role for the public in matters that might affect them (Gibson 2002; Diduck and Sinclair 2002; Abele 1983). These processes may also encourage debate about ways of maximizing the benefits while mitigating the negative effects of resource extraction. It is a trend that has sparked interest among researchers about the democratic potential of land and resource management institutions (Wiklund 2005; O'Faircheallaigh 2010), including those established under comprehensive land claims agreements in Northern Canada (Fitzpatrick, Sinclair, and Mitchell 2008). These latter institutions are especially significant in Nunavut because their structures are intended to provide opportunities for the direct participation of the public and Inuit rights holders in resource governance.

The public role that environmental and social economy organizations can play in resource governance is well established in the literature (Noble 2015). As public participants, they can offer fact-based information, help to define potential problems, identify acceptable solutions, or minimize the risk of legal challenge or other delays. In a process seeking to maximize the benefits of any project or policy while minimizing the risks, they can support decision-making by offering their own assessments of the costs and benefits of specific proposals. Our questions concern the democratic potential of these organizations in resource governance regimes where the lands and livelihoods of Indigenous peoples are at stake. In these cases, we need a way of thinking about environmental and social economy organizations and their relationship to contemporary Indigenous societies.

The formal system of resource governance in Nunavut is the product of a long history of Inuit-Qallunaat relations in the Central and Eastern Arctic. In this way, it shares its origins with the emergence of the voluntary and non-profit sector. From the 1950s onward, world demand for resources in what was then the Northwest Territories led to the expansion of state legal and administrative forms. These supported the growth of extractive industries based on the assertion of Crown sovereignty and the creation of a system of free entry mining across lands historically governed by Inuit. Traditional sources of authority in Inuit society were eroded as federal mining legislation allowed miners to enter onto Inuit lands, stake claims, and lease, produce, and export minerals without requiring Inuit consent or the payment of compensation to them (Bankes and Sharvit 1998; see also Tunngavik Federation of Nunavut 1989). These conditions caused surprise and dismay when Inuit learned they lacked power to control the damaging effects of industrial activities on the land, water, and wildlife they depended on to sustain their way of life (Amagoalik 2007). Circumstances slowly began to change in the 1970s as Indigenous activism transformed the Canadian political and jurisprudential landscape, leading the federal government to recognize the validity of (some) Indigenous claims to lands and resources that had not been formally surrendered to the Crown through treaty (Scholtz 2006). Since 1975, twenty-six comprehensive land claims and self-government agreements have been reached between Indigenous signatories and the federal Crown.

Inuit claims to lands and resources in the Central and Eastern Arctic were settled in 1993 with the signing of the Nunavut Land Claims Agreement. Like other comprehensive land claims agreements, the Nunavut Agreement does not provide Inuit with exclusive ownership or regulatory authority over all lands and resources within the land claim settlement area. Instead, Crown-owned minerals and most Crown land are under federal jurisdiction. Inuit gained proprietary ownership rights over a portion of surface lands and a smaller portion of subsurface resources, as well as the rights to participate in unique environmental management processes. They also agreed to rights to negotiate compensation and benefits derived from resource extraction (Usher 1996; McPherson 2005; Galbraith et al. 2007). These features provide the formal system of Inuit democratic governance with a set of constitutional rights to be exercised by Inuit representational bodies. These include Nunavut Tunngavik Incorporated, regional Inuit associations, and Hunters and Trappers Organizations, all of which are formally empowered under the Nunavut Agreement. They also enable public participation including by third sector environmental and social economy organizations.

Our purpose here is to take an initial step in understanding where and how third sector organizations fit into Nunavut's resource governance system. While the formal system of governance established under the Nunavut Agreement is conditioned by Inuit-specific rights, the system remains an outgrowth of Canada's colonial political and economic systems. The environmental management institutions discussed below are a case in point. Both land-use planning and environmental impact assessment institutions are led by co-management bodies that empower Inuit through shared decision-making at the board level, and which include Inuit-specific provisions such as language rights and rights to legal standing. And yet these institutions are based on and respond to a set of property relations that typically favour industrial developers and landowners over other interests. With power concentrated in these relationships, other actors, such as third sector organizations or Inuit modes of decision-making based on Inuit social relations, may be rendered less visible in formal resource governance discourse. In other words, Inuit concerns are often disregarded or overlooked because of the priorities of these institutions.

Complicating this picture are the complex ways in which Inuit forms of governance and third sector organizations may be related to each other. In a recently published polemic aimed at Qallunaat, Sandra Inutiq, an Inuit rights advocate and an Inuk thought leader, decried the state of Inuit-Qallunaat relations in the territory (Inutiq 2019). She called on Qallunaat to resist the behaviours that "maintain an impoverished Inuit population," including a lack of recognition of Inuit forms of volunteerism. Although Inutiq does not cite evidence supporting her observation that "Inuit do not volunteer or volunteer as much, for boards and committees," it is an observation that suggests the possibility of deep structural divisions within Nunavut's civil society. Since these observations are consistent with the authors' own, we believe it is worth considering whether and how these divisions might influence the character of third sector participation in resource governance. Do the linguistic and cultural divisions that privilege Qallunaat forms of association, authority, and social provision have repercussions for the scope of issues addressed in land-use planning and environmental assessment processes?

How can the resource governance regime evolve with respect to the unrecognized forms of association, authority, and social provision that continue to make democracy a goal for a vibrant and

distinct Inuit society? We address these questions through an exploration of the role that social economy organizations played in resource
governance in Nunavut between 2008-18.

CIVIL SOCIETY AND THE SOCIAL ECONOMY IN NUNAVUT

Civil society has been defined as the "terrain of human association"
that is "distinct from the body politic ... with moral claims independent
of, and sometimes opposed to, the state's authority" (Wood 1995).
More specifically, it is "neither public nor private or perhaps both at
once, embodying not only a whole range of social interactions apart
from the private sphere of the household and the public sphere of the
state, but more specifically the network of distinctively economic relations, the sphere of the marketplace ..." (239). This definition is
grounded in Western social and political thought, and it shares its
heritage with the development of liberal democracy.

Our view of resource politics in Nunavut is premised on the insight
that social economy organizations reflect the complex and contra-
dictory nature of northern civil society insofar as they exist in relation
to two historical developments. The first is the northward expansion
of Canadian social and cultural forms that occurred as missionaries,
traders, and police established their presence in the Central and
Eastern Arctic in the early twentieth century. This process continued
through the administrative period after World War II as settlement,
market formation, and colonial control of political institutions pro-
ceeded under the auspices of the Canadian welfare state. The second
is the continuity of Inuit social and cultural forms, most notably the
mixed economy and the political agency of Inuit in response to colonial domination.

The sphere of a rapidly changing civil society has been one site of
colonial domination in which the terrain of human association is
fraught with geographic, material, linguistic, and normative (ideo-
logical and religious) divisions between Inuit and Qallunaat (Brody
1975). It was in this sphere, through the actions of the Qallunaat who
benefitted most from the processes mentioned above, that many of
the worst "disadvantages" experienced by Inuit could be remedied,
either through the implementation of welfare state programs or the
efforts of early social economy organizations such as churches and
co-operatives (Tapardjuk 2013; Lyall 2013). The ways in which

northern social economy organizations can serve community goals today reflects this history, and the changing structure of civil society in the North.

In a territory where Inuit make up 86 per cent of the population, association between Inuit and Qallunaat must constantly confront differences in language and culture that penetrate the social spheres of school, work, family, and community. Linguistically, although Inuktut use is declining in Nunavut (Martin 2017), it continues to be the primary language spoken by many Inuit families at home and remains central to processes of Inuit democratic governance. And yet formal institutions, including the resource governance boards as well as the labour market and the education system, create substantial pressure for Inuit to function in English.

This use of English benefits the majority of Qallunaat living in Nunavut, many of whom speak little Inuktut and for whom English is their mother tongue. Inuit youth experience considerable pressure to speak both languages (Johnston 2013), despite the lack of support for bilingual Inuit education within the public school system (Berger 2006). The pressure on Inuit to bridge the linguistic divide daily far exceeds the pressure on Qallunaat to do the same, and it helps to explain the desire to revitalize Inuit language and expand the spaces in which Inuit can associate without Qallunaat interference (Inutiq 2019), including in the sphere of the northern social economy.

The way that social provision is structured also contributes to divisions in civil society. Under contemporary conditions of post-industrial capitalism, the vast majority of Qallunaat experience the labour market as the primary institution through which material needs are met, supported by household labour and welfare state programming. Within this system, social needs that are not fully satisfied by these institutions are often addressed through the work of "conventional" social economy organizations, supported by private philanthropy, government grants, and volunteerism. In Nunavut, while the labour market and government programs loom large in people's lives, Inuit forms of social provision centred on land-based production and kinship networks continue to structure association distinct from the market. Often unseen by Qallunaat and unrecognized by the formal structures of the state and the constitutional mechanisms created by land claims, these Inuit-specific structures are mobilized to address any number of issues affecting Inuit lives.

In her letter to Qallunaat, Inutiq (2019) highlights this division clearly:

White people do not seem to understand why Inuit do not volunteer or volunteer as much, for boards and committees, etc. I can promise that Inuit are expending more hours and energy helping each other out than the superstar white volunteer who gets all the volunteer awards. Inuit also deal with a lot of loss or trauma. We help those grieving. The invisible "volunteering" we do is giving rides, visiting each other, helping with application forms, feeding and looking after kids, taking someone shopping or to run errands, helping with sewing projects, fixing snow machines, fundraising for funerals or kids' sports. This is just the tip of the iceberg.

Although Inutiq is focused here specifically on volunteerism, which is only one aspect of the social economy, her message should complicate any assessment of the value social economy organizations present as vehicles for advancing Inuit interests in resource governance.

Recent scholarship on the northern social economy has gone some way to conceptualizing the divisions Inutiq is pointing out, seeking to identify a space or spaces within civil society in which individuals come together to improve the social, cultural, and environmental conditions within northern communities (Southcott et al. 2010). In one sense, the social economy refers to "bottom-up" organizations (Southcott et al. 2010) "whose primary purpose is to serve social goals in the community and whose structures are based on participatory democratic principles" (Abele and Southcott 2016). These organizations are often described as a sector that includes not-for-profit and voluntary organizations, charities, foundations, co-operatives, credit unions, and other social enterprises, as well as Indigenous organizations (see for example Southcott et al. 2010). In this account, social economy organizations could be understood as the products of freely associating individuals acting on their own moral authority to advance claims vis-à-vis the state and large corporations.

A parallel conception of the social economy is given by scholars who have sought to account for Indigenous social formations in Northern Canada since the 1970s. In this sense, the social economy has been understood as the blend of market and non-market forms of social provision known as the mixed economy, and the Indigenous social institutions that continue to underpin them (Abele 2009; Natcher 2009; Kuokkanen 2010). Within the mixed economy, families rely on a combination of wages, social transfers, and the products of harvesting wildlife. The specific means of producing, distributing, and consuming the harvest have come to be known as subsistence, a term that often carries stigma (Reimer 2006; Natcher 2009) reflecting the subordinate position of non-market forms of Indigenous social relations in the northern economy. In this latter account, the social economy emerges not exclusively from the Western social formations and liberal democratic protections alone (although these protections no doubt shape its expression). It also grows out of the continuity of Indigenous social relations that remain distinct from the market and the formal institutions designed to support it.

The relationship between these two conceptions of the northern social economy is not well explored. Functionally, Abele (2009) suggests they may be brought together in an uneasy relationship in which "conventional," or formal, social economy institutions may substitute for Indigenous place-based "mixed economy" institutions depending on the extent to which Indigenous northerners are reliant on the labour market. Less is said in the literature about the political dimensions of Inuit-specific forms of social provision ,and in particular the way they are structured to address the challenges posed by resource extraction.

Contemporary divisions in civil society between Inuit and Qallunaat may result in Inuit choosing to limit their participation in the forms of civil society that are currently recognizable by the state. In that case, how are Inuit able to advance their moral claims and interests in resource governance and resist, if necessary, the preferred outcomes of the extractive industry? Below we discuss third sector participation in resource governance in Nunavut and examine how grassroots Inuit advocacy organizations appear to be one avenue through which Inuit claims are advanced.

OVERVIEW OF SOCIAL ECONOMY ORGANIZATIONS IN NUNAVUT

Research conducted as part of the Social Economy Research Network of Northern Canada (SERNNOCA) between 2006-11 provided a portrait and analysis of the northern social economy. During this period, SERNNOCA researchers identified around 300 social economy organizations of all types operating in the territory. These included long-standing institutions like the Arctic Co-operatives located in each of Nunavut's twenty-five communities, sports associations, early childhood education centres, thrift shops, and Elders' societies to name a few. (NRI 2009; Southcott and Abele 2016; Southcott 2015; Southcott et al. 2010; Abele and Kennedy 2011). Ten years on, many of these organizations are thriving. Of course, some have ceased operation while new ones have emerged in response to community-identified needs and priorities. However, it is not possible to know the extent of the changes to the social economy landscape in Nunavut over the last decade in the absence of a systematic follow-up study of registered Nunavut non-profits.

In this section, we describe in general terms what we know about contemporary environmental and social economy organizations in Nunavut. As such, we have relied on a national registry of charitable organizations, complemented by our own observations and the published research mentioned earlier.

We acknowledge that focusing our description of the sector primarily on those organizations with charitable status imposes some limitations, namely that it excludes a significant proportion of the social economy organizations operating in the territory. However, we believe there are insights to be gained through this approach. Charitable organizations are unique in the social license afforded them by society and the state. As a result of their formal status, they have legitimacy and access to financial and administrative resources to apply in the pursuit of their charitable purposes. For these reasons, their role and influence in resource governance may be substantial. Conversely, it would also be notable if charitable organizations based in Nunavut, particularly those with social service or environmental mandates, did not appear as participants in resource governance processes.

The Canada Revenue Agency lists forty-three charities currently registered in Nunavut (see Table 11.3 below). The majority of these are oriented towards providing religious or social services. Notably, there are currently *no* environmental charities or philanthropic foundations registered in Nunavut. Over half of the charities registered in Nunavut are based in Iqaluit, the territory's capital. The others are spread across the three regions. The organizations located outside of Iqaluit include over a dozen churches, two heritage societies, a childcare centre, and a small number of education-related initiatives, to name a few.

With a few exceptions, the Nunavut-based charities, particularly those located in Iqaluit, closely resemble the types of social organizations that might be found in other places across Canada. These include organizations like food banks, greenhouse societies, the YWCA, Rotary Club, Humane Society, and Habitat for Humanity. While these organizations are "place- based" in that they are registered in Nunavut, with Nunavummiut staff, they are primarily southern transplants, often created by southerners living and working in the territory. This is not to say that the individual organizations are not responding to local needs or that they are not tailored to the communities they serve; rather, that they are predominantly settler institutional forms of representation conceived of elsewhere and, moreover, that the approach of these transplanted structures to local needs is influenced by their respective histories. As a result, they may not reflect how Inuit would choose to respond to those needs (e.g., Inutiq 2019).

In addition to these social service organizations, there is also a set of arts and culture organizations listed in the CRA charities registry. These include a small number of community- based heritage, archival, and museum societies, as well as two arts and entertainment organizations: Alianait and the Qaggiavuut Performing Arts Society, and the Ilitaqsiniq–Nunavut Literacy Council. (The latter supports literacy initiatives in the territory's official languages within a community capacity building framework.) The literature on the northern social economy indicates that, in addition to these charitable organizations, there are also many non-profit and informal entities whose focus is Inuit cultural and/or linguistic continuity. These include the Arviat Film Society and community-based early language learning programs, to name a few. Some of these organizations, such as the Inullariit Society (the driving force behind the renowned Igloolik Oral History Project) in Igloolik for example, have gone through periods where they have been formally registered either as societies or charities, and other periods during which they have lost their "status." This does not necessarily mean, however, that they are inactive (Kennedy and Abele 2010; MacDonald and Wachowich 2019; Kennedy Dalseg 2018).

What these organizations share is that they are largely dedicated to documenting, preserving, and promoting Inuit traditional knowledge, history, and cultural continuity in various ways, and they are most often created and run primarily by Inuit. While community-based museums and heritage societies are generally settler imports, they are forms that seem to resonate with Inuit and have been chosen and adapted by Inuit across Nunavut to serve local purposes.

✳

RESOURCE GOVERNANCE AND SOCIAL ECONOMY
ORGANIZATIONS IN NUNAVUT

In this next section, we briefly describe the land-use planning and environmental assessment institutions established under the Nunavut Agreement before turning to a discussion of the civil society organizations that *have* participated in resource governance in Nunavut over the course of the last decade. See table 11.1 for a summary of organizations that participated in various aspects of the environmental assessment process in Nunavut between 2009-19.

Land-Use Planning and Environmental Assessment in Nunavut

The land-use planning process administered by the Nunavut Planning Commission (NPC) and the environmental assessment process administered by the Nunavut Impact Review Board (NIRB) are the products of careful negotiation between Inuit and the Crown. As constitutionally protected bodies created by the Nunavut Agreement, they exert significant influence over non- renewable resource extraction across the Nunavut settlement area. The structure of these institutions creates what Usher (1996) has described as a "permanent, institutionalized relationship between governments and representative Aboriginal bodies." Together, these so- called Institutions of Public Government and the processes they oversee are meant to "protect and promote the existing and future well-being of the residents and communities of the Nunavut Settlement Area, taking into account the interests of all Canadians ..." (NLCA 11.3.2).

The design of these two resource management processes is consistent with increasing recognition of Indigenous rights and interests by the state, as well as a trend toward greater openness, transparency, and public participation. This has been the case across Canadian land and resource management institutions since the 1970s. For example, the NPC must conduct public hearings on draft land use plans "that give weighty consideration to the tradition of Inuit oral communication and decision-making" (NLCA 11.4.17(a)) and that "allow standing at all hearings to a [Designated Inuit Organization]" (NLCA 11.4.17(b)). The NIRB screens and reviews project proposals and determines whether reviews are required based in part on the likelihood that the project will cause public concern. In cases where development projects must be reviewed (i.e., fully assessed for their environmental and socio-economic effects), the NIRB may conduct reviews by correspondence, public hearings, or other appropriate procedures. Public hearings must emphasize flexibility and informality while respecting the "broad application of the principles of natural justice and procedural fairness" (NLCA 12.2.24(a)).

In addition, the NIRB must take all necessary steps to promote public awareness of and participation in public hearings (NLCA 12.2.27), which must be conducted in the Inuktut language upon request (NLCA 12.2.26). Together, these processes are intended to create opportunities for the public to submit knowledge and value claims to the NPC and NIRB. Those recommendations must be acted upon by responsible government authorities. Often, these claims are presented by social economy organizations on behalf of the interests they represent.

Social Economy Organizations in Nunavut

The list of social economy organizations participating in resource governance processes for the extractive sector in Nunavut is short. Table 11.1 shows that between 2008-18, nine social economy organizations have participated in the screenings or reviews of the seven mining projects that have fallen under NIRB's Part 5 review process. Table 11.2 shows that since 2011, seven social economy organizations have participated in the land-use planning process.

Our analysis of these organizations has led us to identify three types of environmental and social economy organizations participating in resource governance in Nunavut over the last ten years: 1) national or international environmental NGOS; 2) northern environmental NGOS with a conservation focus, and 3) grassroots Inuit advocacy organizations. Each of these categories is discussed below.

The social economy organizations listed for each project were taken from final hearing reports, or screening decision reports in cases where final hearing reports were not issued. The social economy organizations listed for each project are a subset of the interveners participating in each assessment. Other interveners included government agencies, designated Inuit organizations, and First Nations, Métis, and Inuit organizations from outside Nunavut.

Table 11.1
List of resource extraction projects screened and recommended for review under Part 5 of the Nunavut Land Claims Agreement

Resource Extraction Project (2008-18)	Report and Year Issued	Social Economy Organizations Participating as Intervenors in Environmental Assessment
Hackett River Project	Screening decision report (2008)	Canadian Arctic Resources Committee
Mary River Project	Final hearing report (2012)	Dr. Zacharias Kunuk (Digital Indigenous Democracy); World Wildlife Fund
Izok Corridor Project	Screening decision report (2012)	Beverly Qamanirjuaq Caribou Management Board; Canadian Arctic Resources Committee; Canadian Parks and Wilderness Society - Northwest Territories Chapter; Wildlife Conservation Society Canada; MiningWatch Canada; World Wildlife Fund Canada
Meliadine Project	Final hearing report (2014)	N/A
Kiggavik Project	Final hearing report (2015)	Beverly Qamanirjuaq Caribou Management Board; Nunavut Makitagunarningit
Back River Project	Final hearing report (2017)	N/A
Hope Bay Belt Project Phase 2	Final hearing report (2018)	N/A

Source: www.nirb.ca.The social economy organizations listed for each project were taken from final hearing reports or screening decision reports in cases where final hearing reports were not issued. The social economy organizations listed for each project are a subset of the intervenors participating in each assessment. Other interveners included government agencies, designated Inuit organizations, and First Nations, Métis, and Inuit organizations from outside Nunavut.

Table 11.2
List of social economy organizations participating in land use planning processes between 2011 and 2016.

Land Use Planning Stage	Social Economy Organization Participating
Draft Land Use Plan (2011-16)	Beverly Qamanirjuaq Caribou Management Board, World Wildlife Fund, Ecojustice, MiningWatch Canada, Makita, Canada, Oceans North, Van Horne Institute

Source: http://lupit.nunavut.ca/portal/registry.php?public=docs

1. *National and International Environmental* NGOS

The first category of organizations participating in resource governance in Nunavut includes a small number of national or multinational environmental NGOS (ENGOS). All but one of these are registered charities. These are: World Wildlife Fund, MiningWatch Canada, WCS Canada, CPAWS, and Ecojustice Canada. In general, these larger ENGOS are focused on environmental protection and conservation, except MiningWatch Canada, which has a broader global mandate related to mining.

While none of the large ENGOS are registered in Nunavut, the WWF has an Arctic program which is headquartered in Iqaluit. Mostly directed and staffed by non-Inuit "experts" and funded by sponsors and donors in the south, the work of these ENGOS is not driven by northern priorities alone, although they do sometimes align. And while this set of organizations is not directly accountable to northerners, they do seem to acknowledge that their ability to do their work successfully in the territory depends heavily on public perception, which in turn is bound up with the extent to which they prioritize, or at the very least, recognize Inuit as local resource users and rights holders in the region.

Partnerships between these larger ENGOS and northern organizations offer mutual benefits for both parties. On the one hand, they offer one way through which ENGOS can achieve their purposes and at the same time demonstrate their acknowledgment of Inuit as rights holders, and their commitment to Inuit self-determination vis-à-vis conservation and environmental protection. On the other hand, partnerships also offer opportunities for northern organizations, including land claims organizations, to access information and resources that may otherwise be unavailable.

Relationships (in the form of partnerships with Inuit and Inuit organizations) are clearly important to the success of ENGOS' work in Nunavut, but we also observed a distance from "community" despite these partnerships. The ENGOS, even the ones operating from the North, are not claimed by Inuit.

2. *Northern Environmental* NGOS *with a Conservation Focus*

The second category of organizations participating in resource governance includes three organizations that are environmental NGOS. These are northern-focused but not Nunavut-based: the Canadian Arctic Resources Committee (CARC), the Beverly and

Qamanirjuaq Caribou Management Board (BQCMB), and Oceans North. CARC and BQCMB are both charities, registered in Alberta and Manitoba respectively, while Oceans North is a non-profit organization with offices across Canada, including one in Iqaluit and one in Nain.

This subset of environmental NGOS shares some characteristics with the larger ENGOS listed above. Their work is dedicated to conservation and environmental protection, and they are funded primarily by southern donors and sponsors. However, they differ in terms of their focus and structure. Unlike the larger ENGOS above, all three of these organizations have mandates focused almost exclusively on the North or the Arctic. In all three cases, Inuit hold leadership positions within the organization, influencing the strategic direction and priorities of each organization. Each has adapted aspects of their structure and focus to align with Indigenous- identified goals and priorities. For its part, BQCMB is one of the earliest organizations to use a co-management structure which has become common across land claim regions in Northern Canada. The model is premised on power sharing between Indigenous and Western scientific knowledge holders, who work together at the board level to offer advice on caribou management policy. While CARC is not as active as it once was, the organization has a long history working to support the land claims movement, and thus developed long-standing committed partnerships with Indigenous organizations across the North to support and advocate on issues that mattered to Indigenous communities. The approach of Oceans North is heavily informed by its partnerships with Inuit Nunangat communities and the Indigenous organizations and governments that serve them.

3. Grassroots Inuit Advocacy Organizations

The third category of organizations participating in resource governance is, what we are calling, grassroots Inuit advocacy organizations. This category includes two entities: 1) the Digital Indigenous Democracy (DID) project led by Isuma's Zacharias Kunuk, created in response to the Baffinland Iron Mines project on northern Baffin Island; and 2) Makita, led by Joan Scottie and others out of Baker Lake in response to the Kiggavik uranium mine project. Makita's concerns were shaped by the implications of resource development, specifically uranium mining, for the liveli-hoods of people in the region, while DID was a response to the structural and communication deficits of resource development decision-making. Both represent a direct response by Inuit to the state of resource governance in Nunavut.

The Digital Indigenous Democracy (DID) project launched in 2012 during the environmental review process for Baffinland Iron Mines' $6 billion Mary River Project on northern Baffin Island. DID was a direct response by Kunuk, his team, and supporters to concerns that the Baffinland review was not allowing sufficient and effective Inuit par-ticipation in decision-making. Using the internet, community radio, local television, and social media, DID set out to "empower Inuit trad-itional knowledge and create new tools and networks to help voice individual and collective views" (DID 2019). The DID project challenged the decision-making processes and practices, set up by the land claim, and offered Inuit communities an alternative framework through which to "adapt their traditional democratic process of "aajiiqatigiingniq" [deciding together] by providing information in a language they under-stand, facilitating open discussion and helping communities reach col-lective decisions" (DID 2019). Put another way, DID offered an Inuit alternative to decision-making absent the authority over lands and resources that might fully empower Inuit democracy.

In July 2012, Zacharias Kunuk and human rights lawyer Lloyd Lipsett made a formal submission to the Nunavut Impact Review Board (NIRB). Kunuk's submission, called *Ataatama Nunanga, My Father's Land,* used digital media to communicate Inuit concerns about the potential impacts of the Mary River Project on communities, lands, and waters in the region. *Ataatama Nunanga, My Father's Land* dif-fered significantly from conventional NIRB submissions, and had two main parts: (1) a set of written documents including Kunuk's statement in English with Inuktitut and French summaries and two technical analyses provided by two academic partners on matters relating to the potential socio-economic and environmental impacts of the pro-ject; and (2) a video of Kunuk giving his statement in Inuktitut. This was coupled with hours of previously recorded interviews with Inuit Elders and community members; recordings of the radio call-in shows conducted as part of the DID project; and other information that provided context for Kunuk's personal statement. *Ataatama Nunanga, My Father's Land* represents an analysis of the regulatory process itself, tying it to human rights discourse and the potential impacts of the Mary River Project. Through his submission and through the creation of DID itself, Kunuk and his supporters have articulated a vision for how mining companies and regulatory agencies can meaningfully consult and engage with Inuit. This would create the conditions for Inuit to be more actively involved in decision-making about resource development in their homelands.

Nunavummiut Makitagunarningit ("the people of Nunavut can rise up")—or Makita for short—was formed in 2010 in response to AREVA's proposed uranium mine near Baker Lake, Nunavut. It was created by a small number of Nunavummiut including Joan Scottie from Baker Lake and Jack Hicks from then-Iqaluit (both established anti-uranium mining activists) and an emerging set of young Inuit and non-Inuit activists including Sandra Inutiq (now a prominent leader in her own right), Hilu Tagoona, Warren Bernauer, and others. Makita was a response to serious concerns raised by community members and allies about the proposed mine's proximity to the community's water supply and to a sensitive caribou habitat.

One of the group's earliest acts was to submit a petition to the Government of Nunavut(GN) in summer 2010 requesting a public inquiry into the overall issue of uranium mining in the territory. Makita claimed that an inquiry would be more democratic than the regulatory process and would give Nunavummiut a more meaningful opportunity to understand and assess decisions that "were being made without their participation" (Bernauer 2012). A public inquiry was never established but the GN did set up a series of public forums, the background materials for which were publicly criticized by both Makita and MiningWatch Canada for being "factually inaccurate and biased towards industry" (Bernauer 2012). Like the DID project team, Makita expressed deep concern about the lack of information available in Inuktitut, a concern shared publicly with the Baker Lake Hunters and Trappers Organization. Both organizations requested that the review process be halted until the Environmental Impact Statement, which was available only in English, be translated into Inuktitut. NIRB denied these requests (Bernauer 2012).

Like the DID project, Makita was critical of the environmental assessment process, and the extent to which Inuit (and citizens') concerns, knowledge, and interests were included and considered in resource development decision-making. Like Kunuk and his team, Makita's founders saw that Inuit were excluded from decision-making through a lack of access to information needed to adequately assess the scope and potential impacts of the proposed mine, and through a lack of opportunities to participate fully and in meaningful ways.

DISCUSSION

Our review of Nunavut social economy organizations and recent environmental assessment and land-use planning processes in Nunavut has led us to identify two findings that may be symptomatic of the challenges of integrating social economy organizations into resource governance in Nunavut.

First, the social economy organizations participating in these processes represent a small subset of the Nunavut-based social economy organizations discussed above. Broadly speaking, these Nunavut-based organizations—many of which, as we've already pointed out, are focused on social service provision or poverty relief—are working to fill needs created by an economic system in which social inequalities are reproduced or compounded by resource development. Many of these organizations are responding daily to the needs of individuals and families impacted by the industry, as well as to the limits of government and land claim organization service provision. Yet, despite the opportunities presented by a regime that explicitly seeks to integrate environmental and social considerations, none of those organizations with social service, cultural, or educational mandates is actively participating in land-use planning or environmental assessment processes.

Our second finding is that most social economy organizations that have formally intervened in land-use planning or EA between 2008-18 have been the national and international environmental NGOs (Type 1) and the conservation-focused northern NGOs (Type 2). Our desktop review of EA and land-use planning documentation found that a number of these Type 1 and 2 organizations have participated in more than one process. At the same time, we found only two grassroots Inuit advocacy organizations (Type 3) participating during this period, with each intervening in only one environmental assessment process for one specific mining project.

How do we interpret the paucity of Nunavut-based social economy organizations' participation in resource governance? What does it mean for the realization of Inuit democratic empowerment that so few Nunavut-based social economy organizations are participating? To answer these questions fully would require more time and careful consideration than we are able to provide at this stage in our research. Here, our goal is to consider how the embeddedness of resource governance institutions in Nunavut civil society contributes to an explanation.

The literature on northern social economy suggests that there is a way to carve out space for democratic community development to happen within a capitalist economy, through the wide range of social economy organizations that exist in the region. However, Inutiq (2019) explains that many Inuit may choose not to participate in those forms and decide rather, to prioritize their own social relations. This would differ from the ways settler society wants to organize things to address the social and economic issues facing Nunavut communities. If Inutiq's claims are generally true within the territory (and we believe they likely could be), then it would follow that these social divisions, specifically the relative invisibility of Inuit forms of association, would undermine any presumption that conventional social economy organizations advance the moral claims of Inuit civil society. Yet the absence of these organizations from decision-making—including those engaged in social service delivery and education—may also mean that the information they have about their clients' needs, or the impacts of extractive sector development on their ability to deliver services, does not enter directly into deliberations.

One reason why so few Nunavut-based social economy organizations are participating in resource governance may be because Inuit have chosen not to adopt them as vehicles for this purpose. Rather, the choice has been, at least for the most part, to rely on public or land claim institutions to advance their claims. Those institutions include territorial and municipal governments and Inuit land claim organizations. In other cases, however, the choice is to create alternative, though somewhat ephemeral, organizational forms such as Digital Indigenous Democracy and Makita. By their very existence, these latter organizations represent an implied critique of resource governance institutions. They challenge resource development projects while explicitly and implicitly challenging the structures set up by the NLCA. The founders and supporters of each organizational form sought to create alternative ways to express Inuit rights and interests separate from the land claim.

The participation of national or international environmental NGOS and conservation- focused northern NGOS in Nunavut's resource governance regime reflects a differently mediated relationship between Inuit civil society and the formal processes of deliberation and decision-making established under the land claim. As we have discussed, the national or international NGOs are relatively well resourced and accountable to members and donors outside the region for carrying out their conservation mandates. Increasingly, these organizations are working in partnership with local Inuit organizations and community groups, and they take care to align their activities in the region with Inuit concerns and interests. Of these two kinds of organizations (Type 1 and Type 2), it is perhaps the BQCMB that represents the deepest expression of this type of co-operation. Its long-term success bringing Inuit and Western scientific knowledge holders together in a co-management process has gained it a certain legitimacy within the resource management community that other Type 1 and Type 2 organizations may lack on their own. More generally, however, the moral claims made by conventional social economy organizations—whether they are locally, regionally, or nationally organized—may not necessarily reflect Inuit values or priorities. And that is true even if they are conceived locally and even if they are oriented toward social or environmental improvements. This is because of the formally unrecognized character of Inuit social provision and the possibility of Inuit exclusion from the conventional channels of volunteerism within the social economy.

CONCLUSION

As comprehensive land claims agreements have given Indigenous peoples a measure of control over the extractive sector, questions of sustainability necessarily shift to the way democratic control is exercised for the benefit of Indigenous societies. As we have seen, social economy organizations play a role in extractive sector governance in Nunavut, taking advantage of participatory processes such as land-use planning and environmental assessment to advance their mandates and shape the outcomes of extractive sector decision-making. But we have also found that these organizations represent only a small subset of Nunavut-based social economy organizations. As well, only a fraction of those that do participate are either locally based or are the direct expressions of Inuit forms of association. These facts lead us to two tentative suggestions—one pragmatic and one conceptual—that might contribute to enhanced community control through democratic means and at the same time help to shape priorities for future research in this area.

First, communities would benefit from long-term support aimed at strengthening the role of civil society in land and resource governance. It is essential to build local institutions through which Inuit forms of association might lead to more varied expressions of rights and interests vis-à-vis the extractive sector. This would not only create greater opportunities for knowledge sharing within the existing regime, it might also help planners, policy-makers, and project developers anticipate and avoid issues farther in advance of key decisions. It would also help Inuit communities exercise their own forms of authority over lands and resources in ways perhaps not contemplated by the land claim. Where this support should go and how it might be accessed is a quandary though, especially when Inuit forms of association and opinion formation are repressed or likely to be rendered invisible. Solutions must be imagined carefully, perhaps drawing on the experience of or building directly upon earlier attempts to organize, such as through DID and Makita.

Government programs aimed at supporting Indigenous participation in resource governance outside the context of specific project reviews may be part of the answer. Of course, that assumes that funding is available to those who need it, and it is of sufficient quantity and duration for meaningful institution building to occur. Another partial answer may lie in finding ways to adjust CRA's charitable status rules to make it easier for these new institutions to benefit from them.

Whatever practical solutions materialize, they must flow from the recognition that institutional change in Nunavut is far from complete. In the wake of the NLCA and the establishment of Nunavut as a territory, Inuit forms of authority, association, and opinion formation remain the foundation of a political terrain that continues to seek expression in land and resource governance. If it is here that the democratic potential of resource governance takes root, researchers and policymakers would do well to continue learning its contours to understand how sustainability may be achieved.

Table 11.3
List of Registered Charities in Nunavut (October 18, 2018)

Name of Registered Charity	City	Effective Date
Alianait Entertainment Group	Iqaluit	2010-04-01
All Saints Anglican Church	Arctic Bay	1973-01-01
All Saints Anglican Parish	Arctic Bay	1996-01-01
Arctic Children and Youth Foundation	Iqaluit	2005-01-01
Arctic for Christ Ministries	Iqaluit	2012-04-05
Arviat Alliance Church	Arviat	2010-06-15
Cambridge Bay Childcare Society	Cambridge Bay	1996-03-24
Church of the Holy Comforter	Arctic Bay	1993-01-01
Habitat for Humanity Iqaluit	Iqaluit	2006-01-01
Holy Trinity Anglican Church	Arctic Bay	1967-01-01
Ilisaqsivik Society	Clyde River	1999-04-01
Iqaluit Christian Fellowship	Iqaluit	2018-03-13
Iqaluit Congregation of Jehovah's Witnesses	Iqaluit	2010-12-16
Iqaluit District Education Authority	Iqaluit	2005-03-03
Iqaluit Humane Society	Iqaluit	2013-04-19
Iqaluit Pentecostal Church	Iqaluit	1989-01-01
Isaksimagit Inuusirmi Katujjiqatigiit Embrace Life Council	Iqaluit	2006-04-01
Kitikmeot Heritage Society	Cambridge Bay	1996-03-06
Kivalliq Science Educators Community	Rankin Inlet	1995-03-10
Kuut Uumanaqtut Tuksiarviit Fellowship	Iqaluit	2011-01-14
Niqinik Nuatsivik Nunavut Food Bank	Iqaluit	2004-05-17
Northern Lighthouse Ministries	Iqaluit	2008-08-15
Nunavut Arctic College	Iqaluit	1995-01-01
Nunavut Elders' Pension Trust	Rankin Inlet	1994-10-15
Nunavut Kamatsiaqtut Help Line	Iqaluit	2000-11-02
Nunavut Literacy Council	Rankin Inlet	1999-07-22
Nunavut Harvesters Support Program Incorporated	Iqaluit	2016-07-14
Nunavut Wildlife Research Trust	Iqaluit	1996-04-01
Pond Inlet Ukpirtuuqatigiit	Pond Inlet	2018-05-04
Pulaarvik Kablu Friendship Centre	Rankin Inlet	1993-08-31
Qaujisaqtiit Society	Iqaluit	2004-10-08
Qayuqtuvik Society (Food Bank)	Iqaluit	2016-09-21
SKYE (Science for Kitikmeot Youth and Educators)	Kugaaruk	2008-12-12
St. Andrew's Anglican Parish	Arctic Bay	1967-01-01
St. David's Church Anglican Church of Canada	Arctic Bay	1967-01-01
St. George's Anglican Church	Arctic Bay	1994-01-01
St. John's Anglican Church	Arctic Bay	1967-01-01
The Iqaluit Music Society - Iqalunninihhausijarnimut katujjijit	Iqaluit	1999-01-01

The Nunavut National Parks Scholarship Trust	Iqaluit	2000-01-01
The Parish of St. Simon and St. Jude	Arctic Bay	1967-01-01
The Pond Inlet Library and Archives Society	Pond Inlet	2007-07-01
The Rotary Club of Iqaluit Charitable Society	Iqaluit	2007-07-01
YWCA Agvvik Nunavut	Iqaluit	1993-02-14

Source: Canada Revenue Agency Charities Database. https://apps.cra-arc.gc.ca/ebci/hacc/srch/pub/dsplyBscSrch?request_locale=en

REFERENCES

Abele, Frances. 1983. *The Berger Inquiry and the Politics of Transformation in the Mackenzie Valley.* PhD dissertation, University of Alberta.
–2009. "The State and the Northern Social Economy: Research Prospects." *The Northern Review* 30 (Spring): 37-56.
Abele, Frances and Chris Southcott, eds. 2016. *Care, Cooperation and Activism in Canada's Northern Social Economy.* Edmonton: University of Alberta Press.
Amagoalik, John. 2007. *Changing the face of Canada: The Life Story of John Amagoalik.* Iqaluit: Nunavut Arctic College.
Bankes, Nigel, and Cheryl Sharvit. 1998. *Aboriginal Title and Free Entry Mining Regimes in Northern Canada.* Prepared for the Canadian Arctic Resources Committee.
Berger, Thomas. 2006. Conciliator's Final Report: "The Nunavut Project." https://www.tunngavik.com/documents/publications/2006- 0301%20Thomas%20Berger%20Final%20Report%20ENG.pdf.
Bernauer, Warren. 2012. "Uranium Controversy in Baker Lake." *Canadian Dimension.* 3 February. https://canadiandimension.com/articles/view/the-uranium controversy-in-baker-lake.
Brody, Hugh. 1975. *The People's Land: Eskimos and Whites in the Eastern Arctic.* Penguin Books.
Bryant, Raymond L. and Sinéad Bailey. 1997. *Third World Political Ecology.* London: Routledge.
Cohn, Norman and Zacharias Kunuk. 2012. "Our Baffinland: Digital Indigenous Democracy." *Northern Public Affairs Magazine* 1, no.1 (Spring): 50–52.
–2014. *My Father's Land (Ataatama Nunanga).* Documentary film. Kingulliit Productions. http://www.isuma.tv/myfathersland/film.
Diduck, A., and A. John Sinclair. 2002. "Public Involvement in Environmental Assessment: The Case of the Nonparticipant." *Environmental Management* 29, no. 4: 578-588.

Fitzpatrick, P., A. John Sinclair, and B. Mitchell. 2008. "Environmental Impact Assessment Under the Mackenzie Valley Resource Management Act: Deliberative Democracy in Canada's North?" *Environmental Management* 42, no. 1: 1-18.

Galbraith, Lindsay, Ben Bradshaw, and Murray B. Rutherford. 2007. "Towards a New Supraregulatory Approach to Environmental Assessment in Northern Canada." *Impact Assessment and Project Appraisal* 25, no. 1 (March): 27-41.

Gibson, Robert. 2002. "From Wreck Cove to Voisey's Bay: The Evolution of Federal Environmental Assessment in Canada." *Impact Assessment and Project Appraisal* 20, no. 3 (September): 151-159.

–2012. "In Full Retreat: The Canadian Government's New Environmental Assessment Law Undoes Decades of Progress." *Impact Assessment and Project Appraisal* 30, no. 3: 179-188.

Inutiq, Sandra. 2019. "Dear Qallunaat: Racism, Public Government and Inuit Nunanga." *Yellowhead Institute*, 7 Feb 2019. https://yellowheadin-stitute.org/2019/02/07/dear-qallunaat/.

IsumaTV. 2019. *Digital Indigenous Democracy*. Retrieved from http://www.isuma.tv/isuma/did.

Johnston, Aviaq. 2013. "Society: My Mother Tongue." *Northern Public Affairs Magazine* 1, no. 3 (Spring): 42 44. Retrieved from: http://www.northernpublicaffairs.ca/index/wp content/uploads/2016/02/Johnston.pdf.

Kennedy Dalseg, Sheena. 2018. *Seeing Like a Community: Education, Citizenship, and Social Change in the Eastern Arctic*. PhD dissertation, Carleton University.

Kunuk, Zacharias. "My Father's Land: Ataatama Nunanga." Submission to the Nunavut Impact Review Board Mary River hearings, June 8, 2012. Retrieved from: http://www.isuma.tv/en/did/zacharias-kunuk-for-mal-intervention-to-nirb-written submission-june-8-2012.

Kuokannen, Rauna. 2011. "Indigenous Economies, Theories of Subsistence, and Women: Exploring the Social Economy Model for Indigenous Governance." *American Indian Quarterly* 35, no. 2: 215-240.

Linders, Sarah. 2018. "Recognizing Inuit Governance in Canada." *Queen's Gazette*. 9 March 2018. https://www.queensu.ca/gazette/stories/recognizing-inuit-governance-canada.

Lyall, William. 2013. *Helping Ourselves by Helping Each Other: The Life Story of William Lyall*. Iqaluit: Nunavut Arctic College.

MacDonald, John and Nancy Wachowich, eds. 2019. *The Hands' Measure: Essays Honouring Leah Aksaajuq Otak's Contribution to Arctic Science*. Iqaluit: NAC Media.

Martin, Ian. 2017. *Inuit Language Loss in Nunavut: Analysis, Forecast, and Recommendations*. Submission to the Government of Nunavut.

7 March 2017. https://bill37.tunngavik.com/files/2017/03/Inuit-Language-Loss in-Nunavut-Martin-status-report- Mar-7-2017-v3.pdf.

McPherson, Robert. 2003. *New Owners in Their Own Land: Minerals and Inuit Land Claims*. Calgary: University of Calgary Press.

Merritt, John. 1985. "In Search of Common Ground: Ottawa Rethinks Its Approach to Comprehensive Claims." *Northern Perspectives* 15, no. 1.

Natcher, David. 2009. "Subsistence and the Social Economy of Canada's Aboriginal North." *The Northern Review* 30 (Spring): 83-98.

Nunavut Land Claims Agreement Act, SC 1993, c 29.

Nunavut Research Institute. 2009. https://www.nri.nu.ca/sites/default/files/public/files/nunavutsocialeconomy eng.pdf.

O'Faircheallaigh, Ciaran. 2010. "Public Participation and Environmental Impact Assessment: Purposes, Implications, and Lessons for Public Policy Making." *Environmental Impact Assessment Review* 30, no. 1: 19-27.

Reimer, Bill. 2006. "The Informal Economy in Non-Metropolitan Canada." *Canadian Review of Sociology* 43, no. 1: 23-49.

Scholtz, Christa. 2006. *Negotiating Claims: The Emergence of Indigenous Land Claim Negotiation Policies in Australia, Canada, New Zealand, and the United States*. New York: Routledge.

Southcott, Chris, ed. 2015. *Northern Communities Working Together: The Social Economy of Canada's North*. Toronto: University of Toronto Press.

Southcott, Chris, Frances Abele, Dave Natcher, and Brenda Parlee. 2018. "Beyond the Berger Inquiry: Can Extractive Resource Development Help the Sustainability of Canada's Arctic Communities?" *Arctic* 71, no. 4 (December): 393-406.

Southcott, Chris, Valoree Walker, Jennifer Wilman, Carrie Spavor and Karen MacKenzie. 2010. *The Social Economy and Nunavut: Barriers and Opportunities*. Social Economy Research Network of Northern Canada. Research Report Series RR#1-2010.

Tapardjuk, Louis. 2013. *Fighting for Our Rights: The Life Story of Louis Tapardjuk*. Iqaluit: Nunavut Arctic College.

Tunngavik Federation of Nunavut. 1989. "An Inuit Response." *Northern Perspectives* 17, no. 1.

Usher, Peter. 1996. *Contemporary Aboriginal Land, Resource, and Environment Regimes: Origins, Problems, and Prospects*. Report prepared for the Royal Commission on Aboriginal Peoples.

Wiklund, Hans. 2005. "In Search of Arenas for Democratic Deliberation: a Habermasian Review of Environmental Assessment." *Impact Assessment and Project Appraisal* 23, no. 4: 281-292.

Wood, Ellen Meiksins. 1995. *Democracy Against Capitalism: Renewing Historical Materialism*. New York: Cambridge University Press.

12

Impacts of Mining on Well-Being: A Disconnect between Theory and Practice

Brenda Parlee

The relationship between well-being of communities and large-scale development has been a focus of research in numerous disciplines over the last two centuries. Since the Industrial Revolution, sociologists (as early as Durkheim) have attempted to understand and address the complex social, economic, cultural, and political factors that affect communities, including one we examine here: are communities better or worse off when mining comes to town (Ashton and Sykes 1964; Gramling and Freudenburg 1992; Stedman, Parkins, and Beckley 2004)? Inquiry into this complex relationship has grown significantly in recent decades as mineral resource development has expanded into almost every region of the world, including the Arctic. This trend is likely to continue as a result of rising demand for raw materials, which leads to innovation in mining technologies as well as sustainable mining policy development (Bridge 2004; Clark and Clark 1999; Dubiński 2013; Duhaime and Caron 2006; Hilson and Banchirigah 2009). There is much global variation in how, and to what extent, social impacts are considered in decisions about large-scale mining projects; the question of whose values and whose knowledge is represented in these decisions is of interest in this chapter. More specifically, we look at what extent the values of local and Indigenous communities are considered in social impact assessments.

Overall, Indigenous peoples have benefitted little and born the greatest burden of adverse effects from mining, such as contamination of water sources and loss of access to lands and species valued as traditional food (Ballard and Banks 2003; Gedicks 1994; Howitt

2001; Keeling et al., n.d.; O'Faircheallaigh and Ali 2017). As the drive to produce raw materials grows globally, so too do the adverse impacts for Indigenous Peoples. In the present rush to develop the Amazon in Brazil, for example, there is a tendency for some governments and industry leaders to view questions of Indigenous values and traditional uses of the land (e.g., hunting, fishing) as an impediment to development (Hanna et al. 2014; Triner 2015). In parts of Asia, Africa, and Latin America, human rights abuses experienced by Indigenous peoples have drawn international attention. Questions have been raised about the lack of social responsibility demonstrated by many mining corporations, a good deal of which are based in Canada (Gordon and Webber 2008; Helwege 2015).

In Canada, the human costs of mining are seemingly better accounted for through processes and institutions of social impact assessments. That is a dimension of environmental assessment that: "includes the processes of analysing, monitoring and managing the intended and unintended social consequences, both positive and negative, of planned interventions (policies, programs, plans, projects) and any social change processes invoked by those interventions. The primary purpose is to bring about a more sustainable and equitable biophysical and human environment" (Vanclay 2006). But despite significant advances in some jurisdictions, social impact assessment is still a relatively weak tool when compared to other streams of environmental assessment. Many aspects of these assessments lack rigour, and their legitimacy is often questioned, particularly among Indigenous communities (Howitt 2011; B. Noble and Bronson 2006; O'Faircheallaigh 2011; Stevenson 1996; Suopajärvi 2013; Urkidi and Walter 2011).

Communities in Canada's North need to know whether they have the power to better control extractive industry impacts for their benefit. This cannot be done without a better understanding of the social impacts of these activities on the well-being of these communities. Previous research and analysis suggest that a major barrier to improving social impact assessment is lack of engagement with relevant disciplinary and community expertise. Many assessments are ad hoc, based around a narrow range of conventional indicators, and are weak in implementation (i.e., poor methods), resulting in ambiguous determination of impacts (Esteves, Franks, and Vanclay 2012; M. Lane, Dale, and Taylor 2001; Little and Krannich 1988; Rickson et al. 1990). Building off this observation, this chapter explores the ways in which well-being is defined in theory and how it is used in the practice of social impact assessment. A review of selected impact assessments of mineral resource development and the well-being indicators used in previous assessments are also shared. An example of a community-led project from Łutsël K'é Dene First Nation provides an alternative to conventional and top-down approaches. While there is some variation, the majority of impact assessments reviewed for this chapter considered only a narrow range of general indictors, with little consideration of the deeper theoretical dimensions of well-being explored in the social science literature.

CANADIAN WELL-BEING FRAMEWORKS

There is a long interest in the study of well-being dating back over a millennia; and long before that Aristotle explored the concept of well-being or "eudaimonia," which is the experience of living well or actualizing one's human potential (Deci and Ryan 2008). Well-being has many different meanings across cultures, including those in the Canadian North. Family, living on the land, spirituality, and leadership are among the dimensions of well-being important to Arctic peoples (Kirmayer, Fletcher, and Watt 2009; Kral et al. 2011; Parlee and Furgal 2012; Parlee, O'Neil, and Łutsël K'é Dene First Nation 2007; Poppel 2015).

Those involved in the study of well-being can be divided into quantitative or qualitative camps. While quantitative approaches may provide a more structured understanding of well- being that is more easily compared and contrasted, qualitative studies tend to use a more holistic lens to understanding the nuanced and normative relationships, and often delve into the broader context. As shown in Chapters 13 and 14 of this volume, there are some approaches that use mixed methods to combine the opportunities and strengths of quantitative and qualitative study. At the heart of the qualitative lens is recognition that what is well-being for one person is not necessarily true for others, and that this is particularly true across cultures.

Measures of well-being have also been theorized and utilized in national-level surveys and data management initiatives in Canada (Canadian Index of Wellbeing 2016; Diener 2000; Helliwell 2003). Depending on the indicators used, however, the outcomes and the narrative created around the data can be very different. The Government of Canada's *Community Well-Being Index* (cwb) is based on four indicators: education, labour force activity, income, and housing. This index suggests that well-being is improving nationally (Statistics Canada 2016). But these indicators are limited in scope and mirror narrow kinds of mainstream economic and material values; indeed, they resemble many of the kinds of indicators of development that have been used since the post-WWII era. While any effort to understand the landscape of well-being and how it is changing may be seen as important, the re-packaging of conventional indicators under the banner of "well-being" here (as in other frameworks developed by the OECD) does seem an appropriation of a concept that had/has other dimensions and meanings to local people. The *Canadian Index of Wellbeing* (CIW), created and housed at the University of Waterloo, tells a different story from the CWB of Statistics Canada. Using a broader range of indicators such as mental health, leisure time, public engagement, income equality, and others, the CIW shows well-being has declined over the last ten years relative to GDP growth (Canadian Index of Wellbeing 2016).

Caution is therefore needed where the definition of indicators in social impact assessment is driven by governments, project proponents, and consultants, who are tasked with addressing the question of how project x will affect community y. The larger phenomenological questions of what is meant by well-being and how can it be meaningfully measured are of lesser consideration.

INDIGENOUS PEOPLES AND WELL-BEING

Among those most disadvantaged in the use and misuse of statistics on well-being are Indigenous peoples who, through the comparative tracking with other populations, tend to be defined and measured as unwell. As evidenced in reviews and studies in health and sociology, narratives provided about Indigenous peoples from statistical datasets lean towards being pathological in nature. For example, studies measure things such as number of children in care, unemployment rate, incidence of violent crime, and HIV and STD infection rates. These are often used in assessments in Canada. While these indicators can be useful in determining the degree of vulnerability of a particular demographic, they can also create and perpetuate stereotypes and narratives of deviance and deficiency. Such narratives serve to perpetuate a certain kind of power dynamic that disadvantages Indigenous peoples. "In other words, an image of sick, disorganized communities can be used to justify paternalism and dependency (O'Neil et al. 1998).

A case in point is the descriptive statistics used to depict Inuit women and families in the social impact assessment of the Voisey's Bay nickel mine (Archibald and Crnkovich 1999). Some aspects of these inequities in power relations are performative. For example, the presentation of huge binders of material in assessment hearings is a performance, a demonstration of power that carries meaning even if the data within those binders has little value. Communities that lack their own data or fail to offer similar kinds of performances within the context of a hearing may be considered to have less knowledge and consequently less authority in the same forum. Given the fact that very few social impact assessments involve Indigenous communities and few lead to conclusions of adverse effects, it is not surprising that Indigenous scholars and others have described the enterprise as yet another colonial tool of surveillance aimed at land and resource dispossession and socio-cultural assimilation (Kukutai and Walter 2015; Andersen 2013; O'Neil et al 1998). Given these critiques, it is no surprise that there is ever greater emphasis on community participation or engagement in many aspects of social impact assessment. These efforts include the development of terms of reference, values to be addressed, as well as indicators and metrics of assessment.

THEMES OF COMMUNITY WELL-BEING AND IMPACTS OF MINING: THE VIEW FROM THE LITERATURE

There are multiple domains or thematic areas of community well-being that have been discussed and theorized as affected by mining resource development. These are discussed here in the categories of economy, social cohesion, cultural continuity, and the environment.

Economy and Well-Being

The interactions between well-being and the economic inputs of large-scale mining projects have been researched extensively. A large body of work considers these economic opportunities positively, seeing how large-scale development projects can potentially launch communities into greater economic prosperity and improve quality of life.

The belief in the possibility of economic opportunities stemming from mining began in the post-WWII era, but many scholars have also pointed to negative aspects that may arise, such as the resource curse, staples trap, and dependency theory (Auty 2002; Bebbington et al. 2008). Despite these critiques, pursuit of the economic opportunities of large-scale mineral resource development persists globally and has tended to survive the "fault lines of political upheaval, such as decolonization, the embrace of autarkic models of development, transitions from authoritarianism to formal democracy or the imposition of structural adjustment" (Bridge 2004). Indigenous communities are particularly limited in terms of the economic opportunities that accrue from mining. This has been discussed by the World Bank as well as other organizations linked to the mining industry. "It is widely acknowledged that, because of these processes, the benefits of 'development' projects tended to flow away from Indigenous communities affected by those projects and instead towards provincial or national centres ... studies indicated that these projects not only did not benefit Indigenous groups, but they also often worked to increase poverty and cultural disintegration" (Render 2005).

There are, however, other kinds of economic benefits and human capital development that can, but do not always, occur. In a study of seventy-one regions of Australia, mining activities were shown to lead to increased educational attainment as well as a variety of other markers of economic benefit (Hajkowicz et al. 2011; Hajkowicz, Heyenga, and Moffat 2011). In rural areas of Australia in particular, mining production was "significantly correlated with improved income, housing affordability, communication access, educational levels, and employment" (Hajkowicz, Heyenga, and Moffat 2011). However, many more studies suggest mining has a negative impact on educational attainment, in that natural resource–based industries often rely on low-skilled labour and higher wages, which can lure young people away from school, creating disincentives for education and training (Parlee et al. 2018; Gylfason, 2001).

Economic opportunities, including employment, business development, and rent sharing, are either framed as contributing to or affecting the well-being of local communities. In regions of high unemployment and poverty, mining is often proposed as an "easy-fix." While small-scale local mining activity has economic benefits (although significant environmental risks), large-scale mining has a very poor empirical track record in poverty reduction. A study of communities associated with 110 mining projects in the 1990s showed that they experienced little economic growth, sustained employment,

or improved living conditions (Pegg 2006); this finding aligns with
other research on mining and the resource curse, where mineral
resource abundance and large-scale development is negatively correlated with economic growth (Papyrakis 2017).

There are two primary reasons for these limiting economic benefits.
The capital intensive nature of large-scale mining means the number
of jobs created relative to the revenues generated is quite small
(Horsley et al. 2015). In many regions, there are also numerous deter-
rents to local people gaining employment including cultural and
language barriers, distances, and lack of local mining knowledge and
skills (Brereton and Parmenter 2008; Rodon and Lévesque 2015). In
many cases, high paying jobs go to highly educated and trained outsiders. Another reason relates to the illusion of employment opportunities. Although there can be high paying jobs in the mining sector,
they are few. High paying positions, however, can increase the cost of
living (e.g., drive up prices of housing) and thus exacerbate rates of
poverty, unemployment, and welfare dependence (Lawrie, Tonts, and
Plummer 2011).

Social Cohesion and Well-Being

There have been numerous studies focused on the ways in which the
social fabric of communities is disrupted by large-scale mining. The
social disruption hypothesis, which dates back to Durkheim and his
ideas on anomie in the industrial society, is increasingly linked to
research on psycho-social stress, place-based identity, and attachment (Jacquet and Stedman 2014). The impacts of mining on the
social fabric of communities can be characterized along four lines.
The first is the instability of global markets for mineral exports,
which often creates uncertainties of local employment (Freudenburg
1992). Boom town studies, for example, indicate costs that include
social dislocation and disruption (e.g., marital breakdown, mental
health issues, suicide, and drug and alcohol abuse) (Smith, Krannich,
and Hunter 2001). The second way that the social fabric of communities is impacted relates to rotational work (such as fifo two-
week rotation schedules), and the influx of labour from outside the
region (e.g., migrant labour). This type of work can also break down
the social networks and norms important to well-being. Thirdly,
communities with weakened social relations may be less likely to
control or mediate deviant behaviour, particularly of outsiders.

They also have a diminished capacity to care for vulnerable members of their community, and are generally less likely to know how to care for one another (Freudenburg 1986). Finally, changes in infrastructure important to the social fabric of the community can also occur during the course of a mining boom. However, in addition to the adverse impacts noted above, (e.g., housing shortages), there can be positive changes from dollars earned through employment, business development, and corporate investment in needed community infrastructure (i.e., donations and sponsorships for a recreational facility, road construction).

Cultural Continuity and Well-Being

Cultural continuity is another key dimension of well-being that is impacted by mining activity. It is generally defined as the transmission of the meanings and values centrally characteristic to a culture over time and across generations. Many rural and Indigenous communities have cultural knowledge, practices, beliefs, and languages that are unique and quite different from the mainstream. There, the imposition of a large-scale resource development project can lead to significant changes in many aspects of their well-being. For individuals in some communities already suffering from discontinuity of self, community, and culture due to histories and sociopolitical marginalization, the discontinuity created by the imposition of a short-term boom in mining can be more catastrophic than in other kinds of cultural contexts (Chandler and Lalonde 1998). A root concern is that Indigenous communities have lost connection to the land, customs, culture, modes of self-governance, languages, and ways of life due to policies of colonialism (e.g., residential schools). Some studies evidence how recognition of cultural difference is protective of health and a precursor to sustainable development.

Cultural continuity, for example, is increasingly being recognized as protective of well-being, as indicated by research on many kinds of social illnesses and issues including alcohol addiction and youth suicide (Chandler and Lalonde 1998; Kirmayer et al. 1998; Currie et al. 2011). The notion and evidence that culture is "protective" and important to sustain in the context of development (rather than a problem to be addressed) is a significant departure from historic and conventional frames of reference both in academia and in business practice. This shift is evidenced in Canada in increasing respect for Indigenous peoples, but it is also represented in policy by the World Bank and other global institutions (Duer and Levine 1999). The shift is from viewing culture as the problem to considering it as a solution. This is attributed to a variety of socio-political shifts, as well as the embrace of more pluralistic approaches to the study and treatment of illness and community well-being.

Environment and Well-Being

The interconnections between environment and well-being are perhaps the most clearly defined in the literature. The relationship between environmental losses and degradation and well-being are particularly acute in Indigenous communities. There can be myriad kinds of economic, health, cultural, and spiritual impacts that are interconnected with environmental change and disturbance of ecosystems. A great deal of research related to indirect impacts of mining on well-being deal with issues of land use conflict—when land is being taken up for mining, it is no longer accessible for other uses (G. Hilson 2002). Traditional economy, loss of opportunities, or less access to land can lead to declines in harvesting of wild foods with consequent impacts on food security (Parlee, Sandlos, and Natcher 2018). In the Arctic and other northern regions where access to other food resources is limited, such losses to land and opportunities for harvesting can be particularly devastating. Related cultural and spiritual practices can also be fundamentally affected (Gibson and Klinck 2005; M.B. Lane et al. 2003; Piotrzkowski 2016; Ross 2001). The depth of loss to individuals and communities is significant.

Another important theme of research linking environmental disturbance from mining and well-being relates to the loss of place and attachment to place. As with other kinds of development activity, mining can lead to physical displacement of people from places of value to them (Frantál 2016). Even if the places still exist, they may be transformed in aesthetic, so impacts can be significant. The term "solastalgia" was developed and used to describe the emotional impacts of strip mining in the Hunter Valley of Australia on local communities (Albrecht et al. 2007). It can also be a kind of homesickness for a home that no longer exists.

WELL-BEING AND SOCIAL IMPACT ASSESSMENT

Well-being has been included in a variety of ways in environmental assessments. Although some aspects of well-being (discussed above) are included as determinants in health impact assessments (Bronson and Noble 2006; Noble and Bronson 2005), well-being is considered an endpoint or an outcome in social impact assessment.

Social impact assessment (SIA) is a decision-making tool which is evidenced in a variety of jurisdictions globally, including the Canadian Arctic. It is considered a critical dimension of planning and assessment of large-scale development projects by numerous institutions such as the World Bank (Esteves, Franks, and Vanclay 2012; Vanclay 2003). Despite efforts to standardize and develop rigorous guidelines (US Department of Commerce, National Oceanic and Atmospheric Administration, and National Marine Fisheries Service 1995), SIA remains relatively weak, particularly in respect of community engagement; SIA "experts," like those involved in other dimensions of environmental assessment, have not valued the knowledge and experience of local peoples (D.R. Becker et al. 2003; Vanclay 2006). Critical discussion is needed about how it might be improved and more effectively harnessed, particularly by communities most impacted by large-scale development such as mining (D.R. Becker et al. 2003).

The subjective qualities inherent in the definition and choice of indicators of well-being are a major issue in SIA. It is a common concern that the kinds of indicators chosen tend towards those that would rationalize and justify approvals of development projects. Indeed, many social impact assessments result in conclusions of "no adverse social impacts" or ambiguous and unsubstantiated conclusions of impacts (Esteves, Franks, and Vanclay 2012; Lockie 2001; Mancini and Sala 2018; Noble and Bronson 2006; Noble and Bronson 2005; Suopajärvi 2013).

This problem of scope is compounded by biases in process. Although there are some notable exceptions, social impact assessment is very often framed as something done *to* communities under an illusion of objectivity, rather than a process of community engagement, social learning, and discursive decision-making (Webler et al. 1995; Vanclay 2006). Within such a framework, communities are more often treated as objects of research, rather than knowledgeable participants or citizens with real expertise about their own lives (Suopajärvi 2013). Few social impact assessments are initiated, led, or implemented by communities themselves. Instead, they are proponent-driven (Arce-Gomez, Donovan, and Bedggood 2015; Mayoux and Chambers 2005). This has created inequities and perversions not only in the kinds of indicators chosen and the process undertaken, but also in ownership and access to data. For example, data about community health conditions collected by consultants are the property of the proponent or contracted consultants.

The politics involved in social impact assessment include the question, "whose knowledge matters?" This can compound what might otherwise be considered technical or cultural barriers to community participation. On the grounds of objectivity, oral histories, stories, and personal statements shared by average citizens are commonly dismissed as anecdotal or inconsequential. This pattern has led scholars to question, "whose definition of an impact, an aspiration, a value and a fact is considered legitimate and whose is dismissed as subjective, emotional and irrelevant" (Lockie 2001)? "Technocratic rationality is often favoured by professionals who are trained in engineering or the natural sciences and are uncomfortable with, or sceptical about, the involvement of what they regard to be an ill-informed public" (Lockie 2001). It is in this context that this chapter begins an exploration of the meanings of well-being and how those meanings translate into conceptual frameworks and indicators of measurement.

WELL-BEING IN SOCIAL IMPACT ASSESSMENT: A VIEW FROM PRACTITIONERS

The ways in which well-being has been operationalized in assessments varies significantly by region, the specifics of the project being assessed, as well as by the terms of reference of assessment agencies and their experts. Examples of the indicators used in these assessments are found in table 12.1.

Many assessments use standardized indicators as the basis for measuring well-being. For example, employment, business development, housing, and related infrastructure are commonly highlighted. As well, skill development, training, and education are also highlighted as potential benefits for communities. The degree to which predicted benefits are achieved however varies significantly by region and jurisdiction; where governments set terms and conditions for local hiring and investment in local training, the economic opportunities have tended to be more significant and sustainable.

Table 12.1
Examples of well-being indicators in social impact assessments

Project	Examples of Indicators
Ranger Uranium Mine, Alligator Rivers Region The Alligator Rivers Region (ARR) is located about 220 km east of Darwin in the Northern Territory of Australia. The Ranger Uranium Mine and the associated town of Jabiru is about 230 kilometres east of Darwin, in the Northern Territory, surrounded by the Kakadu National Park. The mine is located on Aboriginal land as defined under the Aboriginal Land Rights Act (1976). The ability of people to maintain their health and a lifestyle that is not detrimental to their well-being (e.g., nutrition and diet, physical and mental health, alcohol and substance abuse) Issues of concern related to uranium mining have been explored in reports published by the Parliament of Australia (Wilson 1997)	• The quality of water resources • Safety and hazard exposure • Substance abuse and related health issues • Health service capacity and viability
Goa Iron Ore Mining, Goa, India Ecosystem approach recognizing that human well-being is a function of both the biophysical and the social domain. A Quality of Life Index was created for assessing and monitoring mining impacts by international organizations (i.e., IDRC). There is recognition of the importance of social cohesion, good governance, participation, and command over goods and services (Conway 2011).	• Impacts on agriculture • Job availability • Alcoholism • Influx of migrant labour • Land access, compensation • Environmental quality • Post closure access • Public health • Effectiveness of administrative systems • Level of individual/ community satisfaction • Perception of impacts on constituents of wellbeing (i.e., health, freedom, and activity) (Noronha 2001) https://www.idrc.ca/en/article/case-study-india-tracking-health-and-well-being-goas-mining-belt

Kittilä Mine Gold Mine

Lapland, Finland

Kittilä Mine, also known as Suurikuusikko Mine, is a gold mine in Kittilä, in the Lapland province of Finland. The mine is owned and operated by Agnico Eagle Mines Limited and is located 36 kilometres (22 mi) north-east of Kittilä. It is the largest gold mine in Europe and reportedly employs 400 people, many of whom live in Kittilä and the province of Lapland (Jantunen et al. 2015).

Well-being may be regarded as an umbrella concept; living conditions and comfort can be seen as factors that create well-being. (Suopajärvi 2013)

- Human health
- Employment
- Distribution of well-being
- Land use
- Cultural heritage
- Infrastructure

McArthur River Uranium Mine

McArthur River, Northern Saskatchewan

The McArthur River Uranium Mine is located 620 air kilometres north of Saskatoon, Saskatchewan and 80 kilometres northeast of the Key Lake mill in the uranium rich Athabasca Basin. Mine construction began in 1997, with production commencing in 1999. The mine achieved full commercial production in November 2000.

Well-being is considered in the assessment according to the WHO definition and as synonymous with quality of life; emphasis is on physical health issues; risks of exposure to radiation by direct and indirect ingestion of contaminated water from sites used for hunting and other land use activities.

- Exposure to radiation
- Cancer risks and mortality
- Employment
- Income
- Education
- Housing
- Lifestyle
- Environment
- Traditional land-use activities

Voisey's Bay Nickel Mine

Voisey's Bay nickel-copper-cobalt mine is located in northern Labrador/Newfoundland, Canada, 35 kilometres from the Indigenous Innu community of Nain. The project includes a mining, processing, and shipping component a via port facilities in nearby Anaktalak Bay. Construction started in 2009 and operations began in 2014.

Project proponents made an explicit promise to health and quality of life issues were integrated into the assessment under the banner of sustainability, with consideration given to the traditional economies, cultural practices, and land uses of Inuit and Innu peoples of the region.

- Demographics
- Employment
- Income
- Education and skills
- Use of land (including water and ice) and resources, including fish and wildlife harvesting
- Housing
- Quality of life
- Health
- Morbidity and mortality
- Diet, including country food
- The interrelations of all of the indicators listed above

Kvanefjeld Uranium Project

The project area is in South Greenland, approximately 10 km from Narsaq and approximately 35 km from Narsarsuaq. The main commodities of interest in the Kvanefjeld ore body are rare earth elements (REES). There are also sufficient levels of uranium and zinc in the ore body to produce commercially viable by-products. The project includes the development of an open pit mine, a processing plant, a port, mine accommodation, tailings facility, and roads connecting the parts of the project (Bureau of Minerals and Petroleum 2009).

Health is a state of complete physical, mental, social, and spiritual well-being, and not merely the absence of disease or infirmity.

- Prevalence and rates of infectious and chronic diseases
- Trends in existing health problems
- Health knowledge, practices and attitudes
- Health and social care services
- Nutrition pattern
- Existing levels of environmental pollution and natural level of radiation
- Housing conditions
- Social problems such as drug use and suicide
- Literacy rates and levels of education
- Employment and unemployment rates
- Existing community concerns and aspirations

Northwest Territories Diamond Mining (Ekati, Diavik, Gahcho Kué)

Diamond Mining

A range of similar indicators were developed for the assessment and monitoring of three diamond mines (Ekati, Diavik and Gahcho Kué).

The studies use conventional measures of employment and targets of hiring northerners and Indigenous persons as well as employment benefits (e.g., apprenticeships and training opportunities), business investment, purchasing statistics, capital operating … "[T]he reports point to economic growth as evidence of progress but fail to situate economic growth within the broader context of community well-being, which is bothersome since these same reports reveal limited connection between economic growth and community well-being" (Fonda and Anderson 2009).

"Well-being in the Aboriginal context is closely related to health and the integrity of the eco-system. One conclusion, therefore, might be that reporting should seek to go beyond simple economic cost benefit analysis and apply instead a holistic approach that clearly links complex interrelationships of social economic, political and cultural determinants to the natural environment" (Fonda and Anderson 2009).

- Employment, labour, participation
- Income assistance
- Education
- Housing
- Indigenous language use
- Life expectancy/mortality
- Suicide rate
- Injuries (incidence)
- Prevalence of sexually transmitted diseases
- Tuberculosis cases
- Teen births
- Single parent families
- People-reported spousal assault
- Children in care receiving services
- Women and children in shelters
- Mental health/addictions

These simple development indicators reinforce the narrative that mining development is wholly positive with few adverse implications. Hajkowicz et al. examined the relationship between mining activity and life expectancy, housing, employment, income, education and internet access, and purported to challenge "widely held beliefs that mining negatively affects people's quality of life" (2009). Although Indigenous communities in the study benefitted little from mining, Hajkowicz et al. argued that communities themselves failed to take advantage of the mining boom in the region (Hajkowicz et al. 2009).

In some jurisdictions (e.g., Nunavut, Northwest Territories, Greenland), social impact assessments include consideration of traditional and cultural uses of lands and resources. For example, in Greenland, the assessment used a range of related indicators to evaluate the Kvanefjeld project. These were drawn from the Survey of Living Conditions in the Arctic (SLICA) assessment including traditional economic activities, social networks, and well-being. The outcome of this assessment was similarly a conclusion that there would be no adverse impacts for the local communities of Southern Greenland.

Mining activity impacts on well-being in the Goa region in India were assessed in relation to well-being using three tools: "(1) a set of environmental and social performance indicators to measure the economic, environmental, and social costs of mining; (2) a "quality of life" instrument to assess the well-being of people in mining areas over time; and (3) an income-accounting tool to gauge the long-term economic viability of mining activities" (Conway 2011). The indicators developed within these assessments were focused on specific stressors of the mining project itself, for example land use (mining versus agriculture and issues of quality of life).

One of the most comprehensive assessments related to well-being in Canada was in respect of the Voisey's Bay nickel mine in Labrador (Noble and Bronson 2005). The review required the proponent to consider aspects of traditional land use of the Inuit and Innu people, quality of life, health, diet, and consumption of traditional foods, as well as the interactions between these variables for different communities and employee groups and by gender. While some have called the assessment ground-breaking, others have suggested that it is skewed. Critics claim that the proponent took advantage of the existing social and health challenges of the Labrador Inuit and Innu and overemphasized the positive impacts of the project for the economy while disregarding known adverse impacts. The review of these case studies of social impact assessment tells different kinds of stories. It seems important to recognize that it is not only the definition of well-being and the associated measures developed for assessment that are important, but also the treatment of indicators and the kinds of meanings extracted "about" communities.

AN ALTERNATIVE: COMMUNITY-BASED APPROACHES TO INDICATOR DEVELOPMENT

Participatory processes of indicator development and assessment provide a different vantage point to understanding impacts of mining activity. By virtue of asking communities what matters to them and how they will know if the well-being of their community is changing, the tool of

social impact assessment can become more refined, with impacts being more meaningfully documented and interpreted.

A community-based project aimed at documenting and developing indicators of well-being with the Łutsël K'é Dene First Nation (formed during the assessment of Ekati, the first diamond mine in Canada's North) came about as a result of concerns that the indicators being used by government and the project proponent were of limited meaning to local community members. While some indicators were of great interest, the vast majority presented a pathological narrative of crime, violence, and social illness. The core questions of the indicator project, "what is community well-being?" and "how do you know if the well-being of the community is changing?" provided a much different set of measures for assessing and monitoring the impacts of the mining project as well as understanding the broader landscape of change in well-being.

Unlike other kinds of indicators used on projects that highlighted more economic measures of development, the indicators developed through this process reflect three domains: self-government, healing, and cultural preservation. These domains reflect a range of social, economic, cultural, spiritual, and political dimensions of the way of life of the community.

The methodology has its limitations. The research was carried out with a community researcher and under the direction and guidance of the chief and council of Łutsël K'é Dene First Nation (see table 12.2). The indicators were developed based on almost 100 qualitative interviews in the community (i.e., a representative from every household). To that extent, the research can be easily framed as a community-based project and the indicators developed as a meaningful reflection of the community's way of life. However, the fact that the lead researcher (lead author) is a non-Indigenous scholar living outside of the community might lead one to question the extent to which the project is completely community-based. Subjectivity is inherent in all social science research, so the development of indicators necessitates recognition of the values, biases, and interests of this author/researcher. Other critical questions that emerged in the review of the list included the extent to which indicators (developed in 1996) reflected phenomenon considered by community members to be core to well-being, and to what extent these indicators are snapshots of what was of concern only at a specific point in time. One also might question whether the final list of indicators reflect dimensions of well-being that were normalized or so fundamental to the community's identity and way of life as to be deemed irrelevant in conversation. For example, having healthy and safe drinking water was not considered an issue important to community well-being. It may not have been mentioned because the ability to drink water straight from Great Slave Lake is so intuitive that its importance is rarely pondered. Other ideas about well-being may have been difficult to write down or communicate with researchers during interviews. Such tacit knowledge of well-being including phenomena that are experienced, felt, and intuitively understood, but that are difficult to articulate verbally or in writing, are thus less represented.

The rationale behind a narrower focus on the social-economic and cultural dimensions of change was methodological. The aim was to create a simple social survey based on the indicators that could be carried out by community researchers in an ongoing monitoring program. Other projects considering the health of the land were subsequently developed and have also become the basis for the development of an environmental monitoring program.

DISCUSSION AND CONCLUSION

Many sociologists have studied the complex social, economic, cultural, and political factors that determine how communities are affected by large-scale resource development (Gramling and Freudenburg 1992; Stedman, Parkins, and Beckley 2004). As mining activity grows around the world, questions also grow about how best to mitigate and manage these impacts, if only for the pursuit of greater economic benefits (Bridge 2004; Duhaime and Caron 2006; G. Hilson and Banchirigah 2009).

Table 12.2 Indicators of the Dene way of life from the Łutsël K'é Dene First Nation

Theme	Indicator
Self-Government	
1. Effectiveness of the Leadership	a. **Knowledgeable** – Number of meetings attended by Chief and Council
	b. **Action** – Number of decisions taken (that benefit the community) (i.e., number of Band Council Resolutions)
	c. **Communication** – Number of meetings held/home-visits in the community to share information
2. Togetherness	a. **Good communication** – respectful and supportive words being spoken; not gossip
	b. **Participation** – volunteerism at community events
	c. **Knowledgeable** – attendance at public meetings; number of questions asked at public meetings
3. Economic Development	a. **Employment** - number of people employed including number employed in the mining sector
	b. **Local Control Over Development** – number of agreements with Band Council that ensure local control over development
	c. **Environmental Sustainability** – limited adverse impacts of development on the environment; perceptions of environmental quality
4. Role of Youth	a. **Youth Participation** – number of youth participating in community meetings; number of youth graduating from high school; engagement in post-secondary education
	b. **Community Support of Youth** – level of investment of the community in youth activities and programs
5. Infrastructure and Services	a. **Health and Safety** – number of safe and healthy places to spend time in the community/on the land
	b. **Safe and Healthy Places for Children** – number of safe and healthy places for children to play/learn
6. Individual Well-Being	a. **Traditional Harvesting and Nutrition** – caribou, moose, and other traditional/country food harvested and consumed
	b. **Quality of Life** – reported symptoms of emotional, spiritual, mental stress/well-being
	c. **Physical Wellness** – reported incidence disease, accidents, injuries and harms
7. Family Well-Being	a. **Knowledge/Access** – level of participation in healthcare and wellness programs
	b. **Parent Support** – parental attendance at youth activities
	c. **Family Activities** – number of family-oriented activities in the community
8. Well-Being of Children	a. **Quality of Life** – number of children visibly laughing, sharing, and respecting one another
	b. **Identify Goals** – children able to make and meet short- and long-term goals

9. Healing Services	a.	**Effectiveness of Healing Services** – participant success in addiction and wellness programs
	b.	**Capacity** – capacity of health services to meet local needs (i.e., responding to requests)
	c.	**Respect for Healing Approaches (for Addictions)** – availability of diversity of healing programs
10. Traditional Knowledge	a.	**Knowledge of Sharing** – opportunities for knowledge sharing
	b.	**Understanding of Respect** – people who understand respect as important
	c.	**Traditional Knowledge and Skills** – number of people with traditional knowledge and skills
11. Cultural Education	a.	**Educating Youth** – opportunities for educating youth
	b.	**Cultural Programs** – success of cultural programs
	c.	**Quantity and Quality of Time** – quality of time spent between Elders and youth
12. Land	a.	**Traditional Harvesting** – number of people/amounts of time spent in traditional harvesting and land use activities
	b.	**Respect for the Land** – respect demonstrated during land use
	c.	**Traditional Land Use Activities** – number of organized events and activities for traditional harvesting and land use
13. Language	a.	**Family Use of Chipewyan** – number of homes where Chipewyan is spoken
	b.	**Language Learning** – opportunities for language learning
	c.	**Formal Use of Language** – use of Chipewyan during meetings and other public events

Source: Parlee et al. 2008.

Frameworks aimed at measuring the social costs of development in Canada and other OECD countries are common with well-being increasingly used in policy frameworks and in assessments on sustainable development. However, the ideas evidenced in these frameworks can be ambiguous relative to the theoretical depth of the social sciences literature.

Comparing the themes and indicators of well-being from the literature and those used by social impact assessment, one can see a disconnect in terms of depth and use of indicators beyond those conventionally associated with development. More often, the focus of these assessments is on material costs and benefits infrastructure (e.g., housing, roads), as well as end-of-the-pipe changes in disease burdens and social illness (e.g., violent crime). Consideration of the front-end causes (e.g., deterioration of social capital and social networks) are a lesser focus. Assessments involving Indigenous communities in Canada do include consideration of traditional economies and land uses (e.g., hunting, trapping) which is consistent with the literature. However, there is also a strong focus on aspects of sickness and social dysfunction in using indicators such as crime, alcoholism, and prevalence of sexually transmitted disease. This observation is consistent with what has been argued in previous studies. In a reflection on the Voisey's Bay nickel mine social impact assessment, community members argued that proponents and their consultants tend to depict communities, including Indigenous ones, as vulnerable and in need of the positive impacts of mining development, rather than identify potential adverse impacts. This is problematic in several ways. In such contexts, the onus is on Indigenous communities and others to produce proof to the contrary in the very short time frame of consultation and assessment. Given impact assessments tend to privilege outside experts rather than inside expertise, combined with the limited resources and capacities of communities to produce the burden of proof about adverse impacts, it is not surprising that social impact assessments are highly criticized by Indigenous organizations as tools of colonization and corporate interest.

Participatory approaches to assessment may address some aspects of these inequities in power (D.R. Becker et al. 2003). Supporting and resourcing communities in developing their own indicators for use in project-specific social and strategic impact assessments and ongoing monitoring is likely to become best practice. There are unique opportunities to develop indicators that are more relevant to Indigenous communities and others through community-led research initiatives such as the work described from the Łutsël K'é Dene First Nation. While there is some overlap between the theories and indicators described in the theory and the practice of social impact assessment (e.g., employment indicators, some land use indicators) the work in Łutsël K'e provides a unique perspective on well-being that is very specific to place and culture. As large-scale resource development continues in the Canadian Arctic and elsewhere, and as social impacts continue to be a concern, the issue for communities and governments is to find ways to assess, monitor, and manage these impacts. Of course, this must address the indicators of value to communities, as well as issues of social, cultural, and ecological sustainability.

REFERENCES

Aikenhead, Glen S. and Masakata Ogawa. 2007. "Indigenous Knowledge and Science Revisited." *Cultural Studies of Science Education* 2, no. 3: 539–620.

Albrecht, Glenn, Gina-Maree Sartore, Linda Connor, Nick Higginbotham, Sonia Freeman, Brian Kelly, Helen Stain, Anne Tonna, and Georgia Pollard. 2007. "Solastalgia: The Distress Caused by Environmental Change." *Australasian Psychiatry* 15, suppl 1: S95–S98.

Anand, Paul. 2016. *Happiness, Well-Being and Human Development: The Case for Subjective Measures: UNDP Human Development Report Background Paper.* New York: United Nations Human Development Program. http://hdr.undp.org/sites/default/files/anand_template_rev.pdf.

Aporta, Claudio. 2011. "Shifting Perspectives on Shifting Ice: Documenting and Representing Inuit Use of the Sea Ice." *The Canadian Geographer/Le Géographe canadien* 55, no. 1: 6–19.

Arce-Gomez, Antonio, Jerome D. Donovan, and Rowan E. Bedggood. 2015. "Social Impact Assessments: Developing a Consolidated Conceptual Framework." *Environmental Impact Assessment Review* 50: 85–94.

Ashton, Thomas S. and Joseph Sykes. 1964. *The Coal Industry of the Eighteenth Century.* Manchester: Manchester University Press.

Auty, Richard. 2002. *Sustaining Development in Mineral Economies: The Resource Curse Thesis.* Routledge.

Ballard, Chris, and Glenn Banks. 2003. "Resource Wars: The Anthropology of Mining." *Annual Review of Anthropology* 32, no. 1: 287–313.

Bebbington, Anthony, Leonith Hinojosa, Denise H. Bebbington, Maria L. Burneo, and Ximena Warnaars. 2008. "Contention and Ambiguity: Mining and the Possibilities of Development." *Development and Change* 39, no. 6 (November): 887–914.

Becker, Dennis R., Charles C. Harris, William J. McLaughlin, and Erik A Nielsen. 2003. "A Participatory Approach to Social Impact Assessment: The Interactive Community Forum." *Environmental Impact Assessment Review* 23, no. 3: 367–382.

Becker, Henk A., and Frank Vanclay. 2003. *The International Handbook of Social Impact Assessment: Conceptual and Methodological Advances.* Cheltenham: Edward Elgar.

Brereton, David, and Joni Parmenter. 2008. "Indigenous Employment in the Australian Mining Industry." *Journal of Energy & Natural Resources Law* 26, no. 1: 66–90.

Bridge, Gavin. 2004. "Mapping the Bonanza: Geographies of Mining Investment in an Era of Neoliberal Reform." *The Professional Geographer* 56, no. 3: 406–421.

Bronson, Jackie, and Bram F. Noble. 2006. "Health Determinants in Canadian Northern Environmental Impact Assessment." *Polar Record* 42, no. 4: 315–324.

Bryceson, Deborah, and Danny MacKinnon. 2012. "Eureka and Beyond: Mining's Impact on African Urbanisation." *Journal of Contemporary African Studies* 30, no. 4: 513–537.

Bureau of Minerals and Petroleum. 2009. *Guidelines for Social Impact Assessments for Mining Projects in Greenland.* Nuuk, Greenland: Bureau of Minerals and Petroleum, Government of Greenland. https://www.socialimpactassessment.com/documents/sia_guidelines.pdf.

Canadian Index of Wellbeing. 2016. *How Are Canadians Really Doing? The 2016 CIW National Report.* Waterloo: University of Waterloo.

Chandler, Michael J. and Christopher Lalonde. 1998. "Cultural Continuity as a Hedge against Suicide in Canada's First Nations." *Transcultural Psychiatry* 35, no. 2: 191–219.

Clark, Allen L. and Jennifer Cook Clark. 1999. "The New Reality of Mineral Development: Social and Cultural Issues in Asia and Pacific Nations." *Resources Policy* 25, no. 3: 189–196.

Conway, Kevin. 2011. *Case Study: India — Tracking Health and Well-Being in Goa's Mining Belt. New Tools Promote the Sustainable Development of Mining.* Ottawa: International Development Research Centre. https://www.idrc.ca/en/article/case-study-india-tracking-health-and-well- being-goas-mining-belt.

Cultural Survival. 2016. *The "Canada Brand": Violence and Canadian Mining Companies in Latin America.* Osgoode Legal Studies Research Paper 17.

Cultural Survival: Cambridge MA. https://www.culturalsurvival.org/news/canada-brand- violence-and-canadian-mining-companies-latin-america.

Diener, Ed. 2000. "Subjective Well-Being: The Science of Happiness and a Proposal for a National Index." *American Psychologist* 55, no. 1: 34-43.

Diener, Ed, Eunkook M. Suh, Richard E. Lucas, and Heidi L. Smith. 1999. "Subjective Well-Being: Three Decades of Progress." *Psychological Bulletin* 125, no. 2: 276-302.

Dubiński, Józef. 2013. "Sustainable Development of Mining Mineral Resources." *Journal of Sustainable Mining* 12, no. 1: 1–6.

Duer, Kreszentia and Nancy Levine. 1999. *Culture and Sustainable Development: A Framework for Action.* Washington, DC: World Bank.

Duhaime, Gérard and Andrée Caron. 2006. "The Economy of the Circumpolar Arctic." *The Economy of the North* 17: 17-23.

Duhaime, Gérard, Edmund Searles, Peter J. Usher, Heather Myers, and Pierre Fréchette. 2004. "Social Cohesion and Living Conditions in the Canadian Arctic: From Theory to Measurement." *Social Indicators Research* 66, no. 3: 295–318.

Esteves, Ana Maria, Daniel Franks, and Frank Vanclay. 2012. "Social Impact Assessment: The State of the Art." *Impact Assessment and Project Appraisal* 30, no. 1: 34–42.

Fonda, Marc and Erik Anderson. 2009. *Diamonds in Canada's North: A Lesson in Measuring Socio-Economic Impacts on Well-Being.* Montreal: Canadian Issues.

Frantál, Bohumil. 2016. "Living on Coal: Mined-out Identity, Community Displacement and Forming of Anti-Coal Resistance in the Most Region, Czech Republic." *Resources Policy* 49: 385–393.

Freeman, Donald G. 2009. "The 'Resource Curse'and Regional US Development." *Applied Economics Letters* 16, no. 5: 527–530.

Freudenburg, William R. 1986. "The Density of Acquaintanceship: An Overlooked Variable in Community Research?" *American Journal of Sociology* 92, no. 1: 27–63.

–1992. "Addictive Economies: Extractive Industries and Vulnerable Localities in a Changing World Economy 1." *Rural Sociology* 57, no. 3: 305–332.

Gedicks, Al. 1994. *The New Resource Wars: Native and Environmental Struggles Against Multinational Corporations.* Black Rose Books Ltd.

Gibson, Ginger, and Jason Klinck. 2005. "Canada's Resilient North: The Impact of Mining on Aboriginal Communities." *Pimatisiwin* 3, no. 1: 116–139.

Gifford, Blair, and Andrew Kestler. 2008. "Toward a Theory of Local Legitimacy by MNEs in Developing Nations: Newmont Mining and Health Sustainable Development in Peru." *Journal of International Management* 14, no. 4: 340–352.

Gordon, Todd, and Jeffery R Webber. 2008. "Imperialism and Resistance: Canadian Mining Companies in Latin America." *Third World Quarterly* 29, no. 1: 63–87.

Government of Canada. 1998. *Mackenzie Valley Resource Management Act (s.c. 1998, c. 25).* https://laws-lois.justice.gc.ca/eng/acts/m-0.2/FullText.html.

Gramling, Robert, and William R. Freudenburg. 1992. "Opportunity-Threat, Development, and Adaptation: Toward a Comprehensive Framework for Social Impact Assessment 1." *Rural Sociology* 57, no. 2: 216–234.

Hajkowicz, Stefan A., Sonja Heyenga, and Kieren Moffat. 2011. "The Relationship Between Mining and Socio-Economic Well Being in Australia's Regions." *Resources Policy* 36, no. 1: 30–38.

Hall, Rebecca. 2013. "Diamond Mining in Canada's Northwest Territories: A Colonial Continuity." *Antipode* 45, no. 2 (March): 376-393.

Hanna, Philippe, Frank Vanclay, Esther Jean Langdon, and Jos Arts. 2014. "Improving the Effectiveness of Impact Assessment Pertaining to Indigenous Peoples in the Brazilian Environmental Licensing Procedure." *Environmental Impact Assessment Review* 46: 58–67.

Haslam McKenzie, Fiona M. and Steven Rowley. 2013. "Housing Market Failure in a Booming Economy." *Housing Studies* 28, no. 3: 373–388.

Helliwell, John F. 2003. "How's Life? Combining Individual and National Variables to Explain Subjective Well-Being." *Economic Modelling* 20, no. 2: 331–360.

Helwege, Ann. 2015. "Challenges with Resolving Mining Conflicts in Latin America." *The Extractive Industries and Society* 2, no. 1: 73–84.

Hilson, Gavin. 2002. "An Overview of Land Use Conflicts in Mining Communities." *Land Use Policy* 19, no. 1: 65–73.

–2004. "Structural Adjustment in Ghana: Assessing the Impacts of Mining-Sector Reform." *Africa Today* 51, no. 2: 53–77.

Hilson, Gavin, and Sadia Mohammed Banchirigah. 2009. "Are Alternative Livelihood Projects Alleviating Poverty in Mining Communities? Experiences From Ghana." *The Journal of Development Studies* 45, no. 2: 172–196.

Hitch, Michael, and Courtney Riley Fidler. 2007. "Impact and Benefit Agreements: A Contentious Issue for Environmental and Aboriginal Justice." *Environments Journal* 35, no. 2: 45–69.

Horsley, Julia, Sarah Prout, Matthew Tonts, and Saleem H. Ali. 2015. "Sustainable Livelihoods and Indicators for Regional Development in Mining Economies." *The Extractive Industries and Society* 2, no. 2: 368–380.

Hota, Padmanabha, and Bhagirath Behera. 2016. "Opencast Coal Mining and Sustainable Local Livelihoods in Odisha, India." *Mineral Economics* 29, no. 1: 1–13.

Howitt, Richard. 2001. *Rethinking Resource Management: Justice, Sustainability and Indigenous Peoples*. Psychology Press.

–2011. "Theoretical Foundations." In *New Directions in Social Impact Assessment: Conceptual and Methodological Advances*, edited by Frank Vanclay and Ana Maria Esteves, 78–95. Northampton, MA: Edward Elgar.

Humby, Tracy-Lynn. 2013. "Environmental Justice and Human Rights on the Mining Wastelands of the Witwatersrand Gold Fields." *Revue générale de droit* 43: 67–112.

Ingersoll-Dayton, Berit, Chanpen Saengtienchai, Jiraporn Kespichayawattana, and Yupin Aungsuroch. 2004. "Measuring Psychological Well-Being: Insights from Thai Elders." *The Gerontologist* 44, no. 5: 596–604.

Jacquet, Jeffrey B., and Richard C. Stedman. 2014. "The Risk of Social-Psychological Disruption as an Impact of Energy Development and Environmental Change." *Journal of Environmental Planning and Management* 57, no. 9: 1285–1304.

Jantunen, Jorma, Tommi Kauppila, Marja Liisa Räisänen, and Others. 2015. *Guide: Environmental Impact Assessment Procedure for Mining Projects in Finland*. Helsinki, Finland: Ministry of Employment and the Economy, Finland. http://en.gtk.fi/export/sites/en/mineral_resources/EIA_guidelines_for_mining_projects_in_Finl and_2015.pdf.

Kara, Siddharth. 2018. "Is Your Phone Tainted by the Misery of the 35,000 Children in Congo's Mines?" *The Guardian*: 1.

Keeling, Arn, and John Sandlos. 2015. *Mining and Communities in Northern Canada: History, Politics, and Memory*. Calgary: University of Calgary Press.

Keeling, Arn, John Sandlos, Jean-Sébastien Boutet, and Hereward Longley. "Managing Development? Knowledge, Sustainability and the Environmental Legacies of Resource Development in Northern Canada." Gap Analysis Report #12, RESDA.

Kirmayer, Laurence J., Christopher Fletcher, and Robert Watt. 2009. "Locating the Ecocentric Self: Inuit Concepts of Mental Health and Illness." In *Healing Traditions: The Mental Health of Aboriginal Peoples in Canada*, edited by Laurence J. Kirmayer and Gail Guthrie Valaskakis, 289-314. Vancouver, BC: UBC Press.

Koivurova, Timo, Adam Stepien, and Paula Kankaanpaa. 2014. "Strategic Assessment of Development of the Arctic." Assessment conducted for the European Union.

Kral, Michael J., Lori Idlout, J. Bruce Minore, Ronald J. Dyck, and Laurence J. Kirmayer. 2011. "Unikkaartuit: Meanings of Well-Being, Unhappiness, Health, and Community Change among Inuit in Nunavut, Canada." *American Journal of Community Psychology* 48, no. 3-4: 426–438.

Lane, Marcus B., Allan P. Dale, and Nick Taylor. 2001. "Social Assessment in Natural Resource Management: Promise, Potentiality, and Practice." *Social Assessment in Natural Resource Management Institutions*: 3-12.

Lane, Marcus B., Helen Ross, Allan P. Dale, and Roy E. Rickson. 2003. "Sacred Land, Mineral Wealth, and Biodiversity at Coronation Hill, Northern Australia: Indigenous Knowledge and SIA." *Impact Assessment and Project Appraisal* 21, no. 2: 89–98.

Lawe, L.B., J. Wells, and Mikisew Cree First Nation. 2005. "Cumulative Effects Assessment and EIA Follow-up: A Proposed Community-Based Monitoring Program in the Oil Sands Region, Northeastern Alberta." *Impact Assessment and Project Appraisal* 23, no. 3: 205–209.

Lawrie, Misty, Matthew Tonts, and Paul Plummer. 2011. "Boomtowns, Resource Dependence and Socio- Economic Well-Being." *Australian Geographer* 42, no. 2: 139–164.

Little, Ron L., and Richard S. Krannich. 1988. "A Model for Assessing the Social Impacts of Natural Utilization on Resource-Dependent Communities." *Impact Assessment* 6, no. 2: 21–35.

Lockie, Stewart. 2001. "SIA in Review: Setting the Agenda for Impact Assessment in the 21st Century." *Impact Assessment and Project Appraisal* 19, no. 4: 277–287.

Lockie, Stewart, Maree Franettovich, Vanessa Petkova-Timmer, John Rolfe, and Galina Ivanova. 2009. "Coal Mining and the Resource Community Cycle: A Longitudinal Assessment of the Social Impacts of the Coppabella Coal Mine." *Environmental Impact Assessment Review* 29, no. 5: 330–339.

Lyver, P. O'B., and Łutsël K'é Dene First Nation. 2005. "Monitoring Barren-Ground Caribou Body Condition with Denésǫłıné Traditional Knowledge." *Arctic* 58, no. 1: 44–54.

Mancini, Lucia, and Serenella Sala. 2018. "Social Impact Assessment in the Mining Sector: Review and Comparison of Indicators Frameworks." *Resources Policy* 57: 98–111.

Marais, Lochner, Fiona Haslam McKenzie, Leith Deacon, Etienne Nel, Deirdre van Rooyen, Jan Cloete. 2018. "The Changing Nature of Mining Towns: Reflections from Australia, Canada and South Africa." *Land Use Policy* 76: 779–788.

Mayoux, Linda, and Robert Chambers. 2005. "Reversing the Paradigm: Quantification, Participatory Methods and Pro-poor Impact Assessment." *Journal of International Development* 17, no. 2: 271–298.

Mehlum, Halvor, Karl Ove Moene, and Ragnar Torvik. 2006. "Institutions and the Resource Curse." *The Economic Journal* 116, no. 508: 1–20.

Millennium Ecosystem Assessment. 2005. *Ecosystems and Human Well-Being: Synthesis.* Washington DC: Island Press. https://www.millenniumassessment.org/documents/document.356.aspx.pdf.

Moller, Henrik, Fikret Berkes, Philip O'Brian Lyver, and Mina Kislalioglu. 2004. "Combining Science and Traditional Ecological Knowledge: Monitoring Populations for Co-Management." *Ecology and Society* 9, no. 3: art. 2.

Morphy, Frances. 2016. "Indigenizing Demographic Categories: A Prolegomenon to Indigenous Data Sovereignty." In *Indigenous Data Sovereignty: Toward and Agenda*, edited by Tahu Kukutai and John Taylor, 99-116. Canberra: ANU Press.

Natural Resources Canada. 2017. *The Minerals and Metals Policy of the Government of Canada.* Ottawa: Minister of Public Works and Government Services Canada.

Noble, Bram, and Jackie Bronson. 2005. "Integrating Human Health into Environmental Impact Assessment: Case Studies of Canada's Northern Mining Resource Sector." *Arctic* 58, no. 4: 395–405.

–2006. "Practitioner Survey of the State of Health Integration in Environmental Assessment: The Case of Northern Canada." *Environmental Impact Assessment Review* 26, no. 4: 410–424.

OECD. 2017. *How's Life? 2017: Measuring Well-Being.* Paris: OECD iLibrary, OECD Publishing. https://doi.org/10.1787/23089679.

–2019. *Enhancing Well-Being in Mining Regions: Key Issues and Lessons for Developing Indicators.* Skellefteå, Sweden: OECD Publishing.

O'Faircheallaigh, Ciaran. 2011. "Social Impact Assessment and Indigenous Social Development." In *New Directions in Social Impact Assessment: Conceptual and Methodological Advances,* edited by Frank Vanclay and Ana Maria Esteves. Cheltenham: Edward Elgar.

O'Faircheallaigh, Ciaran, and Saleem Ali. 2017. *Earth Matters: Indigenous Peoples, the Extractive Industries and Corporate Social Responsibility.* London: Routledge.

Panelli, Ruth, and Gail Tipa. 2009. "Beyond Foodscapes: Considering Geographies of Indigenous Well-Being." *Health & Place* 15, no. 2: 455–465.

Papyrakis, E. 2017. The Resource Curse-What Have We Learned From Two Decades of Intensive Research: Introduction to the Special Issue. *The Journal of Development Studies* 53, no. 2: 175-185.

Parlee, Brenda. 2015. "Avoiding the Resource Curse: Indigenous Communities and Canada's Oil Sands." *World Development* 74: 425–436.

–2015. "The Social Economy and Resource Development in Northern Canada." In *Northern Communities Working Together: The Social Economy of Canada's North*, edited by Chris Southcott, 52–73. Toronto: University of Toronto Press.

Parlee, Brenda, and Chris Furgal. 2012. "Well-Being and Environmental Change in the Arctic: A Synthesis of Selected Research from Canada's International Polar Year Program." *Climatic Change* 115, no. 1 (November): 13-34.

Parlee, Brenda, and Łutsël K'é Dene First Nation. 2005. "Understanding and Communicating about Ecological Change." In *Breaking Ice: Renewable Resource and Ocean Management in the Canadian North*, edited by. Fikret Berkes, Rob Huebert, Helen Fast, Micheline Manseau, and Alan Diduck. Calgary: University of Calgary Press.

Parlee, Brenda, John O'Neil, and Łutsël K'é Dene First Nation. 2007. ""The Dene Way of Life": Perspectives on Health from Canada's North." *Journal of Canadian Studies/Revue d'études canadiennes* 41, no. 3: 112–133.

Parlee, Brenda, John Sandlos, and Dave Natcher. 2018. "Undermining Subsistence: Barren-ground Caribou in a "Tragedy of Open Access."" *Science Advances* 4, no. 2 (February). e1701611.

Pegg, Scott. 2006. "Mining and Poverty Reduction: Transforming Rhetoric into Reality." *Journal of Cleaner Production* 14, no. 3-4: 376–387.

Pelders, Jodi, and Gill Nelson. 2019. "Living Conditions of Mine Workers from Eight Mines in South Africa." *Development Southern Africa* 36, no. 3: 265–282.

Piotrzkowski, Maxwell James. 2016. "Exploiting the Sacred: Natural Resource Extraction on Native American Tribal Lands." Honors thesis, Texas State University, San Marcos, Texas.

Piper, L., and John Sandlos. 2007. "A Broken Frontier: Ecological Imperialism in the Canadian North." *Environmental History* 12, no. 4: 759–795.

Poppel, Birger. 2015. *SLICA: Arctic Living Conditions: Living Conditions and Quality of Life Among Inuit, Saami and Indigenous Peoples of Chukotka and the Kola Peninsula.* Nordic Council of Ministers.

Render, Jo M. 2005. *Mining and Indigenous Peoples Issues Review.* London, UK: International Council on Mining & Metals. http://www.ideaspaz.org/tools/download/52001.

Rickson, Roy E., Rabel J. Burdge, Tor Hundloe, and Geoffrey T. McDonald. 1990. "Institutional Constraints to Adoption of Social Impact Assessment as a Decision-Making and Planning Tool." *Environmental Impact Assessment Review* 10, no. 1-2: 233–243.

Rio Tinto Limited. 2020. "Rio Tinto: Sustainability - People." https://www.riotinto.com/sustainability/people.

Roberts, J Timmons. 1995. "Subcontracting and the Omitted Social Dimensions of Large Development Projects: Household Survival at the Carajás Mines in the Brazilian Amazon." *Economic Development and Cultural Change* 43, no. 4: 735–758.

Rocchi, Benedetto, Chiara Landi, Gianluca Stefani, Severino Romano, Mario Cozzi. 2015. "Escaping the Resource Curse in Regional Development: A Case Study on the Allocation of Oil Royalties." *International Journal of Sustainable Development* 18, no. 1-2: 115–138.

Rodon, Thierry, and Francis Lévesque. 2015. "Understanding the Social and Economic Impacts of Mining Development in Inuit Communities: Experiences with Past and Present Mines in Inuit Nunangat." *The Northern Review* 41: 13–39.

Ross, M. 2001. "Extractive Sectors and the Poor." Boston, MA: Oxfam America (28).

Ryff, Carol D. 1989. "Happiness Is Everything, or Is It? Explorations on the Meaning of Psychological Well-Being." *Journal of Personality and Social Psychology* 57, no. 6: 1069-1081.

Sharpe, Andrew, and Jeremy Smith. 2005. *Measuring the Impact of Research on Well-Being: A Survey of Indicators of Well-Being.* Centre for the Study of Living Standards.

Smith, Michael D., Richard S. Krannich, and Lori M. Hunter. 2001. "Growth, Decline, Stability, and Disruption: A Longitudinal Analysis of Social Well-Being in Four Western Rural Communities." *Rural Sociology* 66, no. 3: 425–450.

Statistics Canada. 2016. *National Overview of the Community Well-Being Index, 1981 to 2016.* Ottawa: Government of Canada. https://www.sac-isc.gc.ca/eng/1419864229405/1557324163264.

Stedman, Richard C., John R. Parkins, and Thomas M. Beckley. 2004. "Resource Dependence and Community Well-being in Rural Canada." *Rural Sociology* 69, no. 2: 213–234.

Stevenson, Marc G. 1996. "Indigenous Knowledge in Environmental Assessment." *Arctic* 49, no. 3 (September): 278–291.

Suopajärvi, Leena. 2013. "Social Impact Assessment in Mining Projects in Northern Finland: Comparing Practice to Theory." *Environmental Impact Assessment Review* 42: 25–30.

Suutarinen, Tuomas Kristian. 2015. "Local Natural Resource Curse and Sustainable Socio-Economic Development in a Russian Mining Community of Kovdor." *Fennia-International Journal of Geography* 193, no. 1: 99–116.

Taseko Mines Limited. 2020. "Taseko: New Prosperity." https://www.tasekomines.com/properties/new- prosperity.

Triner, Gail D. 2015. *Mining and the State in Brazilian Development.* Routledge.

United Nations. 2006. *Report of the Meeting on Indigenous Peoples and Indicators of Well-Being.* New York: United Nations Economic and Social Council, E/C.19/2006/CRP.3.

Urkidi, Leire, and Mariana Walter. 2011. "Dimensions of Environmental Justice in Anti-Gold Mining Movements in Latin America." *Geoforum* 42, no. 6: 683–695.

US Department of Commerce, National Oceanic and Atmospheric Administration, and National Marine Fisheries Service. 1995. *Guidelines and Principles for Social Impact Assessment.* Washington DC: US Department of Commerce. https://www.iaia.org/pdf/IAIAMemberDocuments/Publications/Guidelines_Principles/SIA%20Guide.PDF.

Vanclay, Frank. 2003. "International Principles for Social Impact Assessment." *Impact Assessment and Project Appraisal* 21, no. 1: 5–12.

–2006. "Principles for Social Impact Assessment: A Critical Comparison Between the International and US Documents." *Environmental Impact Assessment Review* 26, no. 1: 3–14.

Walter, Maggie. 2016. "Data Politics and Indigenous Representation in Australian Statistics." *Indigenous Data Sovereignty: Toward an Agenda* 38: 79–98.

Walter, Maggie, and Chris Andersen. 2013. *Indigenous Statistics: A Quantitative Research Methodology.* Left Coast Press.

Weber-Fahr, Monika. 2002. "Treasure or Trouble? Mining in Developing Countries." World Bank and International Finance Corporation, Washington DC.

Weitzner, Viviane. 2002. *Through Indigenous Eyes: Toward Appropriate Decision-Making Processes Regarding Mining on or near Ancestral Lands; Final Synthesis Report*. The North-South Institute, Ottawa, ON, CA.

Wilson, Irene. 1997. *Impact of Uranium Mining on Aboriginal Communities in the Northern Territory*. Canberra, Australia: Parliament of Australia. https://www.aph.gov.au/Parliamentary_Business/Committees/Senate/Former_Committees/ura nium/report/c11.

Zorrilla, Carlos. 2009. *Protecting Your Community Against Mining Companies and Other Extractive Industries: A Guide for Community Organizers*. Boulder, Colorado: Cultural Survival; Global Response. https://www.culturalsurvival.org/sites/default/files/guide_for_communities_0.pdf.

13

Measuring the Social and Economic Impacts of Extractive Industry in the Arctic: Developing Baseline Social Indicators for the Inuvialuit Settlement Region

Andrey N. Petrov, Chris Southcott, Philip Cavin, and Bob Simpson

One of the most difficult tasks northern communities have in making decisions about resource development projects is trying to understand what the impacts of these developments are and how they should be measured. The rise of environmental and social impact assessments goes hand in hand with the desire of northern regions to control negative impacts and enhance positive outcomes of resource development. That is why there is an increasing need to measure and monitor the socio-economic change resulting from new projects. There is no consensus on the best way to do this, and as a result, there are different uses of a variety of indicators of change and monitoring systems. Some, such as that described by Parlee in her chapter on well-being (12), look towards enabling communities to develop their own community-based indicators relying on more qualitative methods of data gathering (Fonda and Anderson 2009). Others note the value of having readily available comparative indicators that are more quantitative in nature in addition to community-based indicators and monitoring systems (Kruse et al. 2011; Larsen et al. 2010; Larsen et al. 2015). The desire to have readily available quantitative indicators of social, cultural, and economic changes resulting from resource development led the Inuvialuit Regional Corporation to develop their own set of indicators and to partner with ResDA researchers to link their work with that of the Arctic Council's Arctic Social Indicator Project (Larsen et al. 2010).

The Inuvialuit Settlement Region (ISR) has been affected by several resource boom and bust cycles associated with the resource activities in the Mackenzie Delta and more recently in the Beaufort Sea. The IRC created as a result of the Inuvialuit Comprehensive Land Claim Agreement has been collecting and publishing selected socio-economic data to aid in the decision-making process. It also provides public access to IRC members. Given a growing interest in Arctic resources within the ISR, the IRC engaged in collaboration with a social impacts monitoring team of polar scientists to develop a system of indicators. That system is based on past experiences in the ISR and across the Arctic, local relevance, and data availability.

The goal of this study was to develop a set of measurable, reliable, and accessible indicators to monitor socio-economic conditions in the Inuvialuit Region to trace possible impacts of resource development in the thirty-year period between the late 1980s to the late 2000s. The key question in hand was whether there is evidence that resource development in the Inuvialuit Region resulted in social and economic benefits to local population in comparison to other regions, and to its own pre- and early-development baseline.

The objectives of this research included: (1) to develop nuanced indicators that can reflect shorter-term impacts of resource development on ISR communities; (2) to use these two systems to assess (a) overall well-being of Inuvialuit residents across six social domains as compared to other regions, and to its own pre- and early-development baseline and (b) likely impacts of resource activities on specific socio-economic conditions. Ultimately, upon building an observing system based on existing data sources, the procedures could be designed to enable community-based collection, management, and analysis of data by local actors. This will also facilitate the utilization of the Inuvialuit Baseline Indicators to inform the region's stakeholders as well as aid in the IRC's decision-making and ensure community awareness.

The study involved the development of the Resource Development Impacts Scan (REDIS) indicators, which are represented by a set of socio-economic variables that can be indicative of impacts of resource development on communities. The proposed Inuvialuit Baseline Indicators (IBI) monitoring system includes REDIS indicators as its core component alongside the general social indicators (Petrov 2018). It is thus able to account for both slow and fast variables to capture long-term and shorter-term impacts.

Figure 13.1 Inuvialuit Settlement Region

STUDY AREA

Designated by the Inuvialuit Final Agreement, the Inuvialuit Settlement Region (ISR) is one of the four Inuit regions of Canada spanning across 90,650 square kilometres. Consisting of approximately 6,000 residents, more than 3,000 of them are Inuvialuit living in six communities located in the Mackenzie Delta and on the Beaufort Sea coast (see figure 13.1). The Inuvialuit Regional Corporation (IRC) was established to manage compensation specified in the Final Agreement. The IRC mandate is to work towards improving the economic, social, and cultural well-being of the Inuvialuit. Each community has its own corporation managing local affairs. Part of the IRC's commitment is the monitoring of social-economic conditions. Although the IRC achieved considerable success collecting social data from territorial agencies, a monitoring system is yet to be implemented.

METHODOLOGY

Indicators can be defined as metrics used to determine the current state and, when monitored over time, the trajectory of change for a given phenomenon or process. For example, development indicators allow tracking of trends toward development goals. A considerable amount of new social indicators frameworks were developed in the last decades to measure and monitor various social and economic domains (Michalos 2014; Vanclay 2002; Larsen et al. 2015). Most effective indicators are measurable, based on data that are not too expensive to collect, sensitive (responsive to change), and understandable (Larsen et al. 2010).

In the Arctic context, there has been some progress in developing indicators as pertained to measuring social well-being and impacts of development in Arctic communities (Larsen et al. 2010; Kruse et al. 2011; Ozkan and Schott 2013; Petrov 2018; Impact Economics 2014; Tysiachniouk et al. 2020), although much is yet to be done (Mitchell and Parkins 2011; Wilson and Stammler 2016; Petrov and Vlasova 2021). Leading accomplishments include the Arctic Social Indicators Project (Larsen et al. 2010), the second Arctic Human Development Report (AHDR 2014), the Arctic Observing Network Social Indicators Project (AON-SIP) (Kruse et al. 2011), SLICA (Survey of Living Conditions in the Arctic) (Poppel, 2015), and the Arctic Marine Shipping Assessment (PAME 2009). The ASI/AHDR framework provides the most comprehensive system of indicators across six domains. Case studies using these indicators have been conducted in many regions (Larsen et al. 2015; Vlasova and Volkov 2013; Ozkan and Schott 2013; Petrov 2020). At the same time, a growing volume of case-specific, narrowly constructed methodologies have been used to conduct social impacts assessments for particular (mostly resource-based) projects in the Arctic (Petrov et al. 2013).

The study's approach was to develop a system more nuanced and impact sensitive of Resource Development Impacts Scan indicators. The study heavily relied on Canadian Census and NWT Statistical Bureau datasets and used available Inuvialuit data collected in the IRC database (http://inuvialuitindicators.com/). The purpose of applying a set of standardized indicators was to provide a baseline assessment of socio-economic conditions in the ISR. The study made comparisons between ISR communities and relevant settlements in the NWT, in particular Yellowknife, the NWT capital. In addition, collected and analyzed data could allow further comparisons with communities in other Arctic regions. The project was focused on expanding the system of indicators by incorporating measures that are of particular concern for local residents and beneficiaries. The research team made a field visit and conducted interviews to determine additional indicators to be included in the retrospective or future analysis.

The baseline social indicators developed earlier were designed to provide a baseline assessment of socio-economic conditions experienced by ISR residents in six domains of well-being (Petrov 2018). Periodic analysis of these indicators is valuable in respect to assessing the overall wellness of the Inuvialuit population and to compare it with peer communities in the territories. Most data for this set of indicators are collected and assessed on an annual basis. However, the dynamics of some of the variables are relatively slow (social variables, e.g., language retention, most dimensions of fate control, demographic characteristics), since social change is rarely immediate and demonstrates a tendency to have a delayed, protracted response to external social, economic, and environmental stressors. Therefore, the baseline indicators and trends they exhibit are most useful to analyze on a multiyear basis, such as once every three to five years.

Scan Indicator Selection Principles

Although this section refers to non-renewable resource development indicators, it is important to point out that the nature of development can vary. At different stages of the resource development cycle the impacts may differ, and so will the indicators that are best designed to capture these influences. In the ISR, we primarily deal with implications of resource exploration and associated activities rather than direct resource extraction. It is also important to remember that these effects are complex and intertwined with other drives of economic and social change.

In order to assess more immediate impacts from external stressors, such as resource development, another set of indicators is needed. The ideal measurement and monitoring system should be based on the same guiding principles as ASI/IBI but include fast social variables. These are more sensitive to short-term economic events and, at the same time, representative of the different elements of well-being. In this study we propose to utilize the ASI/IBI framework of six social well-being domains to define a selected group of key indicators measuring impacts of resource development. We propose the following principles for identifying Resource Development Impacts Scan (REDIS) indicators:

- Community relevance
- Contemplation of six well-being domains
- Sensitivity/responsiveness to impacts
- Traceability over time, dynamic nature (fast variables)
- Comparability (in time and across regions; must be comparable with a regional or national baseline)
- Disaggregated nature (ability to measure at fine spatial and temporal scales)
- Little to no cost of measurement

Baseline Community Selection

We suggest that the effects of resource development on communities will be most identifiable if indicators are able to both show the dynamics (change from a previous state) and the relative position of the ISR (compared to territorial or national figures, or to a baseline or benchmark community). In other words, we are interested in the relative change against other parts of the territory or the country. This approach also allows taking into account broader economic changes taking place outside the ISR. The two obvious choices for the baseline are the territorial average and the city of Yellowknife. The advantage of the first is that it compares the ISR with all NWT communities. The disadvantage is that territorial averages are not specific to any particular place and also include the ISR itself. From this perspective, using Yellowknife is more advisable; however, it has to be understood that Yellowknife is a capital city and has different economic, geographic, and population characteristics than the ISR. Below, we use Yellowknife as a baseline for the calculations, unless otherwise is specified.

Material Well-Being

Since most immediate impacts of resource development appear in the material sector and economic variables are dynamic, fast, and responsive to these changes, we propose to use several economic indicators primarily focusing on income and employment (both are typically well-monitored). In addition, we include total employment income and its change rate, the Relative Income Power Index, and a measure of family income disparity (brought up among socially important variables during community interviews). We also use a standard measure of employment: the labour force participation rate. The Relative Income Power Index we define as the ratio between the living cost differential and income differential. In other words, this index accounts for both changes in income (relative to the baseline community/region) and changes in the cost of living (also relative to the same baseline community, e.g., Yellowknife). Considering living costs and income simultaneously is necessary since development, while it may increase personal income, may also inflate cost of living (an influence rarely measured in economic impact studies). The living cost differential can be obtained using figures from the government's isolated posts rates. However, the latter is not dynamic enough to reflect quick changes. Therefore, we propose to use the Consumer Price Index as a proxy of the cost of living. Again, the comparison to the baseline (Yellowknife) is critical to properly assess the changes specific to the ISR.

- Total income
- Total income change rate
- Relative Income Power Index (cost differential /income differential)
- Family income disparity (average household income/median household income)
- Percentage of low-income families (actual and relative to the baseline community)
- Participation rate

Closeness to Nature

Closeness to nature considers the ability to sustain linkages and have access to nature, including traditional economic activities, such as hunting, fishing, and gathering. This is also extremely difficult to measure since little data is typically available outside periodic surveys. For tracking impacts of resource development, we need variables that can be tracked on at least an annual basis. The ideal indicators are either harvest of traditional foods or the percentage of households with more than 50 per cent of their food consumption coming from the land/sea (used by NWT and Statistics Canada). Another indicator is obtainable from payroll and employment records: the percentage of labour force employed part time reflects the likelihood of participation in traditional subsistence activities (assuming that part-time employment is most conducive for active engagement in hunting, trapping, etc., when beneficiaries may have both time and means to be on the land/sea).

- Percentage of adults (15+) who hunt and fish
- Percentage of households with 50 per cent or more food consumption coming from country food
- Percentage of labour force employed part time

Cultural Vitality

The impacts of resource development on culture were mentioned during community consultations. We did not find consensus among Inuvialuit respondents on whether resource development had a largely positive or negative impact on cultural activities. Among positive elements, community members noted availability of financial support for cultural events, functions, and performances, as well as an overall "feeling good" mood that stimulates participation in cultural activities. On the other hand, the Inuvialuit reported a loss of interest and time for cultural activities by many beneficiaries involved with wage sector jobs. The cultural domain is extremely difficult to measure. Language retention used in the ASI/IBI framework is a slow variable and can't be used to detect immediate impacts. Participation in cultural and traditional activities may be an appropriate indicator that can be obtained through surveys (for example, as a part of the Community Based Monitoring Program (CBMP)). However, when survey data are not available, we suggest using the language use/proficiency variable with a particular emphasis on the youth.

- Participation in traditional cultural activities (recommended, but not available)
- Language proficiency/use among youth (15 to 24)

Education

Community members frequently mentioned the impacts of resource development on education, most specifically in respect to school dropout rates. The dropout problem and its links to resource development (as a source of low-skill job opportunities that may attract high school students, leading to them dropping out of school) is well-documented across the Arctic (AHDR 2014). The ASI/IBI monitoring framework focuses on educational attainment, a slow social variable. In the case of short-term impacts, it is more useful to focus on educational attendance, such as the high school dropout rate. This information is supposed to be obtainable from school records. However, these data were unavailable for this study, so we were compelled to use the next best attainment variable: the percentage of people (fifteen and older) with a high school diploma. This variable also allows for assessment of the human capital endowment of the region. It is important to acknowledge, however, that this approach measures only formal education and does not consider informal education, which is of special importance in the Arctic.

- High school dropout rate (recommended, but not available)
- Percentage of 15+ year old residents with a high school diploma

Health and Demography

This domain of human well-being is very complex and multifaceted. Infant mortality and teen birth rates are used in the ASI/IBI monitoring system (Petrov 2018). However, both these indicators have significant lags in responding to other socio-economic changes and are not suitable for assessing impacts of resource development in the short term. Instead, we propose using the data on sexually transmitted infections (STI), premature death (earlier than age of fifty), or the suicide rate (available from local police/hospital records). Premature death and the suicide rates are generic indicators of social cohesion and health. STIS can also serve as an indicator of social problems arising with resource development. If available, we also suggest including both net migration (in minus out migration) and total migration (in plus out migration) per 1,000 residents. The latter variable is designed to show the overall migration burden on the community that can be substantial with the high labour turnover during resource booms.

- Prevalence of sexually transmitted infections
- Premature deaths (including injuries and suicides)
- Total and net migration rate per 1,000 residents (recommended, but not available)

Fate Control and Community Empowerment

The ability of community to define its own destiny is a slow variable. Significant changes in fate control are associated with major shifts in legal, political, and civic spheres, such as land claim agreements, elections, court rulings, etc. Among the components of fate control, economic self-reliance is most sensitive to resource development impacts. The ability of a community to collect more royalties and taxes is an important empowerment mechanism as it allows the community to focus on its own priorities in respect to local development and, most specifically, local spending. We propose to use the per capita municipal budget (a fast variable) as a proxy of the community's fiscal independence. An alternative measure could be the percentage of support income in the total income of the community.

- Per capita budget expenditures (recommended, but not available)
- Percentage of support payments in total income

RESULTS

Below we describe our findings in respect to each domain of well-being.

Material Well-Being

In the 1990s and 2000s, the NWT experienced several waves of resource-related development, including off-shore oil and gas exploration, which is the most relevant for this study. The indicators considered below are guided within that perspective and, where possible and appropriate throughout the text, we attempt to relate the dynamics of the REDIS indicators and resource activities.

The total employment income can be considered a good crude indicator of the overall condition of the wage economy and a proxy of regional GDP. (Note that all figures are in current Canadian dollars, and they are not adjusted for inflation or purchasing power.) In the last decade, the total employment income collected by ISR residents (both IFA beneficiaries and not) has been steadily increasing (see figure 13.2). The rate of change, however, was unsteady, and rapid growth was

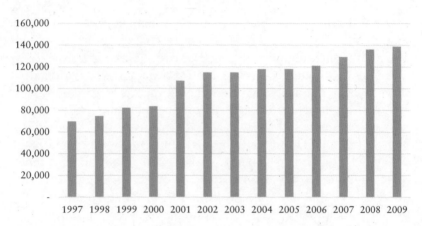

Figure 13.2 Total employment income

associated with active oil and gas exploration phases, especially around 2001 (when it jumped 28 per cent), and much less in 2002-04 and 2006-07 (see figure 13.3). A decline in the total employment income was registered in 2003. As mentioned in the earlier report issues for the IRC (Impact Economics 2014), the gains in benefits from resource exploration seem to stall in the late 2000s. In addition, we can see that the share of the ISR in the NWT total employment income (i.e., ISR weight in the territory's wage sector) has diminished. Still 10 per cent of all wages paid in the NWT are collected in the Inuvialuit Region.

Other studies investigated possible effects of resource-related activities in the ISR on earnings of IFA beneficiaries and Indigenous businesses (Impact Economics 2014). However, a direct approach to understanding the dynamics of incomes and wages may hide important considerations, such as the position of the ISR in respect to other NWT regions and the territorial capital. In selecting REDIS indicators, we advocate the comparative approach when economic variables in the ISR are compared to the territorial baseline. In selecting between the territorial average and Yellowknife, the advantage of the first is that it includes all regions of the NWT and thus compares the ISR with all NWT communities. The disadvantage of the territorial average is that it is not specific to any specific place and includes the ISR itself, which is not desirable. From this perspective, using Yellowknife is more advantageous, although it is indeed very different from the other NWT communities. Nonetheless, in relation to economic prosperity in the wage economy, Yellowknife is the top NWT community, and therefore presents a convenient baseline

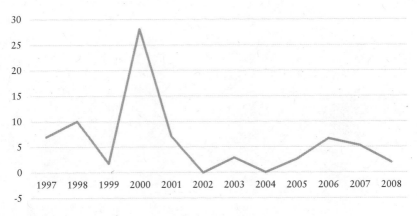

Figure 13.3 Total employment income change rate

for comparisons. Here and below we use Yellowknife as a baseline for most of the comparative assessments.

One of the REDIS key economic impact indicators is the Relative Income Power Index (RIPI) that is derived from comparing personal income and costs of living differentials between the ISR and Yellowknife. This gives us a measure that approximates the "real" income of the ISR population. The measure is also important as it registers both positive impacts of development (increasing incomes) and negative ones (rising consumer prices). Average personal income differential in the ISR has improved between 1997 and 2009, but even in the best year (2001), the average income of an ISR resident was more than 20 per cent below that in Yellowknife. More so, since 2001 we clearly observe a growing gap between the ISR and the capital. In 2009, average income in the ISR was 29 per cent below Yellowknife's average, while the cost of living (food price index) was 50 to 100 per cent higher compared to the territorial capital. In comparing the RIPI we used the food price index for Inuvik, so it is important to understand that the situation in other Inuvialuit communities could be even more troublesome. Figure 13.4 shows that when we consider both income and cost differentials, the relative income of ISR residents grew in the mid-2000s, only to decline later in the decade. Given the rapid rise in consumer prices in the late 2000s, the average relative ("real") income of ISR residents was lower than it had been ten years before. However, it is unlikely that the growth in cost of living in the ISR is directly attributable to the local resource development, but rather to growing fuel prices and other external forces.

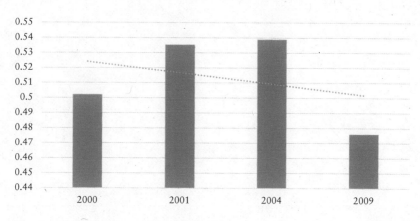

Figure 13.4 Relative Income Power Index

Another key REDIS economic indicator is the participation rate. The latter shows the degree of the population's engagement in the wage economy. Overall, between 1986 and 2009, the participation rate in the ISR remained steady, varying only between 70 and 75 per cent. It is difficult to clearly attribute its spikes to resource-related development impacts, but it appears that in the years of more intensive resource activities, the participation rate was higher (1999-2001, 2004-06), while in the quieter periods, it declined. Still, more than 25 per cent of the ISR labour force remains outside the wage sector, a figure very similar to other non-capital regions in the territory.

As a part of the community input, we included a measure of income inequality. Social justice is a key element of sustainable development, and the distribution of benefits of resource activity among local residents is an important issue for ISR residents. We used data on mean and median family income to calculate two indicators. Family income inequality index is the ratio between mean and median family incomes. A higher ratio indicates higher inequality since the mean is affected by few very high earners, while the median is not. Therefore, a lower ratio shows a more normally spread income distribution. Figure 13.5 demonstrates the dynamics of the family income inequality index between 1997 and 2009. Intensive resource development periods are expected to be associated with growing income inequality and data from the ISR reflects this pattern: the highest income inequality was observed from the early 2000s-2003, and then again in 2008.

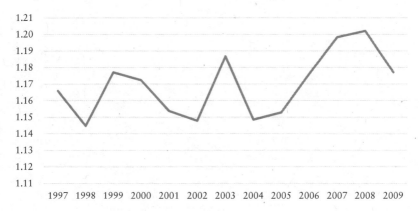

Figure 13.5 Family Income Inequality Index

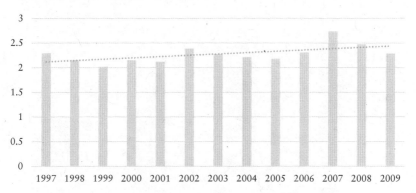

Figure 13.6 Relative percentage of families earning less than $30,000 per year

It is also instrumental to look at low-earning families (with income less than $30,000 per year). Overall, the number of such families dropped, but it cannot be considered a positive trend without adjusting for inflation or comparing with a baseline. Figure 13.6 shows such a comparison: the percentage of low-income families in the ISR vs. Yellowknife. As can be seen, the ISR has relatively more low-income families than Yellowknife, and the gap slightly increased over recent years.

Closeness to Nature

Maintaining closeness to nature, to traditional land and sea-based activities, is an important attribute of well-being of Arctic residents. The REDIS system suggests a number of variables to measure this

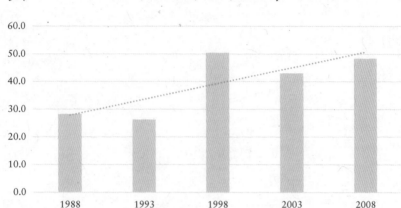

Figure 13.7 a. Percentage of those 15 years and older who hunt and fish

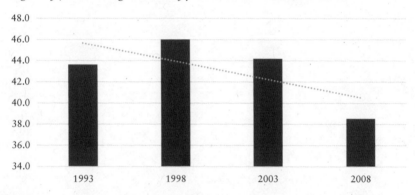

Figure 13.7 b. Percentage of households with 50% or more consumption of country foods

closeness. The main problem with most of the indicators, however, is intermittent, non-continuous data collection. The Aboriginal Peoples Survey and NWT Community Survey data are good sources of information, but they are available for a limited number of years. For this analysis there are few data points collected in 1988, 1993, 2003, and 2008. The first indicator is the percentage of adults (older than fifteen) who hunt and fish (see figure 13.7a). The impacts of resource development on sea and land-based activities are not simple. On one hand, with larger wage employment opportunities, we can expect that the number of people who participate in these activities will decline. On the other hand, availability of financial resources coming from wage employment enables people to purchase equipment and gear for hunting and fishing. In the ISR we observe an increase in the percentage of adults who hunt and fish, although the scarcity of data does not allow for a comparison of this increase to the resource development impacts. We do, however, see some evidence of correlation with employment income. At the same time, fewer families depend on traditional country food in their diets: 46 per cent of ISR households declared that 50 per cent of more of their meat and fish consumption came from country food in 1998 vs. just over 38 per cent in 2009 (see figure 13.7b). This difference, however, may not be an indicator of declined participation in hunting and fishing, but rather result from two processes: 1) an increased reliance on imported foods perhaps due to higher incomes, increased availability of imported foods, and more diverse diets; and 2) the declining role of land and sea-based activities as a source of food. (There is anecdotal evidence that more ISR residents now consider hunting and fishing a recreational undertaking rather than a harvest to feed the family.) Because of these changes, the ISR registers both a diminishing reliance on country food while maintaining high participation in land and sea-based activities.

Another possible variable that can be considered is part-time employment. It can be argued that part-time employment is more conducive to maintaining access to nature while still receiving wage income. In one respect, part-time employment can be advantageous for residents who spend considerable time on the land. In the ISR, the percentage of workers with part-time appointments (less than twenty-six weeks a year) has steadily declined in the study period, indicating that proportionately more wage sector employees are now working full time. This trend, along with other factors, may be responsible for the changing ways in which people utilize their limited time on the land/sea, to participate in hunting, trapping, and other activities more for recreational purposes rather than harvest-oriented ones.

Cultural Vitality

The recommended indicator for cultural vitality is participation in cultural activities and events. However, these data are not routinely collected. Another option, however, is to look at Indigenous language use and proficiency. Although it is a slow variable and data are scarce, the ability of Indigenous people to speak their mother tongue, especially among youth, can provide a valuable insight into cultural vitality. Figure 13.8 suggests that the language retention has

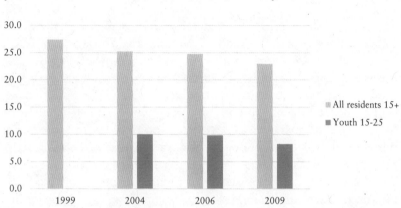

Figure 13.8 Percentage of the population who can speak an Indigenous language

diminished, especially for the younger population, with only 8.2 per cent of people aged fifteen to twenty-four years speaking an Indigenous language in 2009. It is difficult to attribute this change directly to resource development impacts. However, with growing employment in the wage economy, the relevance and use of the Indigenous languages tend to decline.

Education

Resource development has an impact on the education system, and that can be examined in multiple ways. In terms of the effects on enrollment and graduation, the high school dropout rate may be the most telling indicator. However, it requires acquiring and analyzing school records, which is a potentially time-consuming and costly procedure. As an alternative, REDIS uses high school attainment as a measure of student progress that is also reflective of human capital. The percentage of people fifteen years and older with a high school diploma is widely used for this purpose, although this information is only available in the census years. It is important to understand that this parameter is affected by high school graduation, migration, and a changing population age structure (in and out migration of high school graduates and a shrinking older population who tend to have less education). Therefore, over time we generally expect rising educational attainment. Of course, an influx of non-local workers with more than high school education, which is a typical consequence of resource development, will inflate these

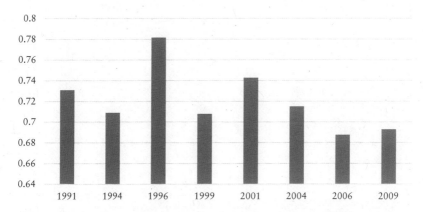

Figure 13.9 Percentage of the population 15 years and older with a high school diploma

attainment figures. In contrast, more frequent dropouts of local students who prefer to seek employment in the resource sector will negatively impact the indicator. In the ISR, 70 to 75 per cent of the population graduated from high school, and over the 2000s, we observe a generally downward trend, although this shift was not very conclusive (see figure 13.9). Still, in 2009 the proportion of high school-educated people in the region dropped from 2001, which was contrary to the territorial and national trends. It is possible to infer that this resulted from a combined influence of youth out migration and resource development, although more data must be analyzed to make a conclusion.

Health and Demography

A variety of health characteristics can be indicative of resource development impacts. Among "fast" ones are the occurrence of sexually transmitted infections (STI) and premature death. The first is reflective of social processes often associated with resource development. The second characterizes the overall health, alcohol abuse, violent crime, and traumatism as well as unnatural deaths (including suicide), but also can be used to assess the state of the health care system. Figure 13.10 reports diagnosed STI per 10,000 residents (based on three-year averages). The general upward trend is concerning, although other regions of the NWT also experienced such growth. In fact, the STI rate in the ISR almost doubled between 2001 and 2011. To assess relative

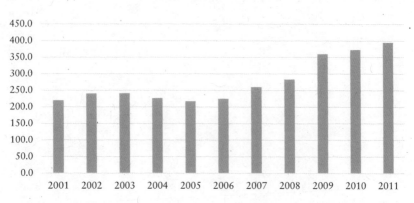

Figure 13.10 Sexually transmitted infections over a three-year period

change in STI mobility, compares STI in the ISR and in Yellowknife. Higher relative rates of infection were detected in the early 2000s (2001-04), but then stabilized at lower levels (though they were growing faster in Yellowknife than in the ISR). Still, levels of STI reported in the ISR were almost two and a half times higher than in the capital. It is likely that the increase in STI may be attributable to population change and turnover during the resource development cycles, as well as to improvement of diagnostics.

One of the ways to measure the prevalence of premature death is analyzing the potential years of life lost. This is a demographic indicator that represents a combined number of years of life that have not been lived by people due to premature death (earlier than predicted by age-specific mortality rates). If people tend to die early in their life, the years lost will accumulate. Figures 13.11a & b plots this variable in the ISR and comparatively in Yellowknife. We do observe the spike in 2001-03 that might be caused by an increased premature mortality due to injuries, suicides, and other unnatural causes. Unfortunately, we do not have data to verify that. Although the gap between the ISR and Yellowknife seems to be closing mostly due to improvements in the Inuvialuit Region, the difference is still staggering: the ISR has 1.6 times more years of life lost than the territorial capital.

Fate Control and Community Empowerment

Resource development may be a positive factor for extending fate control, if community can capitalize on resource activity and has a say in its progress. One means of doing this is fiscal self-reliance of community and its residents. In other words, an improved local budget

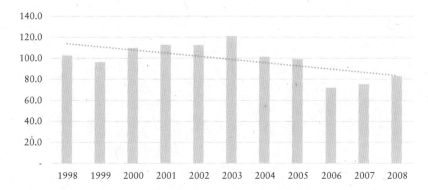

Figure 13.11 a. Potential years of life lost per 1,000 residents

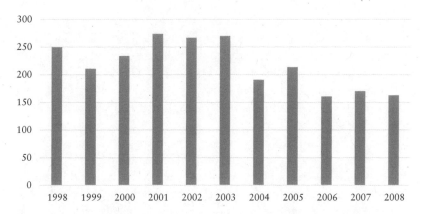

Figure 13.11 b. Relative years of life lost: ISR compared to Yellowknife

situation can be an important instrument to achieve desired spending on community-relevant needs instead of relying on transfers coming from outside, such as income support payments. Similarly, the ability of ISR residents to have their own means to pursue their life goals is key to personal empowerment. Both of these are economic elements of fate control that are sensitive to impacts of resource development.

In terms of dependency on income support payments, figure 13.12 shows the percentage of income support in total income received by ISR residents. The proportion has declined. By the end of the 2000s, over 2 per cent of the total income comes from such payments. This said, there are still many residents who depend on income support. The number of cases rapidly declined in the early 2000s, possibly due to improved job opportunities, but again increased after 2006. The per capita amount

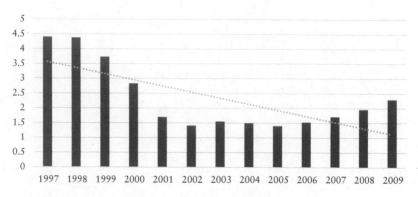

Figure 13.12 Income support payments as a percentage of total income

of support grew substantially from $300 in 2004 to $583 in 2011. It can be argued that although the overall dependency of ISR residents on support income diminished, the reliance of the neediest community members on income transfers may have in fact increased.

DISCUSSION AND CONCLUSIONS

Earlier studies utilized the ASI-based set of measures to compute slow variables of socio-economic dynamics and undertake regional comparison of social well-being in the ISR (Petrov 2018). It was concluded that the ISR experienced some considerable internal differences (Inuvik vs. other communities) and was generally better off than other non-capital NWT regions and Nunavut settlements (except in the area of language retention). The ISR made significant gains in well-being since the settlement of the land claim, however many socio-economic problems have persisted. At the same time, the well-being gap between the ISR and southern Canada has not been closed. This study developed into the implemented set of Resource Development Impact scan (ReDIS) indicators. These indicators and the monitoring system are based on the guiding principles established by ASI/IBI, but they include fast social variables which are sensitive to short-term economic events and, at the same time, are representative of the different elements of well-being.

This approach is also well-aligned with available data and utilizes various measures continuously collected by the NWT Bureau of Statistics and the IRC. It is, however, important to remember that most of the indicators do not measure impacts directly, and thus should be interpreted with this limitation in mind.

The overall analysis of the REDIS produces a complicated and mixed picture of socio-economic conditions in the ISR. Some trends reflect possible positive derivatives of (resource) development, while others suggest that development benefits have not led to improving social conditions or may even have caused negative externalities. Generally, the ISR has been moving toward improving key social-economic conditions. However, it is possible to conclude that resource development has not resulted in substantial and equitable social and economic improvements in the ISR. Below, we focus on the most concerning findings.

First of all, resource-related development did not result in commensurable improvement of material well-being. The percentage of part-time workers has declined so that proportionally more wage sector employees were working full time by the end of the study period. At the same time, more than 25 per cent of the ISR labour force remained outside the wage sector. Despite growing nominal incomes, we observe the rapid rise in consumer prices in the late 2000s. As a result, the average relative ("real") income of ISR residents declined. Intensive resource development periods were associated with growing income inequality: the highest income inequality was observed in the early 2000s-2003, and then again in 2008. The ISR had more low-income families than Yellowknife, and the gap slightly increased over the study period. Although the overall dependency of ISR residents on support income diminished, the reliance of the neediest community members on income transfers may have in fact increased.

This evidence suggests that despite the influx of jobs at certain points of the resource exploration cycle, the benefits to local community remained modest at best and largely confined to the early 2000s. Negative externalities of resource development, such as income inequality and rising costs of living, however, were clearly present. In essence, the ISR was unable to capitalize on development opportunities in the long run. Typical characteristics of the resource curse, including the leakage of capital, labour force turnover, lack of institutionalized channels to secure economic benefits, and dependency on external market forces were likely at play. This said, it is important to reiterate that in comparison to other NWT non-capital regions, the level of material well-being of ISR residents has been mostly higher.

Over time, fewer ISR families depended on traditional country food in their diets, which indicates a possible growing dependency on imported foods. This, however, may not be an indicator of a declined participation in hunting and fishing, but may suggest a changing nature of these activities (i.e., shift to more recreational activities). Declining part-time employment may also be detrimental for the traditional sector. Language retention has diminished, especially for the younger population, with only 8.2 per cent of people aged fifteen to twenty-four years speaking an Indigenous language in 2009. It is difficult to attribute this change directly to resource development impacts. However, with growing employment in the wage economy, the relevance and use of the native language tends to decline.

Finally, we do not see the improvement in human capital: although 70 to 75 per cent of the population graduated from high school, in the 2000s we observe a slight downward trend. Human capital is likely leaking from the ISR communities to other regions and/or schooling could be affected by resource development itself (students drop out of school to pursue employment). The general upward trend in STI is also concerning, although other regions of the NWT also experienced such growth. Finally, while the difference in premature deaths in the ISR and Yellowknife seems to be closing in the 2000s (mostly due to improvements in the Inuvialuit Region), the difference remains considerable. These results are interesting to compare with Rodon et al. and Schott et al. in this volume (Chapter 3), who examine case studies of resource project impacts and benefit distribution in the Eastern Canadian Arctic.

LIMITATIONS AND FUTURE DIRECTIONS

The main limitation of this study is its reliance on existing secondary data. We had no option but to use a "BAD" (Best Available Data) approach. As a result, most proposed and used indicators are only proxies and even these proxies are limited in their utility by the nature and quality of the data in hand. However, the indicators utilized here still provide a snapshot of socio-economic conditions and could be useful tools of impact assessment if interpreted carefully. Another primary concern is the lack of direct linkages between the socio-economic variables available and resource development. Albeit most of the proposed indicators have been demonstrably related to social and economic impacts in other regions, we did not have an opportunity to confirm these relationships in the ISR explicitly. It is preferable that the Resource Development Impacts Scan is complimented by in-depth community-based data collection and economic and social data analysis from primary sources. Recommended future directions include: (1) developing procedures that will enable community-based collection, management, and analysis of data by local actors; (2) collecting necessary data and expanding the IRC database; (3) developing and implementing an integrated Inuvialuit indicators monitoring system to inform the region's stakeholders and aid in the IRC's decision-making and ensure community awareness.

ACKNOWLEDGMENTS

The authors are grateful to the Inuvialuit Regional Corporation for their long-term co-operation and support, and to the Social Sciences and Humanities Research Council (SSHRC) and RESDA for funding this project. We are also indebted to fellow colleagues Joan Nymand Larsen, Gail Fondahl, Peter Schweitzer, and Simon Routh for their advice and assistance. Thanks are owing, as well, to community members of the region for their support, and especially to Bob Simpson for his valuable input and recommendations.

REFERENCES

AHDR. 2014. *Arctic Human Development Report II: Regional Processes and Global Linkages*, edited by Joan Larsen and Gail Fondahl. Nordic Council of Ministers: Copenhagen.

AMSA. 2009. *Arctic Marine Shipping Assessment*. Report. Arctic Council.

Fonda, Marc and Erik Anderson. 2009. "Diamonds in Canada's North: A Lesson in Measuring Socio-Economic Impacts on Well-Being." *Canadian Issues:* 107.

Impact Economics. 2014. *Measuring the Effects of Major Projects in the Inuvialuit Settlement Region*. Yellowknife.

Kruse, Jack, Marie Lowe, Sharman Haley, Ginny Fay, Larry Hamilton, and Matthew Berman. 2011. "Arctic Observing Network Social Indicators Project: Overview." *Polar Geography* 34, no. 1 (March): 1-8.

Larsen, Joan N., Andrey N. Petrov, and Peter Schweitzer, eds. 2015. *Arctic Social Indicators: Implementation*. TemaNord. Nordic Council of Ministers: Copenhagen.

Larsen, Joan N., Peter Schweitzer, and Gail Fondahl, eds. 2010. *Arctic Social Indicators. ASI II: Implementation.* Stefansson Arctic Institute, Akureyri, Iceland.

Michalos, Alex C., ed. 2014. *Encyclopedia of Quality of Life and Well-being Research.* Dordrecht: Springer Netherlands.

Mitchell, Ross E. and John R. Parkins. 2011. "The Challenge of Developing Social Indicators for Cumulative Effects Assessment and Land Use Planning." *Ecology and Society* 16, no. 2 (June): art. 29.

Ozkan, Umut. R. and Stephan Schott. 2013. "Sustainable Development and Capabilities for the Polar Region." *Social Indicators Research* 114, no. 3 (December): 1259-1283.

Petrov, Andrey N. 2014. *Measuring Impacts: A Review of Frameworks, Methodologies and Indicators for Assessing Socio-Economic Impacts of Resource Activity in the Arctic.* Draft Gap Analysis Report #3. Whitehorse: RESDA.

–2018. "Inuvialuit Social Indicators: Applying Arctic Social Indicators Framework to Study Well-Being in the Inuvialuit Communities." *The Northern Review* 47: 167-185.

–2020. "Fate Control and Sustainability in Arctic Cities." In *Urban Sustainability in the Arctic: Measuring Progress in Circumpolar Cities,* edited by Robert Orttung, 163. Berghahn Books.

Petrov, Andrey N. and Tatiana Vlasova. 2021. "Towards an Arctic Sustainability Monitoring Framework." *Sustainability* 13, no. 9: 4800. https://doi.org/10.3390/su13094800.

Poppel, Birger, ed. 2015. *SLICA: Arctic Living Conditions: Living Conditions and Quality of Life Among Inuit, Saami and Indigenous Peoples of Chukotka and the Kola Peninsula.* Nordic Council of Ministers.

Tysiachniouk, Maria, Andrey N. Petrov, and Violetta Gassiy, eds. 2020. *Benefit Sharing in the Arctic: Extractive Industries and Arctic People.* MDPI. https://doi.org/10.3390/books978-3-03936-165-6.

Vanclay, Frank. "Conceptualising Social Impacts." *Environmental Impact Assessment Review* 22, no. 3 (May): 183-211.

Vlasova, Tatiana, and Sergey Volkov. 2013. "Methodology of Socially-Oriented Observations and the Possibilities of their Implementation in the Arctic Resilience Assessment." *Polar Record* 49, no. 3 (July): 248-253.

Wilson, E., & Stammler, F. 2016. "Beyond Extractivism and Alternative Cosmologies: Arctic Communities and Extractive Industries in Uncertain Times. *The Extractive Industries and Society,* 3, no. 1: 1-8.

14

Measuring Social Impacts: Building Tools for Understanding Community Well-Being

Todd Godfrey and Bruno Wichmann

The world demand for minerals has been growing steadily and has doubled in the last thirty years (World Mining Data 2019). New technologies allow companies to extract resources from areas that were previously unfeasible for mining, generating markets for new types of minerals (Hannington et al. 2017; Alonso et al. 2012). Canada is no exception. The technical hardships associated with mining in the Arctic fade in comparison to the prospects of profitable operations in the resource-rich North. The mining industry has been an important engine of economic growth in Northern Canada. In 2017, the Canadian mining industry accounted for 19 per cent of the value of all Canadian exports. In fact, Canada is one of the leading countries in global mineral markets (Mining Association of Canada 2019).

An important question remains: what are the benefits of mining for local Canadians? This literature is constantly evolving and many findings are discussed in previous chapters. For example, in Chapter 3, Schott et al. find that mining creates employment and business development opportunities. Chapter 4 by Hodgkins discusses the potential for vocational education and training opportunities generated by resource development.

However, mining activities can generate concerns about possible negative social impacts, especially for small nearby communities. For example, mining can generate vulnerabilities in local housing markets due to their impacts on affordability and accessibility (Chapter 8, Freeman and Christensen). Many scholars are also concerned about the shortcomings of regulatory frameworks related to mine reclamation and their negative environmental impacts (Chapter 9, Dance et al.).

In light of these effects of mining, there is increasing demand for studies that inform policy interventions to ensure a fair distribution of benefits and costs of resource development, balancing economic opportunity and local well-being. There is a long documented history of research examining the social impacts of large resource development projects on small rural communities. Papers in this literature find that social disruptions caused by large projects can lead to an array of problems that can have an adverse impact on community well-being (Freudenburg and Jones 1991; Gramling and Freudenburg 1992).

There is a consensus between researchers and the communities that resource development can contribute to social disruptions and negative impacts. This concern is grounded in a literature that typically explores qualitative research based on methods like focus groups, in-depth interviews, and case studies. While qualitative research has flourished, quantitative work has not. Statistical estimates of social impacts based on quantitative methods are rarely available. This is partially due to the empirical challenges associated with obtaining large datasets and developing robust estimates of the various effects of mining. Such estimates, however, can be extremely useful in understanding community well-being and, as a result, have the potential to significantly improve the design of regulatory frameworks.

The chapter discusses methodological challenges related to the statistical identification of social impacts of mining on communities. It builds on the indicator chapter by Petrov et al. (13) in this volume by discussing methods that can be applied to estimate quantitative impacts (and hence develop monitoring systems) using the rich set of social indicators examined by those authors. In the Background section, we briefly review the literature on social impacts from mining and report two major findings: 1) there is overwhelming qualitative evidence of several types of social impacts, but little quantitative work trying to establish causal effects; and 2) the lack of impact estimates has been used as an argument for deferring more stringent and objective regulatory framework regarding social impacts.

In the Methods section, we describe different approaches that can be used to estimate social impacts. We start with a discussion of the limitations of standard estimation approaches, highlighting empirical challenges to identifying causal linkages between mining and its impact on communities. Specifically, we discuss the conditions under which estimates fail to capture social impacts and represent actual correlations between mining and pre-existing local trends. We then discuss methods that can be used to disentangle social change resulting from mining, as opposed to other forces of change. Put differently, we discuss how naïve methods fail to estimate causal impacts, focusing on a set of approaches that have the potential to differentiate pre-existing local trends from significant adverse impacts.

We conclude with a review of an application of matching models to estimate mining effects on alcohol consumption. We also offer remarks about how social impact estimates can be integrated in regulatory frameworks to enhance community well-being and create more sustainable economies in the context of resource development. Finally, we present a brief perspective on the role of quantitative methods in complementing research in non-quantitative disciplines in social sciences.

BACKGROUND

Most of the literature that examines social effects of mining uses qualitative methods. For example, Caxaj et al. (2014) study mining in Guatemala and conclude that mining activities contribute to conflicts that lead to a segregated and militarized community. The resulting social implications include an increase in threats, intimidation, kidnapping, and other violent activities. There were also correlated problems with mental illnesses such as anxiety and depression. Bebbington and Bury (2009) find that similar social conflicts are linked to mining development in Peru. At a more individual level, mining has been associated with increases in drug use, gambling, prostitution, violent crimes, injuries, and mental and physical illness (Shandro et al. 2011; Mactaggart et al. 2016; Gibson and Klinck 2005). Other papers suggest that women are more vulnerable to challenges associated with mining operations, as they are victims of increased violence and do not necessarily accrue economic benefits (Shandro et al. 2011; Sharma 2010).

Indigenous groups throughout the world are also susceptible to the impacts of mining (Acuna 2015). Mining activities disrupt unique Indigenous cultures and their strong connection to the land. For example, Brenda Parlee explains in Chapter 12 of this volume that Indigenous land use can be viewed as a deterrent to development. The struggle between Indigenous groups over mining operations occurs in both developing and developed countries. Examples are abound: bauxite mining on sacred land in Odisha, India (Temper and Martinez-Alier 2013); the conflicts between Wopkaimin communities and the Ok Tedi Mine in Papua New Guinea (Hilson 2002); the approval of the Jabiluka Mine and its consequences for the Mirrar people in Australia (Banerjee 2000); to cite a few. Mining can also have significant negative effects on Indigenous peoples in Northern Canada. For example, regulators have allowed companies to develop mines in areas that are sensitive barren-ground caribou habitat, and Indigenous peoples that spend large amounts of time on the land are concerned about the decline in caribou populations (Parlee et al. 2007; Parlee et al. 2018).

The discussion above is not meant to offer a deep and comprehensive review of the literature associated with social impacts of mining. Instead, our goal is to highlight a trend: most of the literature focuses on qualitative methods (e.g., one-on-one interviews), using small sample sizes. While these methods clearly point to a strong correlation between mining and various social challenges, none of these papers attempt to estimate a causal impact of mining on community well-being.

This knowledge gap exists for many dimensions of social well-being. Take the issue of alcohol consumption in northern communities, for example. Since the 1970s, communities have told decision-makers that resource development has caused an increase in alcohol use and abuse (Berger 1977). There is overwhelming qualitative evidence that mining contributes to excessive alcohol consumption (Bowes-Lyon et al. 2009). However, to date, there are no attempts to estimate a causal impact of mining on alcohol consumption. The literature recognizes this gap. As Gibson and Klinck explain, it is not clear whether alcohol consumption is "[...] exaggerated, subdued, or unaffected by the presence of mining" (Gibson and Klinck 2005).

When it comes to policy, the lack of causal interpretation of findings in this literature has practical and concrete consequences to communities. In Canada, and in several other countries, the impact of resource development projects is evaluated in an Environmental Assessment (EA) process where a regulator (e.g., review board) has the authority to establish the conditions under which projects will be implemented and, in some cases, entirely reject a project. Various economic agents participate in this process, but noticeable groups are local and provincial/territorial government, the proposer of the project (e.g., a mining company), Indigenous peoples and local communities, and environmental groups. Each stakeholder has their own interests, incentives, and resources to allocate to this process. All groups participate in the EA process in an attempt to shape the development project to their desired specification, which may include, for example, even a desire to completely reject the project (which is often the case for local communities). The regulator is in charge of making up or down decisions. For projects that are approved, the regulator also has the power to determine mitigation measures to accompany the development of the project.

The EA process also considers not only environmental impacts but also possible negative social impacts that resource development projects may have on communities. Take, for example, the Dominion Diamond Jay Project, which was the subject of the Environmental Assessment EA1314-01 in the Mackenzie Valley, Northwest Territories, governed by the Mackenzie Valley Environmental Impact Review Board. Among other things, the EA1314-01 evaluated the impact of diamond mining on community health and well-being. As stated in the concluding report of February 1, 2016, the Government of the Northwest Territories (GNWT) "expressed challenges to identifying causal linkages between diamond mining and its impacts on communities" and also argues that "it is important to differentiate negative trends from 'significant adverse impacts'" (EA1314-01, 2016). Moreover, "the GNWT cautioned the Review Board on the interpretation of the Communities and Diamonds Report data, saying it `demonstrates a correlation between resource development activity and community wellness—it does not provide causality or speak to the magnitude of a trend`" (EA1314-01, 2016). In addition, the proposer (Dominion) argues that "it is impossible to identify what effects may be caused by the Jay Project" and that difficulties arise "in disentangling social change resulting from mining specifically, as opposed to from other forces of change" (EA1314-01, 2016). The outcome of this process was that the project was approved with no concrete determination of provisions to compensate communities for possible negative social impacts. In conclusion, the Review Board designed Measure 8-1 that states that "the GNWT will actively investigate and address linkages of diamond mining effects on the health and well-being of affected communities" (EA1314-01, 2016). It is unclear if communities will ever receive specific provisions against negative social impacts.

METHODS

The goal of this section is to discuss how quantitative methods can contribute to the debate surrounding community well-being, and as such continues the discussion started in Chapter 13 by Petrov et al. What can the data tell us about the impacts of resource development on communities? What are the methods that can be used to extract impact information from large datasets? Under what conditions do these estimates reflect causal effects? In other words, what are the empirical challenges we need to overcome to estimate the impact of resource development on social indicators? These are questions that need to be addressed carefully for impact estimates to be interpreted as causal estimates.

Step One: Gather Your Data

In addressing the questions above, our goal is not to offer a formal exposition of empirical methods; rather we aim to review empirical approaches of impact evaluation often used by economists and quantitative researchers. In doing so, we offer some intuition about their statistical properties (i.e., the assumptions required by each approach).

The first step in implementing quantitative research is to gather data on the social outcome of interest. Tailoring and implementing a survey allows the researcher to address specific needs by gathering information of local relevance (e.g., refer to the discussion by Petrov et al., Chapter 13). This process can be very advantageous; however, it is also expensive. Nevertheless, a variety of social economic indicators are available through government sponsored surveys. For example, Statistics Canada developed a survey to collect health-related information from Canadians twelve years of age and over—the Canadian Community Health Survey (CCHS). The CCHS was first implemented in 2001 and covers residents of the ten provinces and three territories, including arctic/subarctic regions. A lot of information is available, including indicators that capture access to health care services, mental health, depression, sleep habits, workplace injury, community-based care, activities of daily living, and food security, among others. These data capture information about important aspects of life and well-being, and all of these aspects can be potentially affected by social disruptions in local communities.

As a motivating example, let us go back to the issue of alcohol consumption, which was connected to mining as early as the 1970s (Berger 1977). The CCHS has a group of questions related to alcohol use. Godfrey (2017) uses the CCHS data to study alcohol consumption in Northern Canada (we will review these results below). Researchers interested in mining communities may want to use the CCHS data to determine whether or not changes in social indicators (e.g., alcohol consumption) can be attributed to mining operations. That is, the interest might be on using the data to separate causal effects from correlations. But what exactly are we referring to when we talk about the need to distinguish causation from correlation?

Correlation Versus Causation

When evaluating potential social impacts of mining activities on social indicators, we need to carefully separate correlations from causal effects. Correlation happens when two phenomena manifest themselves together, but this does not mean that one causes the other. Sometimes the relationship can be spurious or coincidental.

The Challenge of Establishing Causality

Consider the following hypothetical example. A mining company starts operations in a region close to a community. A few years later, we measure alcohol consumption in the community and find an increase. From a statistical standpoint, these two observations alone are not enough to attribute the change in the alcohol consumption trend to the start of the mining operations. The challenge in identifying a causal effect is that it is impossible to know with certainty what would have happened to the community if the mining company had not come to the region. The reason is simple: we cannot observe the community both with and without the mine, at the same time period. In technical terms, we do not have *counterfactual* information. In this example, the true social impact of mining operations is equal to the alcohol consumption after the start of mining minus the counterfactual effect (the level of alcohol consumption of the community at the same period in the hypothetical situation that the mine did not come to the region). We now have a theoretical construction of what the causal impact is, as illustrated by figure 14.1.

Figure 14.1 Measuring causal impacts

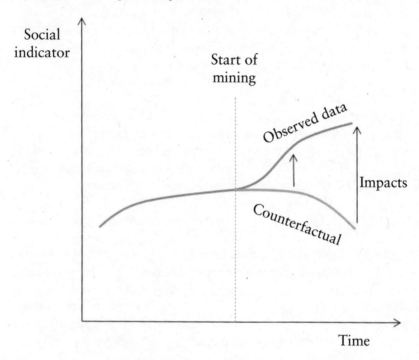

Figure 14.2 The counterfactual and the causal impact

When we work with real-world data, we can assess trends and visualize the overall path of a social indicator by plotting its levels over time. A time plot is also useful in understanding how the concept of a counterfactual delivers a measure of causal impact. Figure 14.2 shows the plot of a social indicator (measured using observed data) over time. Assume that at some point in time a mine starts operations. While we continue to observe how the social indicator behaves after the start of mining, we cannot make any statements about mining effect without a counterfactual.

Suppose information about the counterfactual is available, even though in real life it never is. Figure 14.2 shows the difference between the levels of the community indicator observed in the data (that is, under the influence of a mine), and the counterfactual (the time path of the community indicator had the mine never come to the region). At this point, this is a theoretical exercise, but a valuable one as it allows us to visualize the mining impact at any moment in time. Note that in this hypothetical example, the impacts are initially small, but get larger with time as the observed data diverge from the counterfactual.

Randomization

Randomization is the gold standard for statistical identification of causal impacts of interventions. The idea is to randomly divide the sample into two groups: treatment and control. The intervention is assigned only to individuals in the treatment group. The researcher collects post-intervention data from both groups. The control group is used as the counterfactual for the treatment group. Because intervention was assigned randomly, on average, we expect that individuals in the control group are similar to those in the treatment group; hence the randomly generated counterfactual is expected to be a robust one.

The goal of this experimentation is to eliminate selection bias related to the intervention. This is the strongest case that can be made for a counterfactual because under proper randomization the control group is identical to the treatment group on average. As a result, differences on the outcome indicators can be attributed to the intervention. The randomization design breaks correlations between the treatment and the outcome. This eliminates selectivity bias (or self-selection bias), where the treatment is allocated to especial individuals, i.e., ones that are not representative of the general population.

While experimental approaches hold desirable statistical problems, clearly they are not suited to the evaluation of interventions that can cause negative effects and harm individuals. Moreover, it would be practically impossible to design an experiment to inform impacts of large development projects. For example, how would one randomly assign resource projects to different regions? Therefore the estimation of causal social impacts from mining and other resource development projects must face the challenge of using observational data.

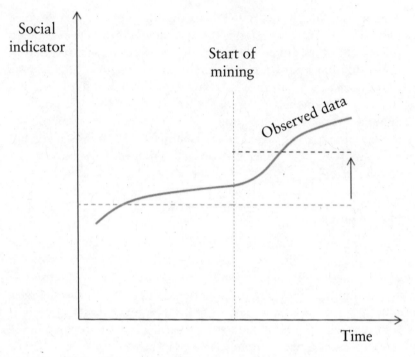

Figure 14.3 Using "before" information as an estimate of the counterfactual

Before and After Comparisons

The counterfactual is a theoretical concept. The empirical problem of establishing causality is therefore one of estimating a credible counterfactual. With no actual counterfactual available for us to compare to our estimate, the quality of impact estimates depends on the assumptions that are necessary for the proposed counterfactual to be a valid one. Typically, an estimate is considered weak if its consistency relies on strong (often unreasonable) assumptions. On the other hand, robust estimates maintain their causal interpretation under fewer and weaker assumptions.

Often differences between the levels of the social indicator before and after the intervention (for our purpose, the start of mining) are used as a measure of impact. Note that, in the context of our discussion, a before and after comparison is an approach that measures impact using before information as an estimate of the counterfactual. Figure 14.3 illustrates this approach. The green dotted line represents the average of the indicator before the start of mining, and it is extended to the after-mining region of the graph representing the fact that this measure serves as an estimate of the counterfactual.

A before versus after comparison is considered by most to be a naïve estimator because causal interpretation relies on the strong assumption that, in the absence of treatment, behaviour in the future (without treatment) and in the past are identical. In real life, too many things can change for this assumption to be reasonable. Take the consumption of alcohol for example. Economic research has shown that the demand for alcohol varies with various factors including, just to name a few, prices, accessibility, regulations, advertisement regulations, factors regarding complements (e.g., cigarette regulation), and even the weather (Baltagi and Goel 1990; Manning et al. 1995; Nayga 1996; Saffer and Dave 2002; Silm and Ahas 2005; Picone et al. 2010). Therefore, it is unlikely that past behaviour can make for a good counterfactual estimate, especially considering possible interactions between all these factors and mining.

The Difference-in-Differences Estimator

Consider two similar communities, however, one is situated in a region with mineral deposits (mining community) and the other is not. At some point in time, a mining company starts operations nearby the mining community (the mining community is the treatment community). There is no mining in the other community. This second community is the control community.

The difference-in-differences (DD) approach establishes a counterfactual based on the control group. The idea is to not only compare changes in the indicator before and after mining, but also to compare after-mining levels of the indicator between the two communities. The DD estimator accounts for pre-existing differences between the mining and control communities. The power of the estimator comes from the assumption of parallel trends; that is, in the absence of mining in the treated community, the time paths of the two communities are identical. This allows us to estimate impacts controlling for both observable and unobservable pre-existing differences in the two communities. Under parallel trends, the counterfactual is consistently estimated by simply adjusting the control data to the levels of the treated community through a parallel shift (see figure 14.4).

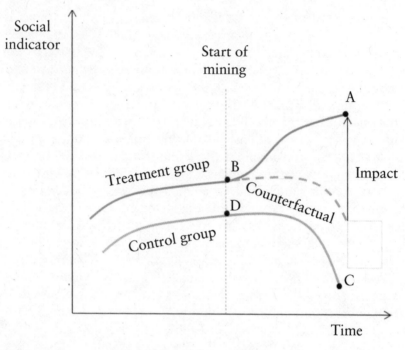

Figure 14.4 Using a control group as the counterfactual

Note that the longer the time frame is between the evaluation moment (after mining) and the baseline (before mining), the more vulnerable or unreasonable the assumption of parallel trends becomes.

Matching

The implementation of a difference-in-differences strategy relies on having a proper control group. However, the task of finding a good control group can be a difficult one. It is possible for some of the observations in the control group not to be similar to observations in the treatment group. Conversely (but different), an observation in the treatment group may not be similar to any of the observations in the control group. In either one (or both) of these situations, the assumption of parallel trends between the mining community and the control community may not hold.

There could be several reasons for such a mismatch, depending on each specific evaluation. Let us go back to the evaluation of mining impacts on alcohol consumption. Parkins and Angell (2011) find that transient workers residing in small communities lack social support and resort to alcohol as a means of coping with feelings of isolation and loneliness. As a result, it may be questionable to use control data from a community with a small share of transient workers to build a counterfactual to a mining community with a high share of transient workers. In other words, the time path of alcohol consumption in these communities may not be parallel.

To address issues of poor specifications of control and treatment groups, King and Nielsen (2019) suggest the use of matching models to "prune" observations. The goal is to enhance the quality of the match between individuals in the treatment and control groups. The approach keeps "good" data and discards "bad" data, hence the terminology "data pruning." The estimation of difference-in-differences impacts using a matched sample improves statistical robustness and makes for a stronger case of identification of causal impacts. That is because it makes the parallel trend assumption more reasonable. Of course, that is the case in some instances more than others.

Several matching models are available, and, while the technical aspects of these methods can be daunting, their goal is very simple: finding good matches between observations in the control and treatment groups. King and Nielsen (2019) classify matching models into two classes, namely Mahalanobis distance matching and coarsened exact matching. Mahalanobis matching simply calculates an index that represents a weighted difference between observable data of individuals in the treatment and control groups. Observations with differences greater than a pre-specified level (the caliper) are pruned. For instance, when evaluating mining and alcohol consumption, age can be a matching criterion. If young individuals are more likely to be flexible and move to find work, then pairs of individuals with significant age difference should be discarded and not be part of the matched sample. Coarsened exact matching does not rely on the specification of a caliper. Instead, the observable characteristics of individuals in the control and treatment group must equal some pre-specified coerced value. For instance, variables may be naturally coarsened at obvious breakpoints, such as high school attainment in the case of education. A matched sample can then be constructed imposing exact matches for high school degree, marital status, gender, and homeownership.

What does this mean in practice? It means that, when estimating mining impacts, we should only use observations of a group (treatment or control) if a reasonable match exists in the other group. For example, suppose we are interested in evaluating the impact of mining on alcohol consumption. Variables such as education, marital status, gender, and homeownership are essential determinants of alcohol consumption trends. In a coarsened matching procedure, individuals are compared with one another based on these observables. For instance, if an individual in the treatment group has not completed high school, and is single, male, and not a homeowner, then that individual should be included in the analysis only if he can be matched to an individual in the control group with the same exact characteristics.

When this process is followed for all individuals, the resulting sample is a pruned or matched sample, where the assumption of parallel trend is more likely to hold. The causal estimate of the mining impact on the indicator would simply be the difference-in-differences estimate applied to the matched sample.

USING MATCHING MODELS TO EVALUATE MINING IMPACTS

As discussed in previous chapters, mining is often associated with local social impacts, one of which is excessive alcohol consumption in northern communities. While the relationship between alcohol use and abuse and mining has been extensively discussed in the literature (Berger 1977, Bowes-Lyon et al. 2009), most papers examining this matter use qualitative approaches. As part of a research project funded by RESDA, Godfrey (2017) contributes to this literature by applying a matching approach to examine the issue of differentiated alcohol consumption between mining and non-mining communities.

The research collected socio-economic data from the Canadian Community Health Survey (CCHS) and mining data from the GEOSCAN Database from Natural Resources Canada to develop a quantitative matching model and obtain estimates of the impact of mining on alcohol consumption. Information from the CCHS allowed for the construction of alcohol consumption indicators. The survey elicits information on alcohol consumption for each one of the seven days leading to the respondent interview. Several indicators of alcohol consumption can be developed, but a simple and powerful one is the number of alcoholic drinks the respondent consumed over the week.

Table 14.1
Alcohol consumption compared to distance from mine

	10 km	20 km	30 km	40 km	50 km	100 km
Impact (# of drinks)	2.010***	2.010***	1.382*	0.866	0.594	0.527
	(0.691)	(0.691)	(0.654)	(0.586)	(0.509)	(0.417)

Sample Size: N=3928. *** P < 0:01, ** P < 0:05, *P < 0:1. Standard errors are reported in parentheses.
Source: Table 5.1, Godfrey (2017)

The sample of Godfrey (2017) consists of 7180 individuals in the territories of Yukon, the Northwest Territories, Nunavut, and the Nunavik region of Quebec and Labrador. Godfrey reports an average of three alcoholic drinks consumed over a week, with a standard deviation of 7.7, which suggests a large variation in alcohol consumption in the North.

Godfrey (2017) uses the Statistics Canada's Postal Code Conversion File to approximate the Global Positioning System (GPS) coordinates of the place of residence (communities) of individuals in the sample. By cross-referencing the position of individuals against GPS coordinates of mining sites, Godfrey is able to determine whether or not the alcohol consumption elicited in the CCHS is from an individual in a mining community.

The geographical location data was used to assign individuals into treatment (all alcohol consumption close to mining activities) and control groups (far from mining activities). Godfrey (2017) uses six classifications of proximity that vary the distance threshold between the residence of respondents and the mining site. These are: 10 km, 20 km, 30 km, 40 km, 50 km, and 100 km.

The matching of individuals of the treatment and control groups was based on variables that capture demographics, wealth and employment, and human capital. Once the sample is matched (with propensity score matching), the control group serves as the contrafactual for comparison of alcohol consumption in mining communities. The estimated impacts, reported in table 14.1 below, show that individuals in mining communities (within 20 km from a mining site) consume on average two more drinks than individuals in control communities. This effect decreases and becomes statistically insignificant as distance between location of residence and mining increases.

DISCUSSION

Valuable resources such as oil, natural gas, and coal are naturally formed as heat and pressure is applied to organic matter over a period of millions and millions of years. But the fact that they are not replenished in an economically feasible time frame does not mean that depletion is the inevitable outcome. Several economic models of optimal extraction predict asymptotic depletion, that is, the resource rate of extraction is such that the reserves are exhausted as time goes to infinity. This is true even ignoring the possibility of exploration to increase the resource stock. Hence, questions surrounding the sustainability of economies that rely on non-renewable resource extraction are not questions about the consequences of depending on a natural resource with a finite stock. Instead, the key element to sustainability is to understand incentives faced by different economic agents, and how their interactions generate benefits and costs, including external effects, or externalities.

Economic sustainability refers to the ability of a community to maintain economic surplus for an indefinite period of time. Even though costs of extraction can increase as resources are depleted, indefinite surpluses are still possible through price effects. That is, as the natural resource gets scarcer, prices tend to increase reflecting the fact that the resource becomes more valuable. However, if non-renewable resource extraction generates negative social effects to communities, it is likely that this development model will not be sustainable because social costs increase with time and cause increasingly disproportionate changes in the distribution of benefits. If a resource development project is causing social externalities, communities may not receive a significant share of benefits and may incur a substantial share of costs (including social costs).

Negative externalities arise when an economic agent acts to maximize its own benefits and in doing so generates external effects that decrease the well-being of other economic agents. The classical economic approach to ensure optimal benefits for the entire economy (according to firms and communities) is to build a regulatory framework. While the theoretical solution to such a problem was conceived at the beginning of the 20th century, and simply involves internalizing the external effects through what is called a Pigouvian tax (Pigou 1920), in practice the issue is significantly more challenging.

Bringing the issue of negative social effects to the realm of regulatory economics involves, at a minimum, two important steps. The first is the challenging task of showing empirically that a negative social effect can indeed be attributed to mining activities. Efforts should not concentrate on showing that a certain social trend is prevalent and has become more prevalent as mining operations expand. That is an argument for correlations, and one that has been dismissed by regulators (see discussion above). The research effort must rely on establishing causal effects. Causal estimates, if confirmed, can and should be interpreted as the elasticities of the social indicator to mining operations and represent the size of the external effect. In summary, step one is perhaps the most important (and extremely challenging) step in designing a policy framework: making the case that one is warranted.

Step two is also challenging: the valuation of social damages. Consider the following conceptual economy. Suppose a society is made up of two economic agents: a mining firm and a community. Economically optimal resource allocation is the one that maximizes joint benefits. If the firm operations generate profits but also impose harm to the community due to a negative externality, there are adjustments that can be made so that harms are mitigated and joint surplus is maximized. Note that, in this example, doing nothing maximizes benefits to the mining firm, but probably not to the economy as a whole. Conversely, if individuals from the community do not participate in production, then there are no significant economic benefits or important social impacts to the community. Maximizing the benefits to the community involves shutting down mining operations, which is also unlikely to be the economically efficient solution. The optimal solution is somewhere in the middle, but where?

Designing a regulatory framework to address external effects involves having economic measures of the trade-offs between the benefits to the firm and social damages to the community. This might seem like a bizarre idea, but our society regularly makes conceptually similar decisions regarding trade-offs. These valuations are embedded in our decisions regarding speed limits, values of speeding tickets, alcohol taxes, and cigarette regulations, to name a few. Hence the final step in designing regulation of external effects is to assign value. In practice, implementation may involve disincentivizing mining, for example through a targeted tax. The result is that the production of harm is mitigated, or internalized, for example by making the mining industry responsible for the cost of the infrastructure that would be required to offer proper compensation services (e.g., medical and social assistance).

In summary, when it comes to northern communities, sustainability involves understanding impacts of resource development projects. If estimates of development impacts are essential, so is data. It takes data, and the effort of statisticians and empirical/quantitative researchers to produce new knowledge to complement the large body of existing literature regarding negative social effects. The trajectory towards this new body of empirical evidence will not be easy. In fact, one would be hard-pressed to think of a more challenging social context to estimate causal impacts than the one involving northern communities and mining.

Estimates may not be perfect, but a discussion around the validity of a causal estimate designed to control for confounding factors is one that is several steps closer to a fair regulatory framework than a discussion relying on local trends that can be easily undermined from a statistical standpoint. Aside from political pressures, it will take strong technical support for regulators to assign the responsibility of certain social problems in local communities to mining operations. Often communities are on the unfortunate side of the struggle. Large financial benefits are easy to measure, as are other economic benefits such as employment and wages. But the community at large seems to carry the burden of the proof when it comes to social disruptions, as they are the stakeholders that must present measurable effects and argue for causal social impacts.

DIRECTIONS FOR THE FUTURE

Most of us have heard of the term "big data revolution," but what does it really mean and how does it affect the work of social scientists? The expression is often associated with the increasing availability of data in social sciences and in many other disciplines. But data alone is not enough. The revolution refers not only to the availability of data but also and perhaps most importantly to the current boom of quantitative methods that are allowing us to extract more and more value from the data.

The uptake of quantitative approaches has been heterogeneous across disciplines. In social sciences, some disciplines are naturally more prone to data analysis and traditionally have relied more on empirical approaches. In other disciplines, quantitative analyses are the exception rather than the rule. The useful application of quantitative approaches in some social contexts can be very challenging and it requires interdisciplinary research teams. Nevertheless, this is already a reality in other contexts. For example, a team with researchers drawn from computer science, medicine, applied math and statistics, and public health worked together to develop a machine learning algorithm that computes a real-time early warning score to predict in advance which patients are at risk for septic shock (Henry et al. 2015). If quantitative approaches can complement medical care delivered by physicians, why can't these approaches be useful to analyze social context?

The big data revolution also highlights the need for deeper ethical considerations. When it comes to data, societies face a trade-off between privacy and analytical power. As argued before, the outcomes of this debate and the current state of regulations reflect the valuations of society regarding this trade-off. Also, the valuations of what is the right mix of benefits and costs when it comes to data policies must be considered. One thing that is becoming more apparent is that the opportunity cost of privacy is increasing as quantitative methods evolve to extract more and more useful information from the same amounts of data. This is a very complex discussion and one that is beyond the scope of this chapter. Nevertheless, future work should further our understanding of the power and limits of quantitative work and shed some light on how data analysis both can and cannot be used to assist communities in the Canadian North.

NOTE

1   Refer to the CCHS website for more information. http://www23.statcan.gc.ca/imdb/p2SV.pl?Function=getSurvey&SDDS=3226).

REFERENCES

Acuna, Roger M. 2015. "The Politics of Extractive Governance: Indigenous Peoples and Socio-Environmental Conflicts." *The Extractive Industries and Society* 2, no. 1 (January): 85-92.

Alonso, Elisa, Andrew M. Sherman, Timothy J. Wallington, Mark P. Everson, Frank R. Field, Richard Roth, and Randolph E. Kirchain. 2012. "Evaluating Rare Earth Element Availability: A Case with

Revolutionary Demand from Clean Technologies." *Environmental Science & Technology* 46, no. 6 (March): 3406-3414.

Baltagi, Badi H. and Rajeev K. Goel. 1990. "Quasi-Experimental Price Elasticity of Liquor Demand in the United States: 1960-83." *American Journal of Agricultural Economics* 72, no. 2 (May): 451-454.

Banerjee, Subhabrata B. 2000. "Whose Land is it Anyway? National Interest, Indigenous Stakeholders, and Colonial Discourses: The Case of the Jabiluka Uranium Mine." *Organization & Environment* 13, no. 1 (March): 3-38.

Bebbington, Anthony J. and Jeffrey T. Bury. 2009. "Institutional Challenges for Mining and Sustainability in Peru." *Proceedings of the National Academy of Sciences of the United States of America* 106, no. 41 (October): 17296-17301.

Berger, Thomas R. 1977. *Northern Frontier, Northern Homeland: The Report of the Mackenzie Valley Pipeline Inquiry* Volume 1. Ottawa: Minister of Supply and Services Canada.

Caxaj, C. Susana, Helene Berman, Colleen Varcoe, Susan L. Ray, and Jean-Paul Restoule. 2014. "Gold Mining on Mayan-Mam Territory: Social Unravelling, Discord and Distress in the Western Highlands of Guatemala." *Social Science & Medicine* 111: 50-57.

EA1314-01. Dominion Diamond Ekati Corp. 2016. *Report of Environmental Assessment and Reasons for Decision.* Mackenzie Valley Review Board. February 1, 2016.

Freudenburg, William R. and Robert E. Jones. 1991. "Criminal Behavior and Rapid Community Growth: Examining the Evidence." *Rural Sociology* 56, no. 4 (December): 619-645.

Hannington, Mark, Sven Petersen, and Anna Kratschell. 2017. "Subsea Mining Moves Closer to Shore." *Nature Geoscience* 10, no. 3 (February): 158-159.

Henry, Katharine E., David N. Hager, Peter J. Pronovost, and Suchi Saria. 2015. "A Targeted Real-time Early Warning Score (TREWSCORE) for Septic Shock." *Science Translational Medicine* 7, no. 299 (August 5): 299ra122.

Hilson, Gavin. 2002. "An Overview of Land Use Conflicts in Mining Communities." *Land Use Policy* 19, no. 1: 65-73.

Gibson, Ginger and Jason Klinck. 2005. "Canada's Resilient North: The Impact of Mining on Aboriginal Communities." *Pimatisiwin: A Journal of Aboriginal and Indigenous Community Health* 3, no. 1: 116-139.

Gramling, Robert and William R. Freudenburg. 1992. "Opportunity-threat, Development, and Adaptation: Toward a Comprehensive

Framework for Social Impact Assessment." *Rural Sociology* 57, no. 2: 216-234.

King, Gary, and Richard Nielsen. 2019. "Why Propensity Scores Should Not Be Used for Matching." *Political Analysis*, 27, no. 4 (October): 1-20. doi:10.1017/pan.2019.11.

Mactaggart, Fiona, Liane McDermott, Anna Tynan, and Christian Gericke. 2016. "Examining Health and Well-Being Outcomes Associated with Mining Activity in Rural Communities of High-Income Countries: A Systematic Review." *Australian Journal of Rural Health*. 24, no. 4 (August): 230-237.

Manning, Willard G., Linda Blumberg, and Lawrence H. Moulton. 1995. "The Demand for Alcohol: The Differential Response to Price." *Journal of Health Economics* 14, no. 2: 123-148.

Mining Association of Canada. 2019. *Facts and Figures 2018*. Report of the Mining Association of Canada. March 25, 2019. https://mining.ca/documents/facts-and-figures-2018/.

Nayga, Rodolfo M. 1996. "Sample Selectivity Models for Away from Home Expenditures on Wine and Beer." *Applied Economics* 28, no. 11 (November): 1421-1425.

Parkins, John R. and Angela C. Angell. 2011. "Linking Social Structure, Fragmentation, and Substance Abuse in a Resource-Based Community." *Community, Work & Family*, 14, no. 1 (February): 39-55.

Parlee, Brenda, John D. O'Neil, and Łutsël K'é Dene First Nation. 2007. ""The Dene Way of Life": Perspectives on Health from Canada's North." *Journal of Canadian Studies/Revue d'études canadiennes* 41, no. 3: 112-133.

Parlee, Brenda, John Sandlos, and Dave Natcher. 2018. "Undermining Subsistence: Barren-ground Caribou in a "Tragedy of Open Access."" *Science Advances* 4, no. 2 (February): e1701611.

Picone, Gabriel, Joe MacDougald, Frank Sloan, Alyssa Platt, and Stefan Kertesz. 2010. "The Effects of Residential Proximity to Bars on Alcohol Consumption." *International Journal of Health Care Finance and Economics* 10, no. 4 (December): 347-367.

Pigou, Arthur C. 1920. *The Economics of Welfare*. London: Macmillan.

Saffer, Henry and Dhaval Dave. 2002. "Alcohol Consumption and Alcohol Advertising Bans." *Applied Economics* 34, no. 11 (July): 1325-1334.

Shandro, Janis A., Marcello M. Veiga, Jean Shoveller, Malcolm Scoble, and Mieke Koehoorn. 2011. "Perspectives on Community Health Issues and the Mining Boom-Bust Cycle." *Resources Policy* 36, no. 2: 178-186.

Sharma, Sanjay. 2010. "The Impact of Mining on Women: Lessons from the Coal Mining Bowen Basin of Queensland, Australia." *Impact Assessment and Project Appraisal* 28, no. 3 (September): 201-215.

Silm, Siiri and Rein Ahas. 2005. "Seasonality of Alcohol-related Phenomena in Estonia." *International Journal of Biometeorology* 49, no. 4 (March): 215-223.

Temper, Leah and Joan Martinez Alier. 2013. "The God of the Mountain and Godavarman: Net Present Value, Indigenous Territorial Rights and Sacredness in a Bauxite Mining Conflict in India." *Ecological Economics* 96: 79-87.

World Mining Data. 2019. Federal Ministry Republic of Austria. Volume 34. Vienna, April 2019. https://www.world-mining-data.info/wmd/downloads/PDF/WMD2019.pdf.

15

What Must Happen for Extractive Industry to Help the Sustainability of Northern Communities?

Chris Southcott, Frances Abele, David Natcher, and
Brenda Parlee

The objective of the RESDA research network was to find ways to maximize the benefits of resource development in northern regions while minimizing the social, economic, cultural, and environmental costs experienced by communities. In this concluding chapter we highlight some of what was learned. It may seem self-serving to state that the most important thing that must happen is that more research needs to be conducted. However, given the wickedness of this challenge, this is one of our most pertinent conclusions, as unsatisfactory as that may sound. Notwithstanding the need for ongoing research, we do feel we have reached a point where we can offer some more definitive answers. It is our hope that these answers will prove useful to northern communities, resource industries, and policymakers as they navigate these challenges together in the future.

SUMMARY OF THE MAIN FINDINGS AS FAR AS COMMUNITIES ARE CONCERNED

The chapters in this volume summarize some of the key research results from the RESDA research program. The first section of chapters focused on the economic impacts of northern resource development. In Chapter 2, Rodon et al. note that one of the most important changes that has occurred in over the past forty years has been in northern communities gaining a more equitable share of benefits from the extraction of resources from their territories. While a large share of resource revenues is designated to mitigate the negative impacts on the environment, other revenues are managed at the discretion of communities and are being used to strengthen the social, cultural, and economic institutions of northern communities. Building on some of the findings of the resource curse-related economic research, Rodon and his colleagues explain how some resource revenues are being distributed in ways that help ensure the long-term well-being of these communities. In their research, they examined the costs and benefits of the following means of revenue distribution: making direct payments to individuals, placing those revenues in trust funds, investing in social programs and public services, or spending the funds on physical infrastructure. Their research indicates that there are positive and negative impacts associated with each of these strategies of distributing resource revenues. The choice of one over another must rely to a certain extent on the particular needs of these communities. For the authors, it is important that each community develop their own particular vision of where their needs lie. They must choose to distribute resource revenues in ways that ensure that any lost natural capital is made up for through investments in other types of capital that are important for the long-term sustainability of the community. They also note the importance of treating resource revenues as finite—a temporary source of income that will eventually end. The results stemming from this research provide communities with information that is based on the experience of others and that will enable them to make more informed decisions about how they can best manage resource revenues in the future.

Chapter 3 by Schott et al. summarizes findings from their project which look at business and individual employment benefits as well as human development from mining in the Nunatsiavut and Nunavik regions of the Canadian Arctic. Traditionally, increased wage employment is often seen as the most visible and immediate benefit from extractive resource development in the North. In a region where few employment opportunities exist, jobs are important for the long-term well-being of families. According to established models of economic development, these jobs come not only directly from the extractive industry itself, but also indirectly from services and other activities in the general business development associated with the industry. In areas such as the Canadian North, where job skills associated with industrial development are limited, the benefits of new employment brought on by mining and other extractive activities are not limited to income from wages, but also the development of human capital.

In their analysis, Schott et al. found that these types of benefits exist in both regions that they studied and that for the most part these benefits are increasing. Both regions are benefitting from employment in mining activities. At the same time there are important differences between the Nunavik and the Nunatsiavut regions. Employment turnover remains an issue, especially in Nunavik, but it is starting to decline. Their research indicates that work needs to continue on finding ways to decrease employment turnover. The employment situation can be improved, at least in part, through more flexible work schedules to accommodate the needs of the traditional subsistence economy such as seasonal hunting, fishing, and gathering. As well, communities need to better understand the impacts of new fly-in/fly-out (FIFO) work schedules on families. Local business capacity needs to be increased in order to encourage more indirect employment. More accessible training programs in smaller communities are needed and more needs to be done to enhance the employment of women. On a general level, the researchers note that employment and training benefits could be maximized and negative impacts minimized, by regions developing a comprehensive vision of development according to Inuit priorities. This vision, guided by a more effective system of local data and indicators, would greatly assist the communities to ensure that the temporary benefits from extractive activities are best used to enhance their long-term well-being.

While the research in Chapter 2 concentrated on resource revenues and that of Chapter 3 dealt primarily with employment, the research presented in Chapter 4 concentrates on training and its ability to result in sustainable employment. It has long been recognized that training is an essential element to enable Indigenous peoples in the Canadian North to benefit from employment opportunities. Equally important is the fact that this training can only be beneficial to these communities if it can be done in a way that recognizes, respects, and integrates Indigenous cultures. Hodgkins' research shows that there are important obstacles that limit the ability of northern communities to benefit from extractive industry-related training programs. The first is the pre-existing inequalities in terms of human capital. Current low levels of educational achievement limit the abilities of northerners to access apprenticeships and pass trade exams. Secondly, miscommunication

between companies and communities concerning issues such as recruitment, training, and termination of employment continue to exist and
show the need for "on the ground" supports to avoid potential problems. Thirdly, the lack of a long- term commitment from companies
to the communities, in addition to contributing to miscommunication
issues, creates a number of other problems. These include the absence
of established role models, a lack of succeeding cohorts to aid our
ability to understand how best to deliver programs, and a lack of
confidence by the local community that the training will lead to long-
term, well-paid career opportunities.

Communities in the Canadian North rely on the mixed economy.
These communities continue to exist because they have been successful
in combining the wage economy with income transfers and the traditional subsistence economy. While many in the south believe that
extractive resource development will lead to a decline in subsistence
activities, many Canada's North desire wage employment so that they
have the funds necessary to practice these activities by buying equipment, supplies, etc. Previously published research supported by ReSDA
showed that the two greatest obstacles to the subsistence economy are
the lack of money and the lack of time (Natcher et al. 2016). While
wage employment from extractive activities may provide the money
to help pay for subsistence activities, it also limits the time allowed
for these activities. While there is no evidence of an antagonistic relationship between extractive industry and subsistence activities, the
complexity of the relationship means researchers do not yet fully
understand the conditions under which these activities can be supportive of subsistence practices (Southcott and Natcher 2018). Chapter
5 by Rooke et al. looks at how extractive industries support the subsistence economy. Their research indicates that the most successful
examples are strategies that can mobilize both public and private
resources, and the co-operation of government, industry, and communities. Northern communities must be involved in the development
of these strategies and they must evolve through meaningful and
respectful relationships.

The shift from building resource-dependent communities in the
North to the increasing use of a mobile workforce has meant that
many of the boom and bust problems previously associated with these
resource towns are no longer an issue for the region. At the same time,
many of the infrastructure, business, and employment benefits that
used to come with these towns are no longer available. How do

northern communities adapt to this new situation in order to ensure that they can maximize benefits? Saxinger's research (Chapter 6) shows that there is a willingness in northern communities to participate in the long-distance commuting workforce, but that to do so requires coping strategies and responsible conduct on the part of the resource companies involved so that a safe and supportive environment is created. Strategies must exist to allow families to deal with the disruptions caused by shift absences. Adequate vocational training programs must be in place. There needs to be a more positive environment for female workers. There must be good relations between companies and communities to ensure that problems associated with the fifo social overlap can be dealt with in a creative manner.

The gendered socio-economic impacts of extractive industry in northern Canada have long been a theme in research on resource-dependent communities (Lucas 1971; Luxton 1980). More recent research on Indigenous communities indicates that gender-based differences go beyond employment issues to one of contributing to decisions concerning extractive activities (Archibald and Crnkovich 1999). If extractive resource development hopes to assist in the long-term well-being of these communities, these gender-based differences must be better understood. There needs to be a way of ensuring a more equal ability of both genders to contribute to these decisions. This is especially the case given the fact that women often face the greatest negative impacts as a result of extractive activities. Chapter 7 by Mills et al. summarizes research looking at two case studies of decision-making around extractive projects—one in the Sahtú Region of the Northwest Territories and the other in Nunatsiavut, Labrador. Both cases showed that women had a strong desire to discuss potential impacts beyond basic employment and training questions. Their perspective was centered around a wider-reaching need to create healthy communities based heavily on Indigenous traditions and cultural perspectives. Their perspective highlighted the need to find ways to better mitigate negative impacts such as addictions, gender-based violence, and food insecurity. The findings indicate that giving a greater voice to women in decisions concerning extractive developments will help ensure that impacts that are often ignored will be more properly considered in these decisions.

The next section of chapters summarizes research that looked at the ability of extractive resource development to help overcome long-standing barriers to the long-term sustainability of northern communities. Chapter 8 looks at housing, a long-standing problem in the region, and while this problem may not directly relate to extractive industry development, communities were interested in finding out whether extractive developments could assist in helping resolve current housing inequalities.

The research project led by Freeman and Christensen (Chapter 8) looked at the relationship between housing and resource development. Using the example of Yellowknife, NWT, they found that extractive development has, in the past, contributed to a constrained and precarious housing landscape. At the same time, it is possible that resources associated with these industries could be used in a more focused attempt to deal with housing issues. Their research notes the importance of building relationships between the main actors around the central purpose of providing suitable housing and then holding the key players in the housing market and extractive resource-based economy to account.

In the past, communities in Canada's North have been severely impacted by the environmental damage caused by extractive industry. One of the supposed improvements concerning mining and oil and gas activities in the region over the past forty years has been the introduction of a new environmental regulatory system designed to limit negative environmental impacts through both mitigation and (at the end of the project), remediation. These mitigation and remediation activities are now supposed to represent an important part of extractive industry planning projects. These activities could be used to help the long-term sustainability of these communities if their interests were properly integrated into the planning. The chapter by Dance et al. (9) examined remediation policies in Canada's North with the aim of determining the place of regional concerns in these policies. They note that many problems still exist with the current regulatory system. These policies need to better integrate the voices and concerns of the region in the processes.

Food security is another long-term concern for northern communities. While extractive industry development may not be primarily responsible for current levels of food insecurity, communities were interested in finding ways to ensure that these activities do not make the current situation worse and seeing if they could make the situation better. The chapter by Keske (10) summarizes a research project to determine whether waste from the Muskrat Falls hydroelectric project could be used to help the region increase local food production.

Through several examples the author was able to point out how greater collaboration between industry and community groups could help industry reduce its waste problems and help communities reduce its food security problems. Waste from the project was used to develop new agricultural activities for the region.

Since the 1970s there has been a shift regarding resource development decision-making. It now strives to be more participatory and democratic in all its objectives, including in environmental and other regulatory processes. Given this objective, it may be that new extractive industry projects could play a role in enhancing civil society and public participation, especially given the empowering potential of new land claim agreements in the Canadian North. The chapter by Gladstone and Kennedy Dalseg (11) summarizes an initial investigation to see if this is indeed happening in the region. While they remain optimistic about future possibilities, they note that in Nunavut, participation by locally-based organizations is very limited. Most organizations that do participate are national or international environmental NGOs and northern environmental NGOs with a conservation focus. This may be because the locally-based social economy sector is of a different nature than that of more conventional settler-based organizations and as such it chooses not to engage with the new regulatory regimes, leaving this task to their Nunavut-based public and land claim institutions. In order for extractive industry-related processes to promote a greater degree of democratization and participation, changes may need to be made to allow differing types of associations to access supports and participate in the new regulatory processes.

The question of how best to maximize benefits of extractive resource development to northern regions and communities and minimize the social, economic, cultural, and environmental costs cannot be answered adequately without a system of measuring these benefits and impacts. The introduction of social impact assessment and other aspects of the new regulatory regimes have improved the situation somewhat, but problems remain. Northern communities have expressed a desire for locally based notions of well-being to be integrated into impact indicators. They also ask that indicators be developed that can be gathered and analyzed by the communities themselves without depending on outside experts. The chapter by Parlee (12) summarizes the earlier work on measuring social impacts on well-being. These indicators have tended to portray Indigenous communities as vulnerable and in need of development. To properly measure impacts and benefits that are relevant to these communities, locally defined notions of well-being must serve as the basis for indicator development. This idea is supported by the work of the Inuvialuit Regional Corporation. The chapter by Petrov et al. (13) describes how a set of existing quantitative indicators can be gathered and adjusted to meet the particular needs of the Inuvialuit. At the same time, they note the need for additional more qualitative indicators to give a more in-depth understanding of impacts. Finally, the chapter by Godfrey and Wichmann (14) builds on the previous chapter by discussing a range of problems that need to be dealt with to ensure northern communities have an adequate means of measuring extractive industry activities in order to ensure their long-term sustainability.

These projects have given us a better idea of what needs to happen in order to maximize benefits to northern communities from extractive resource projects and to mitigate negative impacts. In terms of managing new resource revenues, each community must understand what their most pressing needs are and then choose a distribution system that can best meet their specific needs. Employment benefits will increase if more is done to deal with the problems of turnover, such as the development of more flexible hiring programs and work schedules, and with the implementation of more support programs for employees. Support programs need to be established so that communities can better deal with the negative impacts of participation in a mobile workforce. An increase in local business capacity will provide more indirect employment opportunities. Training benefits can increase through a general increase in human capital in the community but also through better communication on the part of the company and through a long-term commitment of the company to the community. Respectful relations are also essential for developing strategies to support the traditional subsistence economy.

Negative impacts can be reduced and new benefits can be developed if a way is found to allow a more meaningful participation of women in both the decision-making process concerning the extractive project and the project's operation. Research on housing, environmental remediation, food security, and the social economy sector has shown that a more active and meaningful participation of the community throughout the project would allow for innovative ways that the extractive activities could help communities deal with these challenges. Finally, communities need to have in place an adequate system of measuring impacts that reflects their own

particular notions of well-being and that can be collected and analyzed by their own people.

Our research explored ways to maximize the benefits of resource
development in northern regions while minimizing the social, economic, cultural, and environmental costs experienced by northern
communities. In contrast to much of the rich and substantial body of
research published over the last two decades, we did not directly assess
or investigate the new institutions of northern governance or their
connections to the evolution of capitalism in the North. Research on
these matters continues even as we go to press, in programs led, for
example, by Stephanie Irlbacher-Fox (https://moderntreaties.tlicho.
ca) and Thierry Rodon (https://www.mineral.ulaval.ca/en/about/
objectives-and-research-themes). Important published discussions in
the academic literature concern the nature and reach of the governance
innovations (Wilson et al. 2020; Hicks and White 2015. One question
is about the extent to which the new institutions of northern governance are empowering northern communities or whether they are
forcing the assimilation of these communities into capitalism, and
latterly into neo-liberal regimes (Nadasdy 2003; Kulchyski 2016;
Coulthard 2014). This discussion, which extends beyond Northern
Canada and dates at least from the last quarter of the twentieth century, considers how capitalist relations of production shape new
economic and political institutions, and in doing so shape the life
chances of northern community residents (Watkins 1977; Asch 1987;
Usher 2003; Newhouse 2011; Kuokkanen 2011).

Our project is clearly informed by certain strands of this theoretical
work, but our focus has been closer to the ground, as we have
attempted to address directly the specific needs of communities as they
deal with immediate choices in present circumstances. In this way, we
hope to inform the larger debates while speaking to the immediate
concerns of northerners. There is an urgent need for research that
considers how institutions can be further adapted to allow for a more
authentic participation from the Indigenous peoples of Canada's
Arctic. Also to be considered is how to broaden the range of choices
facing them.

The findings from the RESDA projects detailed here and findings
from earlier gap analysis research (Southcott et al. 2018) indicate
that while the possibilities for Canada's northern communities benefitting from extractive resource development have increased, important questions remain and more research needs to be done. Possibilities

for mining, oil and gas, and other projects to contribute to the sustainability of these communities has increased because conditions have changed to allow communities to have more control over these projects. In a practical sense, in areas governed by modern comprehensive land claim agreements, communities have the power to say yes or no to a project. While it may or may not be legally possible to establish a mine in Canada's Arctic when communities are opposed to it, in today's political climate it is difficult to imagine this happening. This situation gives communities in these regions of Canada's North a great deal of power to ensure that if a project goes ahead, it will do so under conditions that meet their needs. The ability of a community to say no to a project is a fundamental part of that community being able to ensure extractive industry contributes to the long-term sustainability of the community.

The ability to say no to a project is not a guarantee that such projects will contribute to the long-term well-being of this community. The regulatory regimes and institutions established over the past forty years are still, by and large, "foreign" institutions that are from outside most Indigenous communities. The structures, rules, guidelines, and processes are primarily adapted from Western socio-political and legal traditions. Communities in the North are still struggling to adapt these institutions and processes to their traditions. As a result, they are still at a disadvantage in terms of their ability to participate in these processes to ensure their notions of well-being are properly recognized and that their needs will be met. Much more work needs to be done in order to find out how these institutions and processes can be more effectively adapted to allow for a more authentic participation from the Indigenous peoples of Canada's Arctic.

NOTES

1   It should be noted that for a variety of reasons, two of the projects organized by RESDA do not have their results included here. The results of these projects are partially available on the RESDA website and the final results will be published separately. These two projects are: "Augmenting the utility of IBAs for Northern Aboriginal Communities," a 2017 project headed by Ben Bradshaw of the University of Guelph, and the 2016 project "Finding What Works: Mining in Inuit Nunangat," headed by Frances Abele of Carleton University.

REFERENCES

Archibald, Linda and Mary Crnkovich. 1999. *If Gender Mattered: A Case Study of Inuit Women, Land Claims and the Voisey's Bay Nickel Project*. Ottawa: Status of Women Canada.

Asch, Michael. 1987. "Capital and Economic Development: A Critical Appraisal of the Recommendations of the Mackenzie Valley Pipeline Commission." In *Native Peoples, Native Lands: Canadian Indians, Inuit and Métis*, edited by B.A. Cox. Ottawa: Carleton University Press.

Coulthard, Glen. 2014. *Red Skin, White Masks: Rejecting the Colonial Politics of Recognition*. University of Minnesota Press.

Hicks, J, and G. White. (2015). *Made in Nunavut: An experiment in decentralized government*. UBC Press.

Kulchyski, P. 2016. "Rethinking Inequality in a Northern Indigenous Context: Affluence, Poverty, and the Racial Reconfiguraton and Redistribution of Wealth." *The Northern Review* 42: 97-108.

Kuokkanen, Rauna. 2011. "From Indigenous Economies to Market-Based Self-Governance: A Feminist Political Economy Analysis." *Canadian Journal of Political Science/Revue canadienne de science politique* 44(2): 275-297.

Lucas, Rex. 1971. *Minetown, Milltown, Railtown: Life in Canadian Communities of Single Industry*. Toronto: University of Toronto Press.

Luxton, Meg. 1980. *More Than a Labour of Love: Three Generations of Women's Work in the Home*, vol. 2. Toronto: Canadian Scholars' Press.

Nadasdy, P. 2003. *Hunters and Bureaucrats: Power, Knowledge, and Aboriginal-State Relations in the Southwest Yukon*. UBC Press.

Natcher, Dave, Shea Shirley, Thierry Rodon, and Chris Southcott. 2016. "Constraints to Wildlife Harvesting Among Aboriginal Communities in Alaska and Canada." *Food Security* 8, no. 6: 1153-1167.

Newhouse, David R. 2011. "Resistance is Futile: Aboriginal Peoples Meet the Borg of Capitalism." In *Ethics and Capitalism*. University of Toronto Press. 141-155.

Southcott, Chris, Frances Abele, Dave Natcher, and Brenda Parlee. 2018. "Beyond the Berger Inquiry: Can Extractive Resource Development Help the Sustainability of Canada's Arctic Communities?" *Arctic* 71, no. 4 (December): 393-406.

Southcott, Chris, and Dave Natcher. 2018. "Extractive Industries and Indigenous Subsistence Economies: A Complex and Unresolved Relationship." *Canadian Journal of Development Studies/Revue canadienne d'études du développement* 39, no. 1 (January): 137-154.

Usher, P. J. 2003. "Environment, Race and Nation Reconsidered: Reflections on Aboriginal Land Claims in Canada." *The Canadian Geographer* 47(4): 365-382.

Wilson, Gary N., C. Alcantara, and Thierry Rodon. 2020. *Nested Federalism and Inuit Governance in the Canadian Arctic.* UBC Press.

Contributors

FRANCES ABELE is Chancellor's Professor, School of Public Policy and Administration at Carleton University.

LEON ANDREW is a Shúhtaot'ine elder with the Tulít'a Dene Band and Research Director of the Ɂehdzo Got'ı̨nę Gots'ę́ Nákedı (Sahtú Renewable Resources Board) in Tulít'a, Northwest Territories.

ANTENEH BELAYNEH is a Policy Analyst at the Department of National Defense in Ottawa Ontario and a graduate of the PH.D. program at the School of Public Policy and Administration at Carleton University.

JEAN-SÉBASTIEN BOUTET is a PhD student at KTH Royal Institute of Technology in Stockholm, Sweden and was previously a mining analyst for the Nunatsiavut Government in northern Labrador, Canada.

PHILIP CAVIN was formerly a student at the University of Northern Iowa, and is currently Project Manager at Data Business Equipment, Des Moines, Iowa.

JULIA CHRISTENSEN is Canada Research Chair in Northern Governance and Public Policy at Memorial University, Newfoundland.

ANNE DANCE was a postdoctoral research fellow at Memorial University Newfoundland from 2014 to 2016. She holds a PhD from the University of Stirling.

LISA FREEMAN is a Research and Policy Analyst at the Hospital Employees Union.

JOSH GLADSTONE is a researcher in resource management and public policy based in Ottawa.

TODD GODFREY is an economist working for the Government of British Columbia.

ANDREW HODGKINS is an independent scholar based in Edmonton, Alberta who has done extensive research on education and training in Northern Canada.

ARN KEELING is a Professor of Geography at Memorial University in Newfoundland.

SHEENA KENNEDY Dalseg holds a PhD in Public Policy from Carleton University where she studied governance and institutional development in Northern Canada.

CATHERINE KESKE was formerly a professor at Memorial University of Newfoundland and is currently a professor in the School of Engineering at the University of California Merced.

HARVEY LEMELIN is a Professor in the School of Outdoor Recreation, Parks and Tourism at Lakehead University in Thunder Bay, Ontario.

ISABEL LEMUS-LAUZON is the director of First Nations and Inuit Relations and Northern Development Branch, Department of Education in Quebec City, Quebec.

SUZANNE MILLS is an Associate Professor in the School of Labour Studies at McMaster University in Hamilton, Ontario.

MIRANDA MONOSKY was a Research Assistant at Memorial University from 2019 to 2020 where she completed an MA in Geography.

DAVID NATCHER is professor in the Department of Agricultural and Resource Economics at the University of Saskatchewan.

BRENDA PARLEE is professor in the Department of Resource Economics and Environmental Sociology at the University of Alberta.

ANDREY PETROV is Professor of Geography at the University of Northern Iowa and is Director of UNI's ARCTICenter.

THIERRY RODON is a Professor in the Political Science Department at Université Laval and chairholder of the Northern Sustainable Development Research Chair at Université Laval in Quebec City, Quebec.

REBECCA ROOKE was a Master's student in the School of Outdoor Recreation, Parks and Tourism at Lakehead University in Thunder Bay, Ontario.

JOHN SANDLOS is a Professor of History at Memorial University in Newfoundland.

GERTRUDE SAXINGER, PhD in anthropology, has done extensive research in social dimensions of natural resource extraction in Arctic Russia and Northern Canada. She is a faculty member of the Austrian Polar Research Institute (APRI) and working as a researcher and lecturer at the Department of Political Science/Uni Vienna, Austria as well as she is associate at the Institute for Social Anthropology/Uni Bern, Switzerland.

STEPHAN SCHOTT is a Professor in the School of Public Policy and Administration at Carleton University in Ottawa, Ontario.

JEAN-MARC SÉGUIN is the director for Mining at the Makivik Corporation in Nunavik, Quebec.

DEBORAH SIMMONS has a doctorate in Social and Political Thought from York University and is Executive Director of the Ɂehdzo Got'ı̨nę Gots'ę́ Nákedı (SaɁehdzo Got'ı̨nę Gots'ę́ Nákedı (SahtúRenewable Resources Board) in Tulít'a, Northwest Territories.

BOB SIMPSON is Director of Government Affairs for the Inuvialuit Regional Corporation in Inuvik, Northwest Territories.

CHRIS SOUTHCOTT is professor of sociology at Lakehead University.

JOHANNA TUGLAVINA is Inuk from Nain, Nunatsiavut. She is an experienced researcher and has served as a leader in several community and regional organizations, including the Nain Women's Shelter and the AnânauKatiget Tumingit Regional Women's Association. She has also worked as a Cultural Awareness Facilitator and Logistics Coordinator at Voisey's Bay Mine.

BRUNO WICHMANN is an associate professor of resource and environ¬mental economics at the University of Alberta.